Biometric Authentication

**Prentice Hall Information
and System Sciences Series**
Thomas Kailath, Editor

Biometric Authentication

A Machine Learning Approach

S. Y. Kung
Department of Electrical Engineering
Princeton University

M. W. Mak
Centre for Multimedia Signal Processing
Department of Electronic and Information Engineering
The Hong Kong Polytechnic University

S. H. Lin
NVIDIA Corporation

PRENTICE
HALL
PTR

Prentice Hall Professional Technical Reference
Upper Saddle River, New Jersey 07458
www.phptr.com

Library of Congress Cataloging-in-Publication Data
Kung, S. Y. (Sun Yuan)
    Biometric authentication / S. Y. Kung, M. W. Mak, S. H. Lin.
      p. cm. — (Prentice-Hall information and system science series)
    Includes bibliographical references and index.
    ISBN 0-131-47824-9 (hardcover : alk. paper)
    1. Pattern recognition systems. 2. Identification—Automation. 3.
    Biometric identification. I. Mak, M. W. II. Lin, Shang-Hung, 1968- III.
    Title.

  TK7882.P3K84 2004
  621.389'28—dc22

                2004012612

Prentice Hall PTR offers excellent discounts on this book when ordered in quantity for
bulk purchases or special sales. For more information, please contact: U.S. Corporate and
Government Sales, 1-800-382-3419, corpsales@pearsontechgroup.com. For sales outside of
the U.S., please contact: International Sales,
1-317-581-3793, international@pearsontechgroup.com.

Printed in the United States of America

Text printed on recycled paper
1 2 3 4 5 6 7 8 9 10—CRW—0807060504
First printing, September 2004

ISBN: 0-131-47824-9

CONTENTS

PREFACE

Biometrics has long been an active research field, particularly because of all the attention focused on public and private security systems in recent years. Advances in digital computers, software technologies, and embedded systems have further catalyzed increased interest in commercially available biometric application systems. Biometric authentication can be regarded as a special technical area in the field of pattern classification. Research and development on biometric authentication have focused on two separate fronts: one covering the theoretical aspect of machine learning for pattern classification and the other covering system design and deployment issues of biometric systems. This book is meant to bridge the gap between these two fronts, with a special emphasis on the promising roles of modern machine learning and neural network techniques.

To develop an effective biometric authentication system, it is vital to acquire a thorough understanding of the input *feature space*, then develop proper mapping of such feature space onto the *expert space* and eventually onto the output *classification space*. Unlike the conventional template matching approach, in which learning amounts to storing representative example patterns of a class, the machine learning approach adopts representative statistical models to capture the characteristics of patterns in the feature domain. This book explores the rich synergy between various machine learning models from the perspective of biometric applications. Practically, the machine learning models can be adopted to construct a robust information processing system for biometric authentication and data fusion. It is potentially useful in a broad spectrum of application domains, including but not limited to biometric authentication.

This book is organized into four related parts.

1. Part I—Chapters 1 and 2—provides an overview of the state-of-the-art in face and speaker biometric authentication systems.

2. Part II—Chapters 3, 4, and 5—establishes the theoretical pillars of machine learning methods adopted in the book. To facilitate the development of effective biometric authentication systems, several modern machine learning models are instrumental in handling complex pattern recognition and classification problems. Part II discusses the expectation-maximization (EM)

algorithm (Chapter 3); describes the fundamental theory on Fisher's linear discriminant analysis (LDA) and support vector machines (SVMs) (Chapter 4); and offers comprehensive coverage of multi-layer learning models, in addition to well-known back-propagation (BP) algorithms (Chapter 5).

3. Part III—Chapters 6 and 7—proposes several flexible structural frameworks based on hierarchical and modular neural networks, under which machine learning modules can be incorporated as subsystems. The discussion introduces several expert-based modular networks, such as the so-called hierarchical mixture-of-experts (Chapter 6), as well as interclass learning strategies based on class-based modular networks (Chapter 7).

4. Part IV—Chapters 8, 9, and 10—presents the theoretical foundations behind the learning networks, which can find natural and fruitful applications in biometric authentication systems. The most important authentication application domains are face recognition and speaker verification. Specifically, Chapter 8 presents probabilistic neural networks for face biometrics, while Chapter 9 covers authentication by human voices. Several multicue data-fusion techniques are addressed in Chapter 10.

As suggested by the title, this book covers two main themes: (1) biometric authentication and (2) the machine learning approach. The ultimate objective is to demonstrate how machine learning models can be integrated into a unified and intelligent recognition system for biometric authentication. However, the authors must admit that the book's coverage is far from being comprehensive enough to do justice to either theme. First, the book does not address many important biometric authentication techniques such as signature, fingerprint, iris pattern, palm, DNA, and so on. The focus is placed strictly on visual recognition of faces and audio verification of speakers. Due to space constraints, the book has likewise overlooked many promising machine learning models. To those numerous contributors, who deserve many more credits than are given here, the authors wish to express their most sincere apologies.

In closing, *Biometric Authentication: A Machine Learning Approach* is intended for one-semester graduate school courses in machine learning, neural networks, and biometrics. It is also intended for professional engineers, scientists, and system integrators who want to learn systematic, practical ways of implementing computationally intelligent authentication systems based on the human face and voice.

Acknowledgments

This book is an outgrowth of many years of teaching and research on the subject of neural networks, pattern recognition, and biometric authentication. The authors are very much indebted to many students for their invaluable questions and contribution of examples and exercises. Some parts of the book are extracted from several postgraduate students' dissertations. We wish to thank all of them, in particular M. C. Cheung, K. W. Cho, K. W. Ku, and K. K. Yiu. We also thank our research assistants, including K. Y. Leung, Chad Myers, Xinying Zhang, Yunnan Wu, C. H. Sit, C. L. Tsang, and W. M. Yu for their efforts in performing some of the simulations and proofreading numerous versions of the manuscript.

We have benefited greatly from frequent and enlightening exchanges and collaboration with many colleagues. To name a few: P. C. Ching, Helen Ming, Brian Mak, and M. H. Siu. Their enthusiastic participation in this project has made the prolonged process enjoyable and rewarding. We also thank Professor W. C. Siu, Professor Jin-Shiuh Taur, and Dr. Hong-Jian Zhang for their invaluable suggestions and comments. Our deep gratitude goes to the reviewers for their critical and constructive suggestions on the book's manuscript. Our sincere appreciation goes to the series editor, Professor Thomas Kailath, for many years of inspiration and friendship.

We have been very fortunate to have worked with Mr. Bernard Goodwin at Prentice Hall and production service coordinator Ms. Marilyn Rash, who have provided the highest professional assistance throughout this project. We are grateful to the Department of Electrical Engineering at Princeton University and the Department of Electronic and Information Engineering at The Hong Kong Polytechnic University for making available such a scholarly environment for both teaching and research.

M. W. Mak is with the Center for Multimedia Signal Processing, Department of Electronic and Information Engineering, The Hong Kong Polytechnic University, Hong Kong. S. Y. Kung was on sabbatical from Princeton University and was the Multimedia Distinguished Chair Professor of The Hong Kong Polytechnic University when this project commenced. We are pleased to acknowledge that the work presented in this book was in part supported by the Center for Multimedia Signal Processing, The Hong Kong Polytechnic University (Grant Nos. A442 and A464); the Research Grants Council, Hong Kong Special Administrative Region (Grant Nos. PolyU 5129/01E and PolyU5131/02E); and the Department of Electrical Engineering, Princeton University. This book is part of a series of publications under the leadership of Professor W. C. Siu, Director of the Center for Multimedia Signal Processing, The Hong Kong Polytechnic University.

Without the support and encouragement of many friends and colleagues, it would be impossible to write a book on such a diversified subject. Our heartfelt thanks go to the researchers who have contributed to the field of machine learning, neural networks, and biometrics, whose work provided the foundation for this book. We offer our sincere apologies for inevitably overlooking many critical topics and references due to time and space constraints.

Finally, the authors wish to acknowledge the generous and kind support of their families. This project would not have been completed so smoothly without their full understanding throughout this long process.

Chapter 1

OVERVIEW

1.1 Introduction

In business and personal life today, security protection systems are critical for many application domains: (1) transaction protection (client verification); (2) access control (key or keyless); (3) computer and network security (terminal user verification); and most important, (4) personal and public safety. Since almost all artificial features used in conventional identification techniques can potentially be forged and some of them (e.g., ID cards or passwords) stolen or forgotten,[1] better and more effective identification and authentication methods are now in greater demand. With recent technological advances in audio and visual microelectronic systems, reliable automatic authentication systems have become a commercial and practical reality.

This book focuses on machine learning techniques for biometric identification and the practical application of neural networks to biometric authentication systems. *Biometric systems* are automated methods of verifying or recognizing the identity of a person on the basis of some physiological characteristic, such as a fingerprint or face pattern, and/or some aspect of behavior, such as the spoken voice, handwriting, or keystroke patterns [245, 339]. Since biometric systems do not identify a person by what he or she knows (a code) or possesses (a card), but by a unique characteristic that is difficult for a different individual to reproduce, the possibility of forgery is greatly reduced.

Opportunities of Biometric Technologies

From a biometric application perspective, since the tragic terrorist attacks of September 11, 2001, there has been a greater awareness of security threats and increased acceptance of more intrusive security systems. This will make public security systems with more controlled and less variant biometric features available. It goes without saying that more reliable raw data implies improved sensitivity and specificity performance, making face and speaker recognition a more viable and feasible option for biometric surveillance systems.

---

[1] According to a Bell Lab's finding, a large percentage of passwords chosen by users were easy to decode in a short period of time.

Biometric identification has the potential to virtually eliminate computer fraud, but cheap, reliable, and foolproof systems are still a long way off. This suggests that more effort should be placed on biometric research because there will surely be increasing demand for reliable security systems. From a machine learning technology perspective, it is important to continue to research and develop more flexible and reliable biometric identification systems. Therefore, this book emphasizes a hierarchical, modular, and structural approach to machine learning, which can arguably lead to a more reliable synthesis of different information sources from various experts or sensors. As a more specific example, Chapter 8 presents several image preprocessing procedures commonly used by face recognition systems in an attempt to reduce the influence of change in lighting conditions and head poses. It also illustrates why facial images can only be detected under limited conditions even with those preprocessing procedures. In fact, achieving true illumination and pose invariants for three-dimensional deformable object recognition is still an unsolved topic for computer vision research. Face recognition technology must overcome this obstacle, otherwise it cannot become a truly human aid-free and automatic method for surveillance and security control. For voice biometrics, Chapter 9 introduces several channel compensation techniques to address the variation in transducers' characteristics. Chapter 10 proposes a multicue approach to personal identification in an attempt to enhance the reliability of biometric systems.

1.2   Biometric Authentication Methods

As depicted in Figure 1.1, biometric authentication methods can be divided into two categories.

1. *Behavioral-based authentication methods* perform the identification task by recognizing people's behavioral patterns, such as signatures, keyboard typing, and voice print. The main problem with behavioral methods is that they all have high variations, which are difficult to cope with. To prevent signatures from being altered, the use of laser-engraved, digitally embossed signatures on the card's signature panel has also been considered. On the other hand, while behavioral characteristics can be difficult to measure because of influences such as stress, fatigue, or illness, they are sometimes more acceptable to users and generally cost less to implement. Good examples are speaker recognition systems, as discussed in Chapter 9.

2. *Physiological-based authentication methods* verify a person's identity by means of his or her physiological characteristics such as fingerprint, iris pattern (eye blood vessel pattern), palm geometry, DNA, or facial features. In general, traits used in the physiological category are more stable than methods in the behavioral category because most physiological features are virtually nonalterable without severe damage to the individual.

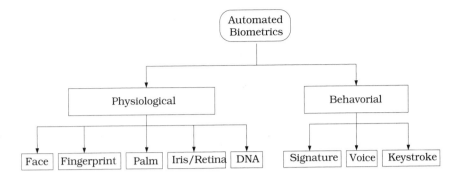

Figure 1.1. Different types of biometric ID methods.

1.3   Face Recognition: Reality and Challenge

Face recognition is probably one of the most *nonintrusive* and *user-friendly* biometric authentication methods available. Machine recognition of human faces from still or video images has attracted a great deal of attention from the communities of psychology, image processing, pattern recognition, neural science, computer security and computer vision. Because of an increasing, omnipresent threat of terrorist attacks, face recognition has now joined the list of many security-related technologies that have attracted widespread media and business attention. The nonintrusive appeal of this seemingly mature surveillance technology has led more and more government and law enforcement agencies to consider it a viable augmentation to enhance security measures in public areas.

In January 2001, face recognition was already in use to scan the faces of people in the crowds attending the Super Bowl in Tampa, Florida. In July 2002, the Virginia Beach Police Department linked Identix Inc.'s FaceIt recognition system to its beachfront cameras [153] (the technology used by FaceIt is described in Chapter 8). In the State of Illinois, the Motor Vehicles Department uses Viisage's Eigenface technology (also described in Chapter 8) on its extensive driver's license face database (15 million images) to combat identity theft [141]. At Sydney's airport, Australian Customs has used Cognitec's FaceVACS system to automatically process the face-to-passport check for the Qantas Air crews since November 2002 [146].[2]

Products for many promising consumer applications, where incorrect decisions are less critical, provide yet another major business opportunity for the application of face recognition techniques. In fact, most face recognition companies (e.g., Viisage, Identix) provide authorization software (or screensaver) for PC and mobile devices, and data-mining tools to search the Internet or image databases. Fuji's dig-

---

[2]Cognitec Systems GmbH is a German company; their FaceVACS system is very similar to Chapter 8's PDBNN face recognition system.

ital photo processing kiosk uses face recognition techniques to detect the presence of people in everyday photos. Once a human face is detected, the kiosk balances the color of the facial region and makes the skin tone in the printed photo more vivid [140]. Automatic red-eye reduction and advanced autofocus are plausible extensions of this application. In the very profitable field of video gaming, face recognition represents a promising and potentially lucrative possibility for yet another add-on feature for video games or toys.

Face Biometrics

After so many years of intense research and development effort, have face recognition technologies really reached a point where large-scale deployment is feasible? The answer to this question must remain cautious and conservative. Take airport security, as an example. Several field studies have suggested that it is still premature to claim reliable performance by any face recognition company, especially in light of the current technology's vulnerability to illumination and pose variation. "It hardly takes a genius of disguise to trick this system," said a security officer. "All a terrorist would have to do, it seems, is put on eyeglasses or turn his head a little to the side."

Such reservations are supported by results obtained from field trials performed at two airports. The two independent reports clearly suggest the inadequacy of current face recognition technology. In 2002, the American Civil Liberties Union (ACLU) tested Identix's face recognition system at the Palm Beach International Airport in Florida [139]. The test was conducted on approximately 5,000 passengers and employees per day at the airport's Concourse C security checkpoint using a test group of 15 airport employees and a database of 250 photographs that, according to press accounts, featured suspected criminals. According to an ACLU press release, the system failed to match volunteer employees who had been entered into the database fully 503 out of 958 times, or 53% of the time. "Even with recent, high-quality photographs and subjects who were not trying to fool the system, the face recognition technology was less accurate than a coin toss," said Barry Steinhardt, director of the National ACLU's Technology and Liberty Program. Similar disappointing results were reported by Boston's Logan International Airport after both Identix and Viisage technologies were tested in 2002 [145]. The system failed to identify positive matches 38% of the time. While false positives based on an operator's decision didn't exceed 1%, machine-generated false positives exceeded 50%.

On the other hand, the need to maintain secure access control and minimize frauds in financial transactions is becoming increasingly important and at the same time, increasingly challenging. Automatic teller machine (ATM) access and building gate control are excellent application examples. From such a commercial application perspective, current face recognition technology already offers a reasonable (or sometimes adequate) performance if "controlled environment" and "user's cooperation" can be expected or ensured. Fortunately, most applications in this category can be conducted in a controlled environment and user's cooperation is often expected. Examples of such controlled application systems include Cognitec's passport checking system and Viisage's Illinois driver's license photo scan.

Germany-based Cognitec seems to have had better success than its competitors Identix and Viisage. Up until November 2003, Cognitec's automatic passport checking system, SmartGate, has successfully handled 4,200 enrolled Qantas Air's crew members in more than 62,000 transactions at the Sydney airport in Australia. The major difference between Sydney's success and Boston's failure is that the face recognition technology was applied to different "applications." Compared to the trial at the Boston airport, where illumination at checkpoints can drastically change from morning to afternoon, from rain to sunshine, and where testers may have been unaware of or unwilling to face the camera, Cognitec's SmartGate usually receives testers'—the registered crew members who want to pass the customs' check—full cooperation. In addition, SmartGate is embedded with auxiliary lights, resulting in better controlled lighting conditions.

In 2002, the U.S. government (DARPA, NIST, DoD) conducted an extensive evaluation program called Face Recognition Vendor Test 2002 (FRVT 2002) to compare the performance of 10 commercial face recognition systems [147]. FRVT 2002 uses a very large database (121,589 operational images of 37,437 people) to compare performance through various indices, including database size, facial expression, head position, lighting condition, demography (age, gender, race), and elapsed time between enrolled images and test images. FRVT 2002 examines the performance on frontal faces under different environmental conditions for three tasks: verification (verifying a tester's claimed ID); identification, or recognition (finding a tester's ID among an existing database); and watch lists (identifying whether a tester belongs to a particular subgroup—the database's "watch list"). The test accurately depict the current status of the face recognition technology, as follows:

- For indoor conditions, error rates of the face recognition systems were reduced 50% since FRVT 2000.

- For indoor conditions, the best face recognition systems had a 90% verification rate at a false acceptance rate of 1%. The three best vendors are Cognitec (90%), Identix (90%), and Eyematic (87%). (In 2003, Eyematic changed its name to Nevengineering, Inc. [150].) Viisage, the top-ranked vendor in FRVT 2000, received a surprisingly lower score of 64%.

- For outdoor conditions, at a false acceptance rate of 1%, the verification rate of the best systems drops to 50%.

- Identification performance decreases linearly with respect to the logarithm of the database size.

- Performance with a smaller watch list is better than performance with a larger watch list.

- The use of morphable models can significantly improve nonfrontal face recognition.

- Demographic characteristics can significantly affect performance. For example, males are easier to recognize than females, and older people are easier to recognize than younger people.

- The change of facial expression severely affects the identification performance.

The FRVT 2002 results show that advances in face recognition technology have been made since 2000. However, the significant performance degradation seen in outdoor environments still greatly limits its range of application. In this respect, the FRVT 2002 results concur with the observations of the ACLU airport trials. Therefore, illumination invariance is a crucial performance criterion and should be a focused research topic for face recognition applications.

Another important research topic is pose invariance. In real-world surveillance cases, people seldom look directly at the camera. FRVT 2002 proposed a preprocessing step to all participating face recognition vendors so that performance for nonfrontal face recognition could be improved. It used a morphing technique by Blanz, Romdhami, and Vetter [28] to transform nonfrontal faces stored in a database into frontal faces. According to the test results, this morphing preprocessing can increase Identix's performance rate from 26% to 84%. In addition to nonfrontal image transformation, such morphable models can be used for facial expression analysis and facial animation. Therefore, the topic of facial image morphing deserves more attention from face recognition researchers.

1.4   Speaker Recognition: Reality and Challenge

The goal of speaker recognition is to verify an individual's identity based on his or her voice. Because voice is one of the most natural forms of communication, identifying people by voice has drawn the attention of lawyers, judges, investigators, and law enforcement agencies [355]. The recent proliferation of home banking has also opened up business opportunities for vendors marketing speaker verification products. The following list summarizes some of the potential applications of speaker recognition.[3]

- *Securing online transactions.* Financial institutions and banks can use speaker verification to enhance their e-banking and phone banking services. Customers' voices can be used together with passwords to verify the identity of individuals before transactions take place. For example, in 1999, speaker verification technologies were used to enhance the user-friendliness and security of BACOB's phone banking systems [144]. Customers register with the system

---

[3]For more potential applications and a list of vendors marketing speaker recognition products, visit *http://www.biometrics.org*.

by uttering three short, random texts or passwords. Before a transaction, the system prompts the customers to utter one of these passwords.

- *Securing critical medical records.* Speaker verification offers a means of verifying the identity of an individual who needs to access his or her own medical records via phone or Internet. Medical personnel can also use this technology to authenticate themselves before accessing the medical records of patients.

- *Preventing benefit fraud.* Speaker verification can be used by governments to track individuals claiming benefits. If the voices of social benefit recipients are stored in a database, any fraudulent attempts to claim benefits twice can be detected.

- *Reset passwords.* A high proportion of phone calls to help desks are requests for resetting passwords. Speaker verification can help automate the password reset process.

- *Voice indexing.* Speaker verification can be applied to create indexes for broadcast news. Given hours of news recordings containing the speech of several news reporters, it is possible to use a small part of the recordings to build a speaker model for each reporter. Once speaker models have been created, the time intervals during which a particular reporter is speaking can be spotted automatically.

Voice Biometrics

To better understand what current technologies can offer, this section examines the results of a recent NIST speaker recognition evaluation [288]. In this evaluation, a system trained on a two-minute cellular phone conversation for each target speaker achieved a false alarm rate (FAR) of 5% and a miss rate (FRR) of 10% given test segments (also cellular phone conversations) of 26 to 35 seconds. It was also reported that handset mismatch plays an important role in degrading performance. For example, in Przybocki and Martin [288], it was reported that at a 5% miss rate, the FAR could increase from 1.5% to 10% when enrollment and verification sessions use different handsets.

More recently, it was found that fusing low-level spectral features with high-level speaker information, such as idiolectal and prosodic information, can dramatically reduce error rates. For example, a recent report shows that with eight conversations (2.5 minutes each) for training a speaker model, and 2.5 minutes of speech for each verification session, the equal error rate can be reduced from 0.7% to 0.22% [42]. Compared to systems that use low-level features only, this represents a dramatic 66% reduction in the equal error rate. While high-level features can significantly improve speaker verification performance, they require long utterances to be effective. This may limit their applicability.

Commercial products have also been evaluated. For example, in 2000, the Centre for Communication Interface Research at the University of Edinburgh performed a large-scale evaluation of the *Nuance Verifier* [143]. The evaluation involved 1,000 participants making calls via the U.K. phone network to simulate phone banking services. The results show that with test utterances consisting of 19 digits, the Nuance Verifier achieved an equal error rate of 0.9%.

Although considerable progress has been made during the last decade, there are still many unsolved problems that prevent voice biometrics from appearing everywhere. In particular, variations in speakers' voices over time could considerably affect system performance (e.g., as a result of changes due to aging [108, 248]). Another challenge is that users may use different devices (e.g., mobile phones, fixed-line handsets, speakerphones) for accessing a system. As different transducers introduce different degrees of distortion to speech signals, it is very difficult to compensate for their effect on speaker characteristics. The increasing popularity of mobile devices introduces another problem—coder distortion. For instance, if a person uses a carbon button handset over a wired network for enrollment and later uses an electret mobile handset over the cellular network for verification, the combination of handset and coder difference is likely to make the system classify him or her as an impostor. Finally, many speakers can alter their voice voluntarily. This ability enables impersonators to attack speaker verification systems. Because of these challenges, it is not surprising to see the following conclusion, which was presented at a recent conference on speech [31]:

> Despite the existence of technological solutions to some constrained applications, at the present time, there is no scientific process that enables one to uniquely characterize a person's voice or to identify with absolute certainty an individual from his or her voice.

1.5   Road Map of the Book

The organization of this book is displayed in Figure 1.2.

1. **Part I** (Chapters 1 and 2): Overview of biometric authentication systems.

2. **Part II** (Chapters 3, 4, and 5): Machine learning models serving as the theoretical pillars of the book.

3. **Part III** (Chapters 6 and 7): Flexible structural frameworks based on hierarchical and modular neural networks, under which machine learning modules can be incorporated as a subsystem.

4. **Part IV** (Chapters 8, 9, and 10): Issues about how to use machine learning technologies to facilitate implementation of practical biometric authentication systems.

Part I provides an overview of state-of-the-art biometric authentication applications, including face and speaker recognition. Chapter 2 contains an overview of the design and system requirements pertaining to biometric authentication systems. It also presents a general pattern classification system in which feature extraction and adaptive classifiers play an important role.

Part II establishes the theoretical foundation for the machine learning techniques advocated in this book. To facilitate development of effective biometric authentication systems, several modern machine learning models are instrumental in handling complex pattern recognition and classification problems. Chapter 3 discusses the expectation-maximization (EM) algorithm useful for pattern representation and classification in an unsupervised training environment. In general, machine learning becomes much more effective under a supervised learning framework. Chapters 4 and 5 cover key learning strategies, taking advantage of a teacher's guidance during the training phase. Note, however, that if a model is overtrained for the sake of achieving nearly perfect training accuracy, it could compromise generalization accuracy. For this reason, the tradeoff between training and generalization accuracies is a focus of this research. The fundamental theory of Fisher's linear discriminant analysis (LDA) and support vector machines (SVM) is presented in Chapter 4. Comprehensive coverage of multi-layer learning models and well-known back-propagation (BP) algorithms is provided in Chapter 5.

Machine learning models can be naturally integrated into hierarchical and modular neural networks to provide a unified framework for biometric authentication. Part III presents several flexible network structures. Chapter 6 elaborates a flexible hierarchical information processing structure comprising many expert- and class-based modules. Prominent expert-based modular networks, such as the mixture-of-experts (MOE) and the hierarchical mixture-of-experts (HME), are presented. Pattern classification and biometric authentication are intimately related, therefore, it is sometimes necessary to adopt interclass learning strategies based on class-based modular networks. In Chapter 7, a probabilistic decision-based neural network (DBNN) is developed based on this design principle.

In Part IV, the learning networks are tailored to special information processing systems for biometric authentication and data fusion. The most important authentication application domains are face recognition and speaker verification. Chapter 8 presents probabilistic neural networks for face-based biometric authentication while Chapter 9 covers voice-based authentication. These two chapters also explore the issue of robustness in the design of practical face or voice recognition systems. Although single modality systems still prevail in current authentication systems, multicue biometric authentication will play a vital role in future security systems. The hierarchical learning model facilitates the adoption of soft-decision strategies, which is vital to several of the multicue data-fusion techniques addressed in Chapter 10.

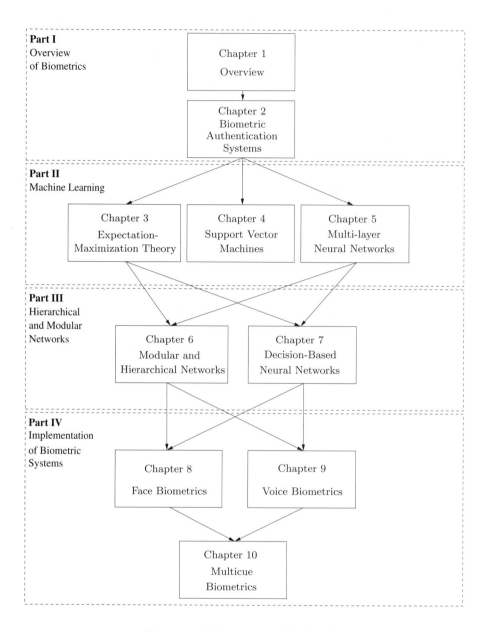

Figure 1.2. Road map of this book.

Chapter 2

BIOMETRIC AUTHENTICATION SYSTEMS

2.1 Introduction

Advances in various information processing technologies, including VLSI devices, digital computers, software applications, and embedded systems have revolutionized the paradigm of information processing. In particular, it has catalyzed rapidly increasing interest in the development of commercial systems for biometric authentication applications. The objective of a commercial system is to satisfy security requirements while incurring minimal implementation cost. This chapter discusses common system deployment requirements as well as critical design tradeoffs pertaining to biometric authentication systems. A general pattern classification system is comprised of feature extraction and adaptive classification. It is vital to acquire a thorough understanding of the input *feature space* before one can properly map such feature space onto the *expert space* and eventually onto the output *classification space*. Therefore, important points in selecting a training strategy are discussed. Finally, several prominent feature extraction algorithms for visual- and audio-based biometric systems are covered.

2.2 Design Tradeoffs

To evaluate a biometric system's accuracy, the most commonly adopted metrics are the *false rejection rate* and the *false acceptance rate*, which respectively correspond to two other popular metrics: *sensitivity* and *specificity*.

1. False rejection rate (FRR), or miss probability, is the percentage of authorized individuals rejected by the system, while *sensitivity*, also known as true positive rate (TPR), is the percentage that an authorized person is admitted. Therefore,

$$\text{FRR} = 1 - \text{Sensitivity} = 1 - \text{TPR}.$$

2. False acceptance rate (FAR), also known as false positive rate (FPR) or impostor pass rate, is the percentage that unauthorized individuals are accepted by the system. On the other hand, *specificity*, also known as the true negative rate (TNR), is the percentage that an unauthorized person is correctly rejected. Therefore,

$$\text{FAR} = \text{FPR} = 1 - \text{Specificity}.$$

A good identification system should have both low FRR (i.e., high sensitivity) and low FAR (i.e., high specificity). Typically, the tradeoff is illustrated by so-called receiver operation characteristic (ROC) curves or by the detection error tradeoff (DET) curves [233] in which FAR is plotted against FRR by varying the threshold. Tradeoff between FAR and FRR is often adjusted by a simple threshold, which generally needs to be adjusted carefully so that the two rates can both satisfy the prescribed security standards. Figure 2.1 shows examples of ROC and DET curves.

2.2.1 Accuracy versus Intrusiveness

While many human biometric characteristics are suitable for identification purposes, different characteristics yield different accuracy rates. Physiological characteristics (e.g., face and fingerprint) generally provide higher recognition accuracy than behavioral features (e.g., voice and signature). While behavioral characteristics can change from day to day because of stress, illness, or mood, physiological characteristics almost always remain unchanged, unless a serious injury occurs. Studies have shown that the average error rate of behavioral biometric methods is 10 to 100 times higher than that of physiological methods [245, 339].

Recall that two other popular metrics for the accuracy of a biometric system are the false rejection rate and the false acceptance rate. A list of error rates (FRR + FAR) for various biometric methods has been created after extensive literature investigation (e.g., [52, 198, 245, 278]); it is shown in Table 2.1. Notice that physiological methods are better, in general, than behavioral methods.

Intrusiveness is another important factor. If a security system makes users feel uncomfortable, either psychologically or physically, then the system is intrusive. In general, a nonintrusive method is more user-friendly and therefore better. For example, in computer network security or access control for areas requiring middle or low security levels (e.g., apartments, hospitals, stores), an intrusive system will annoy users and therefore will discourage them from using it. However, in high security areas, an intrusive system sometimes can turn out to be a benefit, since it may appear to be a highly secure recognition method. This elevated sense of security may in itself discourage intruders.

In biometric methods, a tradeoff exists between recognition accuracy and intrusiveness. Table 2.1 depicts the tradeoffs among various biometric methods. The "intrusiveness" column mainly describes *psychological* intrusiveness (e.g., privacy violation) and the "convenience" column reflects *physical* intrusiveness (e.g., adjusting position to fit within the sensor range). Among the physiological methods,

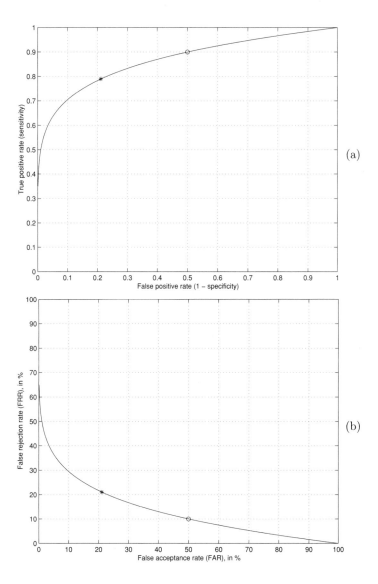

Figure 2.1. (a) ROC and (b) DET graphs illustrating the performance of a biometric recognition system. Both curves show the same information on how the decision threshold affects error rates. Two operating conditions are marked: The asterisk (∗) shows where the FAR equals the FRR (approximately 20% for this biometric system); the circle symbol (∘) marks the condition where the FRR is 10% and FAR is 50%. Lowering the decision threshold will move the operating point toward the right of both curves, which means the system will more probably return a positive (acceptance) response on both true positive and false positive cases.

Table 2.1. Accuracy versus intrusiveness of various biometric methods

	Intrusiveness	**Convenience**	**Error Rate**
Face	No	Good	$10^{-1} \sim 10^{-3}$
Palm	No?	Middle	$< 10^{-3}$
Fingerprint	Yes	Middle	$10^{-2} \sim 10^{-6}$
Iris/Retina	Yes	Very bad	$< 10^{-6}$
Voice	No	Middle	$10^{-1} \sim 10^{-2}$
Signature	No	Bad	$10^{-1} \sim 10^{-3}$

face recognition is the most nonintrusive, but unfortunately to this point it is also the least accurate. At the other end of the spectrum, the iris recognition method provides the highest accuracy (10^{-6} error rate), but it requires that users compromise on convenience.

2.2.2 Recognition versus Verification

The first issue to consider is which means of authentication to use: a recognition system or a verification system. A recognition system captures a biometric sample of an individual then searches a database for an acceptable match. In contrast, a verification system requests the user to claim his or her identity through an *artificial ID* (e.g., magnetic card or password) before the system records the user's biometric sample. The verification system then confirms the claim by comparing the sample with that person's database record.

From the user's point of view, recognition systems are more convenient, but they generally require more computational resources and have higher error rates. On the other hand, due to the assistance of artificial IDs, verification systems require considerably less computational resources but the FRR may increase slightly. This is because the combined false rejection rate for a system that uses both artificial IDs and biometric IDs is

Combined FRR = FRR of artificial ID + FRR of biometric ID.

That is, the system could reject a registered user based either on his or her artificial ID or biometric ID. On the other hand, the combined FAR can be greatly reduced with artificial identities. This is because the combined FAR is

Combined FAR = FAR of artificial ID × FAR of biometric ID.

That is, the system will accept an impostor only if his or her artificial ID and biometric ID are valid. Therefore, requiring an artificial ID can minimize casual attacks to the biometric verification system because random claims can often be rejected as unknown to the database. Moreover, a claim-based verification system requires the impostor to make a conscious false claim and suggest his or her intentions. This artificial claim can be an aid in subsequent investigation and prosecution.

2.2.3   Centralized versus Distributed Systems

Biometric devices have three primary components. The first is an *automatic sensor*, which captures a sequence of samples (e.g., sounds or images) from a person. Second is a *pattern matcher*, which handles compression, processing, storage, and comparison of the captured sample with the stored data. The third is a *central controller* configuring sensor operation and pattern matcher, which can also interface with other application systems. These pieces can be configured in various ways to suit different situations.

There are two types of architectural designs for such biometric systems: *centralized* and *distributed* [84]. Figure 2.2 shows the architecture of a centralized biometric access-control system. During operation, the user approaches a biometric sensor, claims an identity by inserting a card or entering a number, and provides a biometric sample either passively (e.g., waiting for the sensor to take a picture of his or her face) or actively (e.g., placing a finger on a sensor window). The biometric sample and identity claim are then uploaded to the central matcher where the sample is compared to the enrolled templates stored in the database. The central controller then logs the matching result and issues a necessary action back to the security checkpoint. The architecture of a distributed biometric system is depicted in Figure 2.3. As before, a user approaches a biometric sensor, claims an identity, and provides a biometric sample. In the distributed system each checkpoint has a local matcher and a local template database that stores frequently accessed templates. Matching is performed at the checkpoint and an action is taken. Only a logging entry is passed to a central log. A central controller is used to control system operation, adjust distribution of the template database, and continuously monitor system integrity by polling and testing.

Centralized and distributed systems have both advantages and disadvantages; Table 2.2 compares them. Centralized systems tend to consume higher communication bandwidth—the biometric samples must be transferred to the central matcher and the corresponding commands need to be dispatched back to local checkpoints. Distributed systems, on the other hand, duplicate the matcher and database at each checkpoint. Because failures can be isolated to local ports, the distributed systems in general are more robust. The disadvantage of distributed systems is that maintenance and backup are more complex than for centralized systems. It is also more complex to implement individual authority and status changes, since such changes must be propagated to other distributed ports. In short, while the centralized architecture has the advantage of easy maintenance, the distributed architecture has the advantage of lower communication loading and lower risk of systemwide failure. Because network communication and system strength directly affect a system's online performance, increasing numbers of biometric systems have adopted the distributed architecture.

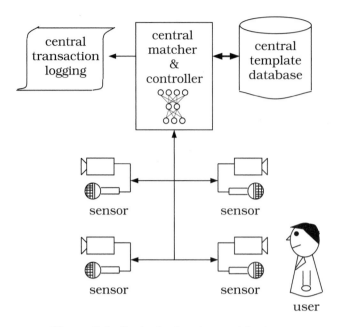

Figure 2.2. Centralized system architecture.

Table 2.2. Distributed biometric system versus centralized biometric system

Distributed System	Centralized System
Less communication loading	More communication loading
Less prone to systemwide failure	Greater risk of systemwide failure
Backup and maintenance are more complex	Easier to supervise and manually override

2.2.4 Processing Speed

Identification time is another important factor in selecting biometric identification (BI) algorithms. If a biometric gateway control system needs one hour to process one entry request, it is useless no matter how accurate it is.

Suppose that the BI algorithm has been chosen. Further suppose that the algorithm has generated a feature extractor F and a pattern classifier C such that F accepts raw data $\mathbf{x}$ as input and produces feature vector $\mathbf{y}$ as output, and that C accepts feature vector $\mathbf{y}$ as input and produces a class ID as output. Let $t(\varphi)$ represent the computation time spent on processing the function φ. Define T as the time needed for the BI model to identify an input pattern $\mathbf{x}$:

$$T = t(F(\mathbf{x})) + t(C(\mathbf{y})). \tag{2.2.1}$$

If the components in $F(\mathbf{x})$ and $C(\mathbf{y})$ cannot be computed in parallel, and if the

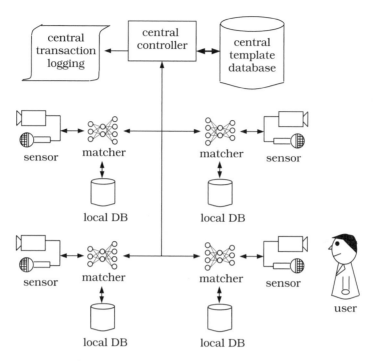

Figure 2.3. Distributed system architecture.

processing time of each component φ_i in function φ is similar, then

$$T \simeq M \cdot t(F_i(\mathbf{x})) + P \cdot t(C_i(\mathbf{y})), \tag{2.2.2}$$

where M is the feature vector dimension and P is the number of processing units in the classifier. The identification time T should be as small as possible. If Eq. 2.2.2 is the time model, then reducing M and P will reduce T.

An example of how processing speed affects the selection of biometric algorithms can be found in fingerprint identification applications. In a fingerprint picture, there are 18 types of fingerprint features (the "minutiae"). It has been found that if all the minutiae are used, the identification error rate could be as low as 10^{-10} [135]. However, to increase processing speed, almost all of the fingerprint feature extraction algorithms sacrifice accuracy (error rate drops to the range of 10^{-2} to 10^{-6}) by detecting only two types of minutiae—ridge ending and ridge bifurcation [198, 295].

Notice in the preceding discussion that it is assumed the chosen BI algorithm can generate the feature extractor and pattern classifier *before* the identification process. Some neural network algorithms, such as multi-layer perceptrons (MLPs) and decision-based neural networks (DBNNs), meet this assumption because they learn the values of the network parameters before the identification phase and those weighting parameters are fixed during the identification phase. If the BI algorithm cannot generate the feature extractor and pattern classifier until the pattern to be identified is available, the identification time will be longer than the time shown in Eq. 2.2.1. Examples of such algorithms are dynamic link architecture and elastic matching algorithms, both of which have been applied to the face recognition problem.

2.2.5 Data Storage Requirements

In addition to the task of transforming original data into a more separable space, the feature extractor serves the purpose of data compression. In most scenarios the size of the raw data is too large for a biometric system to store. Therefore, the system needs to choose a feature extraction algorithm to compress the original pattern into a feature vector with much smaller dimensions. In other words, when selecting a proper BI algorithm, the dimensions of the feature vector M should be as small as possible. For example, in a face recognition problem, Pentland et al. [272] use the eigenspace method to compress a 256×256 facial image to a 20-dimensional feature vector. This 3200:1 compression ratio greatly saves on memory space. In speaker recognition, time-domain speech signals are partitioned into a number of overlapping frames. These frames are transformed to a frequency domain where 12-dimensional spectral vectors (mel-frequency cepstral coefficients) are extracted (see Section 2.6). Typically, each frame contains 30ms of speech, and consecutive frames overlap by 15ms. For speech signals sampled at 8kHz, 16 bits per samples, the conversion to 32-bit floating point spectral vectors represents a compression ratio of $8,000 \times 16 \times 0.015 : 12 \times 32 = 5:1$.

Application types and storage requirements also dictate the biometric system architecture. For example, in an e-banking system, where the number of users is typically large and users can access the system at different locations (office, home, hotel, etc.), it is not desirable to store users' templates locally. Instead, a central database that stores all users' templates is more practical and desirable. On the other hand, mobile phones and PDAs are more personal, so it is practical to store the user's template locally on these devices.

Before implementing a biometric system, it is of vital importance to estimate storage requirements. For example, an e-banking system that augments speaker verification technologies with users' passwords may need to store an M-center Gaussian mixture speaker model for each user (see Chapter 9). More precisely, the number of bytes required to store an M-center GMM is

$$N = N_\mu + N_\sigma + N_\omega,$$

where $N_\mu = M \times 12 \times 4$, $N_\sigma = M \times 12 \times 4$, and $N_\omega = M \times 4$ are the number of bytes for mean vectors, diagonal covariance matrices, and the mixture weights, respectively. For $M = 128$, the storage requirement is 12.5KBytes per user. For a system with one million users, this translates to 12GBytes for the storage of users' models. Note that the assumption here is that the system uses 12-dimensional spectral features and 4 bytes per floating point number. For systems that also use dynamic features, such as delta cepstra, storage requirements will be doubled.

2.2.6   Compatibility between Feature Extractor and Classifier

A total recognition system involves mappings between the following spaces.

- *Instantiation space:* During the instantiation process, a symbol is instantiated into a physical object. The instantiation space contains all actual occurrences of objects. Typically, each symbol (e.g., a character "a", "b", or "c" in the character-recognition application) can have different instantiations.

- *Feature space:* In the feature space, the object is described in terms of a set of primitives (features). The mapping from instantiation space to feature space is called *feature extraction*. Moreover, this mapping represents a data-compression stage. In fact, the power of adaptive classifiers hinges on this representation. This is one of the most essential and/or difficult stages in pattern classification.

- *Symbol space:* Finally, this space contains the symbols representing classes of objects. The mapping from feature space to symbol space is called classification.

As depicted in Figure 2.4, a total recognition system involves two subsystems: a feature extractor and a classifier. To holistically analyze a total system for neural information processing, it is important to clearly identify the role of each of its subsystems.

- *Feature extraction:* The performance of any recognition system hinges on an effective representation of the raw data. In the feature space, an object is described in terms of a set of primitives (features). Feature extraction is represented by a mapping from instantiation space $\mathbf{x}$ to feature space $\mathbf{f}$:

$$\mathbf{x} \rightarrow \mathbf{f}(\mathbf{x}).$$

  Feature representation and extraction are indispensable processes to any adaptive information processing system, dictating the ultimate performance of machine learning and classification systems.

- *Classification:* As noted before, the mapping from feature space to symbol space (representing the classes of objects) is called *classification*. Take an example of a two-class classification: a feature vector pattern $\mathbf{f}$ may belong to either a "positive" class C_+ or a "negative" class C_-. A classifier is characterized by its discriminant function $\varphi_C(\mathbf{f})$ whose sign dictates the final classification result. A perfect classifier is the one whose discriminant function yields correct signs—that is, $\varphi_C(\mathbf{f}) \geq 0$ (resp. $\varphi_C(\mathbf{f}) < 0$) if the feature vector $\mathbf{f}$ is extracted from a pattern originally belonging to C_+ (resp. C_-).

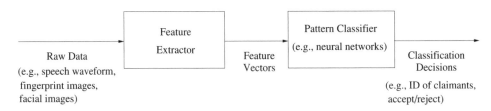

Figure 2.4. A neural processing and retrieving system consists of two subsystems: feature extraction and a neural net.

From a practical point of view, feature extraction and classification algorithms must be tailored to the intended biometric application domain (e.g., face, voice, fingerprint, iris, etc.). The selection of classification algorithms must be made compatible with the corresponding feature extraction algorithms, and vice versa. A feature extractor F and a pattern classifier C are said to be *compatible* if the data distribution for the intended application is φ-separable with respect to the total system function $\varphi_C(\mathbf{f}(\mathbf{x}))$—that is, $\varphi_C(\mathbf{f}(\mathbf{x})) \geq 0$ (resp. $\varphi_C(\mathbf{f}(\mathbf{x})) < 0$) if the input vector $\mathbf{x}$ belongs to C_+ (resp. C_-). They play a complementary role to each other. If the extracted features are not easy to separate, a more powerful pattern classification algorithm becomes imperative. On the other hand, if the designer prefers a simpler classifier, then it is necessary to adopt a more powerful feature extraction algorithm. Therefore, to design an effective biometric recognition system, one needs not only to consider the individual performance of the feature extraction algorithm but also the pattern classification algorithm. The compatibility between the two is, in effect, even more critical to the total system performance.

2.3   Feature Extraction

The power of a recognition system lies in the representation of pattern vectors because it is essential that the representation provide concise, invariant, and/or intelligible information on input patterns. Conversely, the applications intended also dictate the choice of representation. For example, in natural visual systems, it is known that images are preprocessed before being transmitted to the cortex [73]. Similarly, in image and vision analysis, raw image data must be preprocessed to extract vital characteristics (e.g., characteristics that are less dependent on imaging geometry and environment). As another example, in speech signal analysis, there is a wide variation in data rates, from 75bps for parametric text representation to 200,000bps for simple waveform representation [98].

2.3.1   Criteria of Feature Extraction

It is not easy to pinpoint an optimal representation strategy. The following offers some general criteria for the evaluation of feature representations.

   1. *Data compression.* Only vital representations or features are extracted.

2. *Informativeness.* The characteristics embedded in the raw data most essential for the intended applications should be best described.

3. *Invariance.* The dependency of the features on environmental conditions should be minimized.

4. *Ease of processing.* A cost-effective hardware and software implementation should be feasible.

Other factors, such as overall cost, data rates, flexibility of representation, and quality of signals, should also be taken into account. In particular, data compression is an essential process in biometric identification. Depending on the data model, the following two approaches are often adopted to obtain highly compressed representations: (1) dimension reduction by projection onto vector subspace and (2) data clustering to simplify data representation. The subsequent section describes several dimension reduction methodologies. Comprehensive coverage of the data clustering strategy can be found in Chapter 3.

2.3.2 Projection Methods for Dimension Reduction

For convenience of analysis, high-dimensional input vectors are often first reduced into a set of linear combinations of the original raw vector. The reduced dimensionality of the subspace would facilitate many complex analysis tasks. There are two categories of subspace projection methods for dimensionality reduction: unsupervised and supervised.

1. *Unsupervised subspace projection.* Unsupervised projection methods have been widely used in many applications in which teacher values are not readily available. Data patterns can be projected into low-dimensional subspace according to either (1) their second-order moment or (2) their cluster separability. The former leads to the principal component analysis (PCA) approach while the latter leads to the independent component analysis (ICA) approach. PCA and ICA are arguably the most popular unsupervised subspace methods used in biometric applications [77, 159].

2. *Supervised subspace projection.* For many classification applications, including face and speaker recognition, the training data consist of pairs of input/output patterns. In this case, it is more advantageous to adopt a supervised learning strategy to help determine the optimal projection. When class information is known, Fisher's discriminant criterion is commonly used to measure interclass separability. This leads to the so-called linear discriminant analysis (LDA) discussed in Section 4.2. Another supervised projection method very much worth pursuing is the well-known support vector machine (SVM), which is covered in detail in Chapter 4.

Principal Component Analysis

In statistics, a popular approach of extracting representation with maximum information is the *principal component analysis* (PCA). To retain the maximum amount

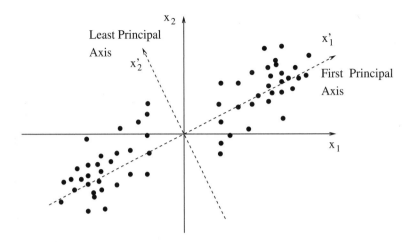

Figure 2.5. The x_1' axis represents the principal component for the pattern space.

of relevant information, the principal components (1) extract the most significant features that can best manifest the original patterns, and at the same time (2) avoid duplication or redundancy among the representations used.

Given an n-dimensional zero-mean wide-sense stationary vector process $\mathbf{x}(t) = [x_1(t), x_2(t), \cdots, x_n(t)]^T$, then its autocorrelation matrix can be derived as

$$\mathbf{R}_x = E\{\mathbf{x}(t)\mathbf{x}^T(t)\}$$

if the probability distribution function of $\{\mathbf{x}(t)\}$ is known a priori. The normalized eigenvectors of autocorrelation matrix $\mathbf{R}_x$ are $\{\mathbf{e}_i\}$, $\|\mathbf{e}_i\| = 1$, corresponding to the eigenvalues $\{\lambda_i\}$, where $\lambda_1 \geq \lambda_2 \geq \cdots \lambda_n \geq 0$. An eigen-component of a stochastic process $\mathbf{x}(t)$, $\mathbf{e}_i^T \mathbf{x}(t)$, is its projection onto the one-dimensional subspace spanned by the eigenvector $\mathbf{e}_i$. The principal component of a signal is the first (and the largest) eigen-component

$$x_1'(t) = \mathbf{e}_1^T \mathbf{x}(t).$$

One such example is illustrated in Figure 2.5—the original pattern space is two-dimensional ($n = 2$) and the two principal axes for the pattern space are shown. Note that the first principal axis x_1' captures most of the distribution variance of the pattern space.

Given an n-dimensional vector process $\mathbf{x}(t) = [x_1(t), x_2(t), \cdots, x_n(t)]^T$, then any m-dimensional representation vector $\mathbf{x}'(t) = [x_1'(t), x_2'(t), \cdots, x_m'(t)]^T$ can be expressed as

$$\mathbf{x}'(t) = \mathbf{W}\mathbf{x}(t),$$

where $\mathbf{W}$ is an $m \times n$ matrix. Mathematically, the PCA is to find a matrix $\mathbf{W}$ such that an optimal estimate $\hat{\mathbf{x}}(t)$ of $\mathbf{x}(t)$ can be reconstructed from $\mathbf{x}'(t)$ in terms of the mean-squared error. The solution $\mathbf{W}$ is formed by the first m eigenvector

$\mathbf{e}_i, i = 1, 2, \ldots, m$. In other words, the first m principal components are the eigen-components corresponding to the largest m eigenvalues: λ_i, $i = 1, 2, \ldots, m$.

The PCA techniques have found many useful applications in signal and image filtering, restoration, pattern classification, and recognition. Specific application examples include biomedical signal processing, speech and image processing, antenna applications, seismic signal processing, geophysical exploration, data compression, high-resolution beam forming, separation of a desired signal from interfering waveforms (e.g., fetal ECGs from maternal ECGs), and noise cancelation [77, 117, 163, 187, 266]. It is worth mentioning that there are already widely available batch-processing methods—with very mature software developments (e.g., LIN-PACK and EISPACK)—for PCA or SVD computations.

Independent Component Analysis

In contrast to PCA, independent component analysis (ICA) extracts components with higher-order statistical independence. A common objective of PCA and ICA is to extract, say, m most representative *linear* features. The main difference lies in their selection criteria (i.e., PCA finds m uncorrelated expression profiles, while ICA finds m statistically independent expression profiles). Note that statistical independence is a stronger condition than uncorrelatedness [61, 159].

The *kurtosis* of a process y is defined as

$$k(y) = \frac{E[y^4]}{E[y^2]^2}.$$

Figure 2.6 depicts the density functions and kurtosis values of three exemplifying processes: (a) a Gaussian random process has a kurtosis value of 3, (b) a uniformly distributed random process has a kurtosis value of 1.8, and (c) a binary-valued random process has a kurtosis value of 1. Note that the lower the kurtosis value, the more separable the data clusters become.

As noted before, the PCA and ICA use very different training criteria. The PCA maximizes the second-order covariance, while ICA minimizes the fourth-order kurtosis. A potentially advantageous feature of using ICA instead of PCA is that the kurtosis function $k(y)$ is scale-invariant. (See Problem 1.)

The problem of finding the single most discriminative independent component can be mathematically formulated as follows: find a vector $\mathbf{w}$ such that $y(t) = \mathbf{w}^T \mathbf{x}(t)$ yields the minimum kurtosis:

$$\min_{\mathbf{w}} \frac{E[\,|\mathbf{w}^T \mathbf{x}(t)|^4]}{E[\,|\mathbf{w}^T \mathbf{x}(t)|^2]^2}. \tag{2.3.1}$$

Similarly, the full ICA problem is to find m linearly combined outputs, represented by a vector process $\mathbf{y}(t) = \mathbf{W}\mathbf{x}(t)$ so that $\mathbf{y}(t)$ extracts the m most discriminative independent components. Here $\mathbf{W}$ is an $m \times n$ matrix formed by m independent row vectors $\mathbf{w}_i$, $i = 1, 2, \ldots, m$, which can be extracted in a sequential manner—assuming that $\mathbf{w}_i$, $i = 1, 2, \ldots, m - 1$ are already extracted and one is

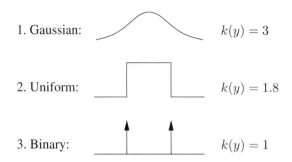

1. Gaussian: $k(y) = 3$

2. Uniform: $k(y) = 1.8$

3. Binary: $k(y) = 1$

Figure 2.6. The density functions and corresponding kurtosis values for three random processes.

set to extract the m-th component. Mathematically, the vector $\mathbf{w}_m$ can be found as the vector such that $y_m(t) = \mathbf{w}_m^T \mathbf{x}(t)$ yields the minimum kurtosis under the orthogonality constraints that $\mathbf{w}_m^T \mathbf{w}_i = 0$, $i = 1, 2, \ldots, m-1$. This constrained extraction scheme can be efficiently implemented via a popular numerical procedure known as the deflation technique. Indeed, the deflation method was incorporated into the *KuicNet* [189] originally designed for sequentially extracting independent components. (See Problem 2.)

2.3.3   Feature Selection

Sometimes, only a few selected features of a high-dimensional vector process would suffice for an intended classification task. For example, among many stocks listed on Hong Kong's stock market, only 33 are selected to calculate the Hang Seng index. This is because they are the most representative of different sectors and industries in Hong Kong. Note that unlike dimension reduction, there is no linear combination of features in the feature selection process.

The mathematical framework established previously can be applied to the preselection of features. Again, there are unsupervised and supervised preselection schemes.

1. *Unsupervised preselection schemes.* Following the same principle behind PCA, the second-order moment for each feature x_i, $i = 1, \ldots, n$ reflects the extent of the pattern spread associated with that feature. The larger the second-order moment, the wider the spread, thus the more likely the feature x_i contains useful information.

Given an n-dimensional zero-mean wide-sense stationary vector process $\mathbf{x}(t) = [x_1(t), x_2(t), \cdots, x_n(t)]^T$, then the feature or features with the highest variance will be selected. Mathematically, the feature index can be identified as:

$$\arg\max_i E[\, |x_i(t)|^2 \,].$$

Similarly, following the ICA-type argument, the fourth-order moment for each feature, x_i, $i = 1, \ldots, n$, reflects how separable the data clusters are along that feature dimension. The lower the fourth-order moment, the more separable are the pattern clusters, thus the more likely it is that the feature contains useful information for the purpose of class discrimination.

Given an n-dimensional zero-mean wide-sense stationary vector process $\mathbf{x}(t) = [x_1(t), x_2(t), \cdots, x_n(t)]^T$, then the feature(s) with the lowest kurtosis value will be selected. Mathematically, its index can be expressed as:

$$\arg\min_i \frac{E[\,|x_i(t)|^4]}{E[\,|x_i(t)|^2]^2}.$$

2. *Supervised preselection schemes.* If class information is known, a preselection method based on the Fisher discriminant analysis becomes very attractive. (Recall that the Fisher discriminant offers a convenient metric to measure the interclass separability embedded in each feature.) Such a preselection approach entails computing Fisher's discriminant $J(x_i)$, $i = 1, \ldots, N$, which represents the ratio of intercluster distance to intracluster variance for each dimension:

$$J(x_i) = \frac{(\mu_{1i} - \mu_{2i})^2}{\sigma_{1i}^2 + \sigma_{2i}^2}, \qquad (2.3.2)$$

where μ_{1i} and μ_{2i} denote the means of the i-th feature x_i for data belonging to C_+ and C_-, respectively. Similarly, σ_{1i} and σ_{2i} are the variances of the feature x_i for data belonging to C_+ and C_-, respectively. The resulting values for each feature can then be used as a basis for feature selection. Ideally, the features selected will correspond to the index (or indices) i, which maximizes the contrast criterion $J(x_i)$. For an application example, see Zhang et al. [404].

2.3.4   Clustering Methods: Gaussian Mixture Models

From a clustering perspective, most biometric data cannot be adequately modeled by a single-cluster Gaussian model. However, they can often be accurately modeled via a Gaussian mixture model (GMM) (i.e., data distribution can be expressed as a mixture of multiple normal distributions).

To estimate the density parameters of a GMM statistic model, several prominent cluster estimation methods (e.g., the K-means or EM algorithms) can be adopted (see Chapter 3). In addition, the clustering task also requires the determination of an optimal number of clusters for the GMM. Such an optimal cluster number is often derived from well-known information theoretic metrics such as AIC or MDL—see [5, 313, 314]. The number of clusters is also tightly coupled with the cluster mass. The larger the number of clusters, the smaller the average cluster mass. From a practical perspective, the choice of cluster mass (and thus the number of clusters) depends heavily on the desired classification objectives—for example, whether the emphasis is placed on training accuracy or generalized performance.

Cluster-Then-Project versus Project-Then-Cluster

For high-dimensional or more complex data patterns, it is also advisable to pursue more sophisticated training strategies such as (1) *cluster-then-project* or (2) *project-then-cluster* methods.

As an example of the cluster-then-project approach, a projection aimed at separating two known classes, each modeled by a GMM, is proposed in Zhang et al. [404]. The proposed scheme makes use of both class membership information and intraclass cluster structure to determine the most desired projection for the purpose of discriminating the two classes.

An example of the project-then-cluster approach is shown in the hierarchical clustering strategy depicted in Figure 2.7 [367, 370]. The following are three subtasks.

1. Project the data set from t-space onto a reduced-dimensional x-space, via say PCA.

2. Perform data clustering in the x-space.

3. Trace the membership information back to the t-space, and use the membership function as the initial condition and further fine-tune the GMM clustering by the EM algorithm in the t-space.

2.4 Adaptive Classifiers

If individual patterns are available, one can build a statistical model for each class of person. A common approach is to model each class by a normal density so that the system estimates the corresponding mean feature vector and covariance matrix for each person. Using a prior distribution of the individuals in the database, the classification task is completed by computing the Bayesian a posteriori probability of each person, conditioned on the observations of the query. If log probability is computed, the classification process can be considered a nearest-neighbor search using the *Mahalanobis* distance metric.

There are other statistical approaches to pattern classification. First, the K-*nearest-neighbor algorithm* [85] determines the class of a test pattern by comparing it with the K-nearest training patterns of known classes. The likelihood of each class is estimated by its relative frequency among the K-nearest training patterns. As training pattern size grows, these relative frequencies converge to the true posterior class probability. Second, the *Parzen windows* [270] estimate the class-conditional densities via a linear superposition of window functions—one for each training pattern. The window function is required to be unimodal and has a unit area under its curve. As training pattern size grows, the linear superposition of window functions for a given class converges to the true class-conditional density.

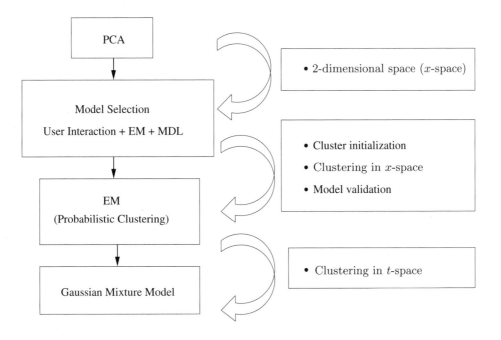

Figure 2.7. An illustration of the project-then-cluster approach. Projection of data from t-space to x-space, then after clustering in the lower-dimension subspace, trace the membership information back to the t-space.

2.4.1 Neural Networks

The ability to learn and adapt is a fundamental trait of any intelligent information processing system, including machine learning models and neural networks. Although the original development of neural networks was primarily motivated by the function of the human brain, modern neural models encompass many similarities with statistical pattern recognition. A neural network is an abstract simulation of a real nervous system that contains a collection of *neuron units* communicating with one other via *axon connections*.

The first fundamental neural model was proposed in 1943 by McCulloch and Pitts [237] in terms of a computational model of "nervous activity." The McCulloch-Pitts neuron is a binary device and each neuron has a fixed threshold, thus performing simple threshold logic. The McCulloch-Pitts model led to the work of John von Neumann, Marvin Minsky, Frank Rosenblatt, and many others. Hebb postulated in his classical book *The Organization of Behavior* [130] that neurons are appropriately interconnected by self-organization and that "an existing pathway strengthens the connections between the neurons." He proposed that the connectivity of the brain is continually changing as an organism learns various functional tasks and that cell assemblies are created by such changes [64]. By embedding a vast number of simple neurons in an interacting nervous system, it is possible to provide computational power for very sophisticated information processing [8].

Several promising neural networks for biometric identification applications are explored in Chapters 5, 6, and 7. Their application to face recognition, speaker verification, and joint audio-visual biometric authentication are investigated in Chapters 8, 9, and 10. In addition to biometric identification applications, neural networks have been applied to a variety of problems such as pattern classification, clustering, function approximation, prediction/forecasting, optimization, pattern completion, and control [128, 186].

2.4.2   Training Strategies

Instead of following a set of *rules* specified by human experts, neural networks appear to learn the underlying rules from a given collection of representative examples. The ability of neural networks to automatically *learn from examples* makes them attractive and, in general, is considered to be one of the major advantages of neural networks over traditional expert systems.

Based on the nature of the training data sets made available to the designers of biometric authentication systems, there are two common categories of learning schemes: supervised learning and unsupervised learning.

1. In supervised learning schemes, the neural network is provided with a correct answer (the "teacher value") for each input pattern. The weighting parameters are determined so that the network can produce answers as close as possible to the teacher values. In many classification applications (e.g., OCR and speaker recognition), the training data consist of pairs of input/output patterns. In this case, adopting a supervised network is advantageous.

2. In contrast, unsupervised learning schemes do not require that a correct answer be associated with each input pattern in the training data set. These schemes explore the underlying correlations between patterns in the data, and patterns are organized into categories according to their correlations. Unsupervised networks have been widely used in applications in which teacher values are not easy to obtain or generate, such as video coding and medical image segmentation.

2.4.3   Criteria on Classifiers

For biometric identification applications, key performance measurements of adaptive learning classifiers include the following.

- *Training and generalization performance.* Two popular metrics to measure the accuracy of a learning algorithm are training accuracy and generalization accuracy. For *training accuracy*, the test patterns are simply samples drawn from the original training patterns, namely memorization. In contrast, for *generalization accuracy*, test patterns are drawn from an independent data set instead of the original training pattern set. In short, the distinction between training and generalization accuracies lies in the test patterns adopted. There is a natural tradeoff between training and generalization accuracies. In other words, high training accuracy does not necessarily yield good generalization accuracy.

- *Invariance and noise resilience.* From the *invariance* perspective, any dependence on environmental conditions should be minimized. For example, in terms of face recognition, the classifier should be made invariant with respect to any geometric transformations, such as coordinate-shift and/or orientation. In addition, the system should be able to tolerate noise corruption of images or speeches because noise is basically inevitable in practical biometric applications.

- *Cost-effective system implementation.* It is important to provide low-cost, high-speed, flexible implementation for adaptive classifiers, thus a cost-effective hardware/software design platform should be considered. To fully harness the advanced VLSI/ULSI and system-on-chip technologies, a processor architecture with distributed and modular structure appears to be the most promising. Since compatibility between feature extractor and adaptive classifier could heavily influence total system performance, emphasis should also be placed on system integration design issues.

2.4.4 Availability of Training Samples

If the chosen biometric identification (BI) algorithms require learning or parameter estimation, then the *availability* of reliable and representative training data is also of critical concern. Training data size differs greatly from one application to another. If the number of training examples is too small or the number of φ functions (the hidden neurons) is too large, the BI algorithm can generate a BI model with spurious decision boundaries or poor interpolation. In other words, the resulting BI model may behave just like a lookup table constructed from the training patterns, and it may possess little *generalization* capability.

Many learning algorithms have mechanisms to control the number of hidden functions (i.e., M and P in Eq. 2.2.2) to prevent the degradation of generalization accuracy. An example of this mechanism is the regularization term in some neural network models [27]. However, for some applications (e.g., criminal identification), where training patterns are difficult to acquire (perhaps one or two fingerprint records or photos for each person), the performance of learning algorithms could be greatly degraded because the *training-by-examples* principle has been violated.

There are two possible solutions to this training example deficiency problem. The first is to perform an intense and thorough analytical study on the nature of the input biometric signals and on the noise models embedded in the data acquisition process. Once knowledge of the true data generative process and noise models becomes available, it will be possible to design a powerful, yet nonlearning feature extraction algorithm, the resulting feature vectors of which are easily separable by simple classifiers (e.g., nearest-neighbor rule). One example of this approach is used in fingerprint problems. For almost 100 years, it has been known that the relative positions between various minutiae are the discriminative features for identifying people [135]. Therefore, there is no need to use examples to tell the system which features should be extracted from sensory images. Most fingerprint researchers detect the locations and orientation of useful minutiae by designing filters using image processing techniques (e.g., edge enhancement and thinning) and by applying graph matching techniques to determine the similarity between the test pattern (a set of points) and the reference pattern. Both feature extraction and pattern classification algorithms can achieve a satisfactory identification rate without applying learning procedures.

Another solution to training data deficiency is to apply the *virtual pattern generation* procedure. To ensure sufficient diversity in the training set, the algorithm should take the acquired data from sensors and transform it to create additional training exemplars (i.e., *virtual training patterns*). The transformations should mimic the variations that might be embedded in the data acquisition process. One example of virtual training pattern generation can be seen in the face recognition experiments (see Chapter 8), where up to 200 virtual training patterns are generated from one original facial image by applying various affine transformations (e.g., rotation, scaling, shifting) and a mirroring process. Robustness of the trained network can be improved via the use of a virtual training set.

2.5   Visual-Based Feature Extraction and Pattern Classification

This section discusses several commonly used feature extraction algorithms for visual-based biometric systems. The most important task of a feature extractor is to extract the discriminant information that is *invariant* to as many variations embedded in the raw data (e.g., scaling, translation, rotation) as possible. Since various biometric methods have different invariant properties (e.g., minutiae positions in fingerprint methods, texture patterns in iris approaches), it is important for the system designer to understand the natural characteristics of the biometric signals and the noise models that could possibly be embedded in the data acquisition process.

Table 2.3 shows several useful features in image-based biometric identification problems. Two main classes of feature extraction techniques apply to the recognition of digital images: geometric feature extraction and template feature extraction. Geometric features are obtained by measuring the geometric characteristics of several feature points from sensory images (e.g., length of the nose, or the distance between two minutiae). Template features are extracted by applying global-level processing to the subimage of interest. The techniques are based on *template matching*, where a two-dimensional array of intensity values (or more complicated features extracted from it) is involved. Figure 2.8 shows a human face as an example of the

Table 2.3. Features used for image-based biometric identification. *Note:* The circles indicate the features that have been used in different BI problems. Various template features and their corresponding feature extraction algorithms are listed here too.

	Template					Geometric Features	Auxiliary Information
	Low-Frequency Component	High-Frequency Component	Principal Component	Texture	High-Level Features		
	Gaussian Pyramid	Laplacian Pyramid—edge filtering	KL SVD APLEX	Gabor Filters—Wavelets	NN		
Face	o	o	o	o	o	o	Color, motion, thermography
Palm		o			o	o	
Fingerprint	o	o		o	o	o	
Iris/Retina		o		o		o	
Signature	o	o				o	Pen movement, pen pressure

difference between these two feature extraction techniques. The nonimage-based biometric approaches (e.g., thermography [287], pen pressure and movement [278], and motion [29]) are not within the scope of this book.

2.5.1 Geometric Features

Geometric feature extraction techniques are widely used in many works of biometric identification. People measure geometric features either manually or automatically to verify identity. In an early work done by Kelly [178], 10 measurements are extracted manually for face recognition, including height, width of head, neck, shoulders, and hips. Kanade [173] uses an automatic approach to extract 16-dimensional vectors of facial features. Cox et al. [65] manually extract 30 facial features for a mixed-distance face recognition approach. Kaufman and Breeding [177] measure geometric features from face profile silhouettes. The ID3D Handkey system [10] automatically measures hand geometric features by an edge-detection technique. Gunn and Nixon [121] use an active contour model "snake" [174] to automatically extract the boundary of the human head. Most automatic fingerprint identification systems (AFIS) detect the location of minutiae (anomalies on a fingerprint map, such as ridge ending and ridge bifurcation) and measure their relative distances [56, 62, 198, 286, 295].

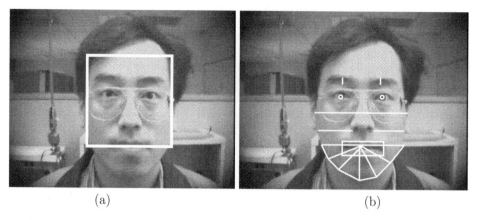

Figure 2.8. Possible template used for a face recognition problem. (a) Image in the white box; (b) geometric measurements indicated in white lines.

Feature point localization is a crucial step for geometric feature extraction. For example, two corners of an eye ("feature points") must be located before the width is measured. People have applied various techniques for automatic feature point localization because locating them manually is both tiresome and impractical. Automatic feature point localization techniques include the Hough transform [263], spatial filters [92, 299, 396], deformable template [126, 399], morphological filters [295, 399], Gabor wavelets [230], knowledge-based approaches [57], and neural networks [207, 323, 343]. The difficulty of feature point localization varies from application to application. For fingerprint approaches, the localizer is easier to design since all feature points are similar (most fingerprint approaches detect no more than two types of minutiae [295]). For face recognition methods, designing localizers is more complex because a much larger variation exists between various facial feature points (e.g., corners of the eye, tip of the nose, bottom of the chin).

One disadvantage to geometric feature extraction is that many feature points must be correctly localized to make a correct measurement. Compared to template-based approaches, which need only locate two to four feature points (to locate the template position) [211, 251], geometric methods often need to locate up to 100 feature points to extract meaningful geometric features.

2.5.2   Template Features

Template-based methods crop a particular subimage (the template) from the original sensory image, and extract features from the template by applying *global-level* processing, without a priori knowledge of the object's structural properties. Compared to geometric feature extraction algorithms, image template approaches need to locate far fewer points to obtain a correct template. For example, in the PDBNN face recognition system (cf. Chapter 8), only two points (left and right eyes) need to be located to extract a facial recognition template. Brunelli and Poggio [38] compared these two types of feature extraction methods on a face recognition problem. They found the template approach to be faster and able to generate more accurate recognition results than the geometric approach.

The template that is directly cropped out of the original image is not suitable for recognition without further processing, for the following reasons: (1) The feature dimensions of the original template are usually too high. A template of 320×240 pixels means a feature vector dimension of 76,800. It is impractical to use such a high-dimensional feature vector for recognition. (2) The original template can be easily corrupted by many types of variation. Table 2.3 lists five types of commonly used feature extraction methods for template-based approaches.

The most commonly used features are *high-frequency components* (or edge features). Although edge information is more prone to the influence of noise, it reflects more of the structural properties of an object. Many fingerprint approaches use directional edge filtering and thinning process to extract fingerprint structures [198, 295]. Rice analyzes the vein structures of human hands by extracting edge information from infrared images [310]. Wood et al. use a modified Laplacian of Gaussian ($\nabla^2 G$) operator to detect retina vessel structures [385].

Low-frequency components (or Gaussian pyramids) are good for removing high-frequency noise as well as for dimension reduction. Sung and Poggio use low-resolution images for face detection [343]. Rahardja et al. use the Gaussian pyramid structure to represent image data for their facial expression classification system [293]. Baldi uses low-pass filtering techniques as the first preprocessing step of his neural network fingerprint recognizer [16]. Golomb and Sejnowski [115] and Turk and Pentland [356] also use low-pass filtering and down-sampling as preprocessing steps for face recognition.

As shown in Table 2.3, high-frequency component features have been widely used for various biometric methods, including face recognition. In the face recognition system described in Lin et al. [211] (see also Chapter 8), low-frequency features were also adopted to assist in recognition.

2.5.3   Texture Features

Another useful feature for biometric identification is *texture information*. Wavelet methods are often used to extract texture information. For example, the FBI has adopted wavelet transform as the compression standard for fingerprint images [34]. Daugman uses Gabor wavelets to transform an iris pattern into a 256-byte "iris code" [67]. He claims the theoretical error rate could be as low as 10^{-6}. Lades et al. also use Gabor wavelets as features in their dynamic link architecture (DLA)—an elastic matching type of neural network—for face recognition [194]. To effectively represent the vast number of facial features and reduce computational complexity, in Zhang and Guo [401], wavelet transform was used to decompose facial images so that the lowest resolution subband coefficients can be used for facial representation.

2.5.4   Subspace Projection for Feature Reduction

Raw image data usually involve very large dimensions. It is therefore necessary to reduce feature dimensions to facilitate an efficient face detector or classifier. There are several prominent subspace methods for dimensionality reduction: (1) unsupervised methods such as principal component analysis and independent component analysis; and (2) supervised methods such as linear discriminant analysis and support vector machines.

If the diversity of input patterns is sufficient to show statistical distribution properties, principal component analysis can be a simple and effective approach to feature extraction. Pentland et al. assume that the set of all possible face patterns occupies a small and parameterized subspace (i.e., the eigenface space) derived from the original high-dimensional input image space [250, 272, 356]. Therefore, they apply the PCA to obtain the eigenfaces—the principal components in the eigenspace—for face recognition.

To reduce the computational burden, facial features are first extracted by the PCA method [87, 356]. With supervised training, the features can be further processed by Fisher's linear discriminant analysis technique to acquire lower-dimensional discriminant patterns. In Er et al. [87], a paradigm was proposed whereby data information is encapsulated in determining the structure and initial parameters of the classifier prior to the actual learning phase.

One of the purposes of feature selection is to support the creation of a face-class model. In Sadeghi et al. [327], a method for face-class modeling in the eigenfaces space using a large-margin classifier, such as SVM, is presented. Two main issues were addressed: (1) the required number of eigenfaces to achieve an effective classification rate and (2) how to train the SVM for effective generalization. Moreover, different strategies for choosing the dimensions of the PCA space and their effectiveness in face-class modeling were analyzed.

It is well known that distribution of facial images, with perceivable variations in viewpoint and illumination or facial expression, is highly nonlinear and complex. Thus, it was no surprise that linear subspace methods, including PCA and LDA, were found to be unable to provide reliable solutions to cope with the complexities of facial variations [217]. The inadequacy of linear reduction can be effectively compensated for by using the nonlinear dimension reduction schemes discussed in the following subsection.

2.5.5   Neural Networks for Feature Selection

Recently, increasing attention has been placed on the use of *neural networks* to extract information for classification. To obtain a nonlinear model, nonlinear hidden-layers are adopted in a multi-layer network. One advantage of neural network feature extractors is their learning ability. To more effectively capture the nonlinear relationship between the compression and output layers and to facilitate the nonlinear dimension reduction effect, Feraund et al. [95] suggested that an additional hidden-layer (with reduced dimensions) be incorporated into the multi-layer networks. The weighting parameters in a neural network feature extractor can be trained so that the network can extract features that contain the most discriminant information for the subsequent classifier.

In fact, most neural network feature extractors are connected to the neural network classifiers so that both modules can be trained together. Examples of neural network feature extractors are the Baldi and Yves convolutional neural network for fingerprint recognition [16]; the Weng, Huang, and Ahuja Cresceptron for face detection [376]; Rowley's convolutional neural network for face detection [323]; and Lades's dynamic link architecture for face recognition. Lawrence et al. [195] combine a self-organizing map [184] and convolutional neural networks for face recognition. More than a 96% recognition rate has been reported on the face database from Olivetti Research Laboratory [329]. The neural network feature extraction approach was also successfully applied to the face recognition system described in Lin et al. [211] (see also Chapter 8), where two neural network components (i.e., face detector and eye localizer) are adopted to locate and normalize the facial templates.

2.6   Audio-Based Feature Extraction and Pattern Classification

Audio-based feature extraction consists of parameterizing speech signals into a sequence of feature vectors, which are less redundant for statistical modeling. Although speech signals are nonstationary, their short-term segments can be considered to be stationary. This means that classical signal processing techniques, such as spectral and cepstral analysis, can be applied to short segments of speech on a frame-by-frame basis.

It is well known that the physiological and behavioral characteristics of individual speakers are different. While the physiological differences (e.g., vocal tract shape) result in the variation of low-level spectral features among speakers, the behavioral differences (e.g., voice source, speaking rate, use of words) contribute to the variation in the high-level, suprasegmental features.

2.6.1   Low-Level Features

Short-term spectral measurements are currently the most common feature extraction method for speaker recognition. In this approach, time-domain speech signals are divided into a number of overlapping frames, with frame sizes of 20ms to 30ms. For each frame, either linear predictive-based cepstral parameters or filter-bank-based cepstral parameters are extracted.

LP-Based Cepstral Parameters

Linear prediction (LP) analysis [228] is based on the assumption that the current sample of speech signals $s(n)$ can be predicted from the past P speech samples;

that is,

$$s(n) \approx \tilde{s}(n) = -\sum_{k=1}^{P} a_k s(n-k),$$

where $\{a_k\}_{k=1}^{P}$ are called the LP coefficients. It is also assumed that the excitation source $Gu(n)$, where G is the gain and $u(n)$ is the normalized excitation, can be separated from the vocal tract. These two assumptions lead to a simple source-filter model where the vocal tract is represented by an IIR filter of the form

$$H(z) = \frac{S(z)}{Gu(n)} = \frac{1}{1 + \sum_{k=1}^{P} a_k z^{-k}}. \tag{2.6.1}$$

In the time-domain, the output $s(n)$ of this IIR filter is a linear regression of its (finite) past output values and the present input $Gu(n)$:

$$s(n) = -\sum_{k=1}^{P} a_k s(n-k) + Gu(n).$$

The goal of LP analysis is to determine a set of LP coefficients $\mathbf{a} = \{a_1, \ldots, a_P\}$ for each frame of speech such that the frequency response of Eq. 2.6.1 is as close to the frequency spectrum of the speech signal as possible (i.e., $|H(\omega)| \approx |S(\omega)|$). This is achieved by minimizing the sum of the prediction error for each frame. Specifically, for speech frame $\{s_n(m)\}_{m=0}^{N-1}$ starting at sample n, compute this:

$$\hat{\mathbf{a}} = \arg\min_{\mathbf{a}} \sum_{m=0}^{N+P-1} e_n^2(m)$$

$$= \arg\min_{\mathbf{a}} \sum_{m=0}^{N+P-1} (s_n(m) - \tilde{s}_n(m))^2$$

$$= \arg\min_{\mathbf{a}} \left\{ \sum_{m=0}^{N+P-1} \left[s_n(m) + \sum_{k=1}^{P} a_k s_n(m-k) \right]^2 \right\}, \tag{2.6.2}$$

where $\hat{\mathbf{a}} = \{\hat{a}_1, \ldots, \hat{a}_P\}$, $m = 0, \ldots, N+P-1$ is the index for accessing the speech samples in the current frame and N is the number of samples per frame. The solution to this minimization problem (see Problem 15) is a matrix equation of the form:

$$\begin{bmatrix} r_n(0) & r_n(1) & \cdots & r_n(P-1) \\ r_n(1) & r_n(0) & \cdots & r_n(P-2) \\ r_n(2) & r_n(1) & \cdots & r_n(P-3) \\ \cdots & \cdots & \cdots & \cdots \\ r_n(P-1) & r_n(P-2) & \cdots & r_n(0) \end{bmatrix} \begin{bmatrix} \hat{a}_1 \\ \hat{a}_2 \\ \cdots \\ \hat{a}_P \end{bmatrix} = \begin{bmatrix} -r_n(1) \\ -r_n(2) \\ \cdots \\ -r_n(P) \end{bmatrix}, \tag{2.6.3}$$

where $r_n(k)$ is an autocorrelation function

$$r_n(k) = \sum_{m=0}^{N-1-k} s_n(m)s_n(m+k).$$

As a numerical example, let $s(0) = 2.0$, $s(1) = -1.0$, $s(2) = 1.0$, and $s(3) = s(4) = \cdots = s(N-1) = 0.0$ and $P = 2$.[1] Thus,

$$r(0) = \sum_{m=0}^{N-1} s^2(m)$$
$$= s^2(0) + s^2(1) + s^2(2) + \cdots + 0 = 4 + 1 + 1 = 6$$
$$r(1) = \sum_{m=0}^{N-2} s(m)s(m+1)$$
$$= s(0)s(1) + s(1)s(2) + s(2)s(3) + 0 \cdots + 0 = 2(-1) + (-1)(1) = -3$$
$$r(2) = \sum_{m=0}^{N-3} s(m)s(m+2)$$
$$= s(0)s(2) + s(1)s(3) + s(2)s(4) + 0 \cdots + 0 = 2(1) + (-1)(0) = 2.$$

Substituting $r(1)$, $r(2)$, and $r(3)$ into Eq. 2.6.3, the following LP coefficients is obtained:

$$\begin{bmatrix} 6 & -3 \\ -3 & 6 \end{bmatrix} \begin{bmatrix} \hat{a}_1 \\ \hat{a}_2 \end{bmatrix} = \begin{bmatrix} 3 \\ -2 \end{bmatrix} \Rightarrow \begin{bmatrix} \hat{a}_1 \\ \hat{a}_2 \end{bmatrix} = \begin{bmatrix} 0.44 \\ -0.11 \end{bmatrix}.$$

Hence, the transfer function of the vocal tract filter is

$$H(z) = \frac{1}{1 + 0.44z^{-1} - 0.11z^{-2}}.$$

Although LP coefficients represent the spectral envelope of the speech signals, it was found that smoothing the LP-based spectral envelopes by cepstral processing can provide a more consistent representation of a speaker's vocal tract characteristics from one utterance to another [13]. The cepstral coefficients c_n can be computed directly from LP coefficients a_k [228] as follows (see also Problem 18):

$$c_0 = \ln G \qquad\qquad\qquad\qquad\qquad\qquad (2.6.4)$$

$$c_n = -a_n - \sum_{k=1}^{n-1} \left(\frac{k}{n}\right) c_k a_{n-k} \qquad 1 \le n \le P \qquad (2.6.5)$$

$$c_n = -\sum_{k=1}^{n-1} \left(\frac{k}{n}\right) c_k a_{n-k} \qquad\qquad n > P, \qquad\qquad (2.6.6)$$

---

[1]For notation simplicity, the subscript n has been dropped.

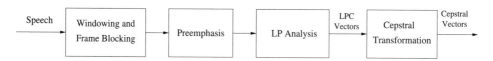

Figure 2.9. Extraction of LPCCs from speech signals.

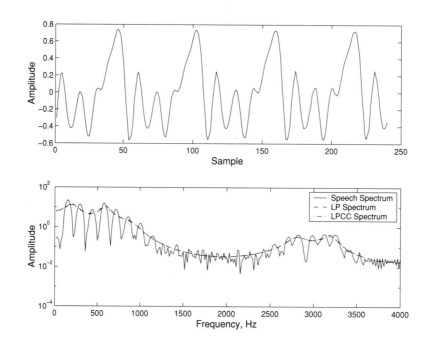

Figure 2.10. Comparison of the spectrum, LP-spectral envelope, and LPCC-spectral envelope of a frame of voiced speech. Top, time-domain speech signal; bottom, frequency spectra.

where G is the estimated model gain and P is the prediction order. Figure 2.9 depicts the process of obtaining the LP-based cepstral parameters. As the coefficients are derived from LP analysis, they are referred to as LP-derived cepstral coefficients (LPCCs) in the literature. Figure 2.10 compares the spectrum, the LP-spectral envelope, and the LPCC-spectral envelope of a voiced speech frame. Evidently, the envelopes created by both the LP coefficients and the LPCCs are able to track the peaks (formants) of the speech spectrum. Because different individuals produce speech with different formants (even for the same class of sounds), LPCCs can be used as a feature for speaker recognition.

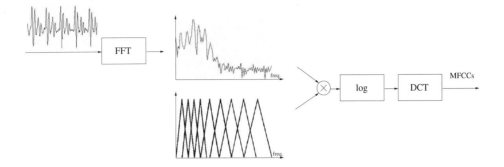

Figure 2.11. Extraction of MFCCs from speech signals.

Filter-Bank-Based Cepstral Parameters

Similar to the LPCCs described before, filter-bank-based cepstral parameters are also derived from speech signals on a frame-by-frame basis. However, unlike LP analysis, a number of triangular filters with different center frequencies and bandwidth are applied to the speech signals. Each of these filters extracts the average of the frequency spectrum in a particular frequency band. Typically, the Mel scale, which is an auditory scale similar to the frequency scale of the human ear, is used for positioning the filters. The outputs of these filters form a spectral envelope, and a discrete cosine transform is applied to the filters' output to obtain the cepstral coefficients [69]:

$$c_n = \sum_{k=1}^{K} X(k) \cos\left[n\left(k - \frac{1}{2}\right)\frac{\pi}{K}\right] \quad n = 1, \dots, P,$$

where $X(k)$ is the logarithm of the k-th filter bank's output, K is the number of filter banks, and P is the number of cepstral coefficients for recognition. Because the Mel scale is used in the processing, the coefficients are referred to as Mel-frequency cepstral coefficients (MFCCs). Figure 2.11 depicts the process of extracting the MFCCs.

Dynamic Parameters

In speech recognition, the first- and second-order LPCCs or MFCCs are also computed by using polynomial approximation as follows [107]:

$$\Delta c_n = \frac{\sum_{k=-L}^{L} k c_{n+k}}{\sum_{k=-L}^{L} |k|} \quad \text{and} \quad \Delta\Delta c_n = \frac{\sum_{k=-L}^{L} k^2 c_{n+k}}{\sum_{k=-L}^{L} k^2},$$

where L is typically equal to 3. Similar dynamic features are also extracted from the log-energy. Together with the cepstral coefficients c_n, $n = 1, \dots, P$, a feature vector will contain 39 ($=12 + 12 + 12 + 1 + 1 + 1$) elements if $P = 12$. The dynamic features are important for speech recognition because the rate of change of spectral shape and energy could be different from phones to phones. Incorporating these parameters into the feature vectors can certainly help improve the performance of large-vocabulary continuous-speech recognition systems, as most of these systems use phone-based hidden Markov models [395].

In speaker recognition, the change in energy and rate of energy change may not be as significant as those in speech recognition. This is especially the case in text-independent speaker recognition, where the energy profile of utterances could be very different, even though they are produced by the same speaker. Therefore, Gaussian mixture models (see Chapter 5), instead of hidden Markov models, are usually used for modeling the vocal tract characteristics of individual speakers, and the order of presenting the feature vectors to the models is irrelevant. On the other hand, energy and energy profile play an important role in text-dependent speaker recognition because it is possible to use phone- or word-based speaker models. During a verification session, the claimant is asked to utter a prompted phrase, and the energy profile can help the spectral features detect the misalignment between the prompted phrase and utterance.

2.6.2   High-Level Features

The low-level features described before assume that every 10ms to 20ms of speech can be treated independently. These features work well when the amount of recorded speech is in the order of seconds. However, if the amount of recorded speech increases to the 10s of seconds, or even in the order of minutes, extracting features that span a longer time scale will become attractive. It is commonly believed that, apart from using spectral contents, humans also recognize speakers based on their speaking style, speaking rate, prosody, rhythm, intonation, accent, word and phrase usage, frequency of occurrence of specific words, and so on. These *high-level* features will be very useful for automatic speaker recognition if they can be extracted reliably.

Early attempts to use high-level features date back to the early 1970s. For example, in Atal [12], pitch contours were used as speaker features. Although it has become clear that using pitch contours alone is not very effective, pitch-based information contained in the MFCCs of the LP-residual signals can be added to the static MFCCs for speaker recognition [290, 292, 398]. The idea comes from speech coding in which the construction of the LP-residual plays an important role in enhancing the quality of coded speech. Other high-level features that have been used for speaker recognition include idiolectal and prosodic information [82, 374]. Building speaker recognition systems that can make use of this information, however, requires a large prosodic database, which may be difficult to achieve in practice.

2.6.3 Robustness in Speech Features

As a result of the proliferation of e-banking and e-commerce, recent research on speaker verification has focused on verifying speakers' identity over the phone. A challenge of phone-based speaker verification is that transducer variability could result in acoustic mismatches between the speech data gathered from different handsets. The recent popularity of mobile and Internet phones further complicates the problem because speech coders in these phones also introduce acoustic distortion into speech signals. The sensitivity to handset variations and speech coding algorithms means that handset compensation techniques are essential for practical speaker verification systems. Therefore, the ability to extract features free from environmental distortion is of primary importance.

Efforts to address reliability issues in speech-based biometrics have centered on two main areas: channel mismatch equalization and background noise compensation.

Channel Mismatch Equalization

In the practical application of speaker recognition, speech is typically collected under different acoustic environments and communication channels, causing mismatches between speech gathered during enrollment and during verification. Three main schools of thought have been used to address this problem. The first looks at the local spectral characteristics of a given frame of speech. Early attempts included cepstral weighting [351] and bandpass *liftering* [172]. These approaches, however, assume that all frames are subject to the same distortion. This assumption has been avoided by more recent proposals, such as adaptive component weighting [11] and pole-zero postfiltering [406]. However, the experiments in these studies have limitations because the speech corpora (KING and TIMIT with channel simulators) they used do not allow for proper examination of handset variability. More effort must be made to investigate whether these approaches are sensitive to handset variation.

The second school exploits the temporal variability of feature vectors. Typical examples include cepstral mean subtraction [13], pole-filtered cepstral mean subtraction [258], delta cepstrum [107], relative spectral processing [134], signal bias removal [294], and the modified-mean cepstral mean normalization with frequency warping [112]. Although these methods have been successfully used in reducing channel mismatches, they have limitations because they assume that the channel effect can be approximated by a linear filter. Most phone handsets, however, exhibit energy-dependent frequency responses [308] for which a linear filter may be a poor approximation. Therefore, a more complex representation of handset characteristics is required.

The third school uses affine transformation to correct the mismatches [229], where speaker models can be trained on clean speech and operated on environmentally distorted speech without retraining. This approach has the advantage that both convolutional distortion and additive noise can be compensated for simultaneously. Additional computation is required during verification, however, to compute the transformation matrices.

Background Noise Compensation

It has been shown that approximately 40% of phone conversations contain competing speech, music, or traffic noise [79]. This figure suggests the importance of background noise compensation in phone-based speaker verification. Early approaches include spectral subtraction [30] and projection-based distortion measures [231]. More recently, statistical-based methods (e.g., the noise integration model [317] and signal bias removal [294]) have been proposed. The advantage of using statistical methods is that clean reference templates are no longer required. This property is particularly important to phone-based applications because clean speech is usually not available.

Joint Additive and Convolutional Bias Compensation

There have been several proposals aimed at addressing the problem of convolutional distortion and additive noise simultaneously. In addition to the affine transformation mentioned before, these proposals include stochastic pattern matching [332], parallel model combination [111], state-based compensation for continuous-density hidden-Markov models [4], and maximum likelihood estimation of channels' autocorrelation functions and noise [405]. Although these techniques have been successful in improving speech recognition performance, caution must be taken when they are applied to speaker recognition. This is because adapting a speaker model to new environments will affect its capability in recognizing speakers [18].

2.6.4   Classification Schemes for Speech-Based Biometrics

Similar to the visual-based biometrics described in Section 2.5, feature extraction in audio-based biometrics is followed by pattern classification in which the characteristics of individual speakers are modeled. For verification applications, a number of decision strategies have been developed to improve system performance.

Speaker Modeling

The choice of speaker models depends mainly on whether the verification is text-dependent or text-independent. In the former, it is possible to compare the claimant's utterance with that of the reference speaker by aligning the two utterances at equivalent points in time using dynamic time warping (DTW) techniques [107]. An alternative is to model the statistical variation in the spectral features. This is known as hidden-Markov modeling (HMM), which has been shown to outperform the DTW-based methods [259]. In text-independent speaker verification, methods that look at long-term speech statistics [232], or consider individual spectral vectors as independent of one another, have been proposed. The latter includes vector quantization (VQ) [342], Gaussian mixture models (GMMs) [301], and neural networks [222, 265].

Decision Strategies

Research has shown that normalizing speaker scores can reduce error rates. Early work includes the likelihood ratio scoring proposed by Higgins et al. [136] and the cohort normalized scoring proposed by Rosenberg et al. [318]. Subsequent work based on likelihood normalization [215, 235] and minimum verification error training [320] also shows that including an impostor model not only improves speaker separability but also allows the decision threshold to be easily set. Rosenberg and Parthasarathy [319] established some principles for constructing impostor models and showed that those with speech closest to the reference speaker's model perform the best. Their result, however, differs from that of Reynolds [302], who found that a gender-balanced, randomly selected impostor model performs better, suggesting that more work is required in this area.

2.7   Concluding Remarks

In this chapter, various biometric identification models are reviewed and key design issues and tradeoffs for biometric authentication systems are explored. As vital background information for later chapters, several commonly used feature extraction algorithms for visual-based and audio-based biometric systems are described. In real-world situations, most people divide the BI model selection problem into optimization of the feature extraction algorithms and then optimization of pattern classification algorithms, under the constraints of compatibility, data storage, speed, and data availability. Various existing feature extraction and pattern classification methods are investigated. The tradeoff between the feature extractor and the pattern classifier in the BI models is also addressed.

Problems

1. A function is scale-invariant if $\phi(\alpha y) = \phi(y)$. Verify that the kurtosis function $k(y)$ is scale-invariant.

2. *Deflation Procedure.* The deflation involves the removal of the already extracted independent component(s) from the original space $\mathbf{x}(t)$. Assuming that the first IC-extracting vector $\mathbf{w}_1$ is already derived from Eq. 2.3.1. Then, let us define $\mathbf{x}'(t) = (I - \mathbf{w}_1\mathbf{w}_1^T)\mathbf{x}(t)$, which in effect spans only a (deflated) $(n-1)$-dimensional subspace. The second most discriminative independent component can be found from the deflated subspace. Mathematically, the aim is to find a vector $\mathbf{w}_2$ such that $y(t) = \mathbf{w}_2^T\mathbf{x}'(t)$ yields the minimum kurtosis:

$$\mathbf{w}_2 = \arg\min_{\mathbf{w}} \frac{E[|\mathbf{w}^T\mathbf{x}'(t)|^4]}{E[|\mathbf{w}^T\mathbf{x}'(t)|^2]^2}. \tag{2.7.1}$$

Furthermore, to better ensure numerical stability, the following SVD approach is proposed in the KuicNet algorithm [189]. By the SVD property, $[\mathbf{W}_{orth}|\mathbf{w}]$ in the following SVD factorization forms a unitary matrix:

$$I - \mathbf{w}_1\mathbf{w}_1^T = \begin{bmatrix} \mathbf{W}_{orth} & | & \mathbf{w} \end{bmatrix} \begin{bmatrix} 1 & & \\ & \ddots & \\ & & 1 \\ & & & 0 \end{bmatrix} \begin{bmatrix} \mathbf{W}_{orth}^T \\ \hline \mathbf{w}^T \end{bmatrix},$$

where $\mathbf{W}_{orth}$ is an $n \times (n-1)$ matrix, corresponding to the $(n-1)$ singular vectors. Now let's define a new $(n-1)$-dimensional vector process: $\mathbf{v}(t) = \mathbf{W}_{orth}^T \mathbf{x}(t)$, and find a vector $\mathbf{w}'_2$ such that $y(t) = \mathbf{w}'_2{}^T \mathbf{v}(t)$ has a minimum kurtosis; that is,

$$\mathbf{w}'_2 = \arg\min_{\mathbf{w}'} \frac{E[|\mathbf{w}'^T\mathbf{v}(t)|^4]}{E[|\mathbf{w}'^T\mathbf{v}(t)|^2]^2}. \tag{2.7.2}$$

(a) Show that the second IC-extracting vector $\mathbf{w}_2$ can be derived from $\mathbf{w}'_2$. In fact, verify that

$$\mathbf{w}_2 = \mathbf{W}_{orth}^T \mathbf{w}'_2.$$

(b) Show that, by a recursive procedure, all of the other components $\{\mathbf{w}_j, j > 2\}$ can be similarly derived.

(c) Write a MatLab program for sequential extraction of the independent components based on the deflation procedure.

3. Among various potential applications of face recognition techniques, significant differences exist in terms of image quality and the availability of a well-defined matching criterion. A small number of exemplar applications are listed here:

(a) Credit card and ID cards (including driver's license, passport, etc.)

(b) Bank security

(c) Photo matching

(d) Public surveillance

(e) Expert/witness identification

Note that the first three applications involve matching one face image to another, while the last two applications involve finding a face resembling the human recollection of a facial image [52]. Among these applications, which is the most advantageous in terms of controlled image, controlled segmentation, and good quality images? Also, which application(s) bears the most disadvantages in terms of

(a) the lack of an existing database

(b) a potentially huge database

    (c) uncontrolled segmentation

    (d) low image quality

4. Among the applications listed in Problem 3, select application(s) where recognition is more desirable than verification.

5. Among the list of applications in Problem 3, select the application(s) where distributed systems are more desirable than centralized systems. That is, which application(s) have the potential to make a best use of a geographically localized search. Explain the situations where data can be stored locally and where data should be stored centrally.

6. Face recognition techniques can be classified into two groups: (1) static matching and (2) dynamic matching, depending on the availability of video images. Among the list itemized in Problem 3, select the application(s) that could effectively make use of the dynamic matching technology. Explain why.

7. Given controlled imaging conditions, segmentation/location of a face is usually straightforward. Among the applications listed in Problem 3, select the application(s) which are potentially easiest for segmentation/location of a face. Which are potentially the most difficult? Explain why.

8. Although imaging conditions are controlled for many key applications, feature extraction and face matching must take into account variations in the face due to aging, hair loss, hair growth, and so on. Suggest plausible solutions to tackle these problems.

9. Image segmentation is important for image analysis and face recognition. The contour of a segmented object offers a more reliable representation than the grey levels of image pixels. Prominent examples for contour representations include chain codes, B-spline representation, and Fourier descriptors (described next). For the Fourier descriptors, a trace function is expressed by a complex function

$$u(n) = x(n) + jy(n), \; n = 0, \ldots, N-1$$

whose discrete Fourier transform representation is

$$a(k) = \sum_{n=0}^{N-1} u(n) \exp\left(\frac{-j2\pi kn}{N}\right), \; 0 \le k \le N-1.$$

The complex coefficients $a(k)$ are called the Fourier descriptors (FDs) [273]. Two shapes are considered similar if their (normalized) FDs have a short distance. The FDs are invariant with respect to many geometric transformations. For example, the FDs can be used to match similar shapes even if they are of different size and orientation. Show that

(a) normalized FDs are invariant to scaling.

(b) the magnitudes of the FDs $|a(k)|$ for all k's except $k = 1$ have invariant properties with respect to starting point, rotation, and reflection.

(c) the magnitudes of FDs are invariant to shift.

10. A face recognition system uses the distances between various facial features to identify people. The feature extractor contains a bank of local feature detection filters to detect the positions of the nose, mouth, chin, and eyes. The feature vector that feeds into the recognizer is formed by the distance measures between those detected points. The recognizer is an MLP with 10 hidden neurons (see Section 5.4 for detailed discussions on MLPs). Suppose the processing time for each local feature detector is the same—that is, $t(F_i(\mathbf{x})) = F$ for all i—and the processing time for each neuron in the MLP is also the same, that is, $t(C_j(\mathbf{x})) = C$ for all j. Assume that C equals to $0.2F$.

(a) If one downsizes the feature extractor to eye and nose detection alone, while keeping the recognizer the same, can the identification time be reduced?

(b) Suppose one needs to increase the number of hidden nodes in the MLP to maintain the same recognition rate. If the number of hidden nodes is linearly proportional to the reciprocal of the dimension of the input feature vectors, can the identification time be reduced with eyes and nose detection only?

11. Template-based visual feature extraction often decomposes the input image into several levels of low- and high-spatial frequency components (e.g., FFT, wavelets).

(a) Describe the pros and cons of using low- or high-frequency information for recognition.

(b) If the object is (1) poorly illuminated or (2) photographed by a low-quality camera, what can be done to maintain image invariance?

12. A centralized speaker verification system uses 1024-center Gaussian mixture speaker models to characterize client speakers and a 1024-center universal background model to characterize impostors. The feature extractor of the system computes 12-dimensional MFCCs and their first derivative every 15ms. How would you reduce the computation requirements when verification is to be done locally on a handheld device?

13. Assuming that an e-banking system has 100,000 users, estimate the storage requirements using your favorite speaker verification technology (please specify) as a means of enhancing access-control security. Repeat the preceding but use your favorite face recognition technology.

14. What is FRR and what is FAR? What is the relationship between FRR and FAR? If a face recognizer claims to have a "100% verification rate," does that mean this face recognizer has achieved perfect performance?

15. The vocal tract can be modeled by an all-pole filter of the form

$$H(z) = \frac{1}{1 - \sum\limits_{k=1}^{P} a_k z^{-k}},$$

where a_k $(k = 1, \ldots, P)$ are the prediction coefficients.

(a) By minimizing the error between the speech signal $s(n)$ and its predicted value, show that

$$\sum_{m} s(m-i) s(m) = \sum_{k=1}^{P} a_k \sum_{m} s(m-i) s(m-k).$$

Hence, show that the LP coefficients $\{a_k; k = 1, \ldots, P\}$ are the solutions of the matrix equation

$$\sum_{k=1}^{p} r_n(|i - k|) a_k = r_n(i), \ 1 \le i \le p,$$

where

$$r(i - k) = \sum_{m=0}^{N-1-(i-k)} s(m)s(m+i-k)$$

is the autocorrelation function of speech signal $s(n)$ at lag $i - k$, and N is the frame size.

(b) Determine the LP coefficients $\{a_k; k = 1, \ldots, P\}$ when $s(0) = -1.0$, $s(1) = 1.0, s(2) = s(3) = \cdots = s(N - 1) = 0.0$, and $P = 2$.

(c) Plot the frequency response of $H(z)$.

(d) Hence, determine the formant frequency (or frequencies) of $s(n)$, assuming that the sampling frequency of $s(n)$ is 8kHz.

(e) Suggest an appropriate value for P so that the first three formant frequencies of speech signals can be accurately modeled.

16. In linear-prediction analysis, a preemphasis filter of the form

$$P(z) = 1 - 0.98z^{-1}$$

is typically applied to speech signals.

(a) Draw the frequency response of $P(z)$.

(b) Describe the effect of $P(z)$ on a frame of voiced speech.

(c) Hence, explain the purpose of the preemphasis filter.

17. Answer the following questions based on a vocal-tract filter of the form

$$H(z) = \frac{1}{(1 + z^{-1} + 0.5z^{-2})(1 + 0.8z^{-1})}.$$

(a) Determine the poles of $H(z)$.

(b) Plot the poles on a z-plane.

(c) Assume that the sampling frequency is 8kHz. Determine the resonance frequencies.

(d) Sketch the frequency response of $H(z)$ (i.e., plot $\left|H(e^{j\omega})\right|$ against ω).

(e) Identify the type of sounds (voiced or unvoiced) whose spectra are best modeled by $\left|H(e^{j\omega})\right|$ and give a brief explanation.

18. In linear-prediction analysis, the transfer function of the inverse filter can be written as

$$A(z) = 1 + \sum_{k=1}^{p} a_k z^{-k},$$

where a_k $(k = 1, \ldots, p)$ are the predictor coefficients. If all the zeros of $A(z)$ are inside the unit circle, $\log A(z)$ can be expressed as:

$$\log A(z) = C(z) = \sum_{n=0}^{\infty} c_n z^{-n},$$

where c_n is the cepstrum of $A(z)$.

(a) By differentiating $\log A(z)$ with respect to z^{-1}, show that

$$\sum_{k=1}^{p} k a_k z^{-k+1} = \left\{ 1 + \sum_{k=1}^{p} a_k z^{-k} \right\} \sum_{n=1}^{\infty} n c_n z^{-n+1}.$$

(b) Hence show that

$$c_n = a_n - \sum_{m=1}^{n-1} \left(\frac{m}{n}\right) c_m a_{n-m} \quad 1 \le n \le p.$$

(c) Hence, express the cepstrum of the vocal-tract filter $H(z)$ in terms of a_n, where $H(z) = 1/A(z)$.

(d) The cepstral coefficients of a speech signal $s(n)$ can be obtained by using the definition of cepstrum:

$$c_n = \frac{1}{2\pi} \int_{-\pi}^{\pi} \log |S(\omega)| \, e^{j\omega n} d\omega.$$

Alternatively, they can be derived from the predictor coefficients a_k ($k = 1, \dots, p$) according to the results in Problem 18b and Problem 18c. Comment on the differences between the cepstral coefficients derived from these two approaches.

19. The spectral similarity measure between two spectra, $S_1(\omega)$ and $S_2(\omega)$, is defined as

$$d(S_1, S_2) = \frac{1}{2\pi} \int_{-\pi}^{\pi} |\log S_1(\omega) - \log S_2(\omega)|^2 d\omega.$$

(a) Show that

$$d(S_1, S_2) = \sum_{n=-\infty}^{\infty} |c_1(n) - c_2(n)|^2,$$

where $c_1(n)$ and $c_2(n)$ are the complex Cepstral coefficients of $S_1(\omega)$ and $S_2(\omega)$, respectively. *Hint:* Use the relationship

$$\left| \sum_{n=-\infty}^{\infty} [c_1(n) - c_2(n)] \right|^2 = \sum_{n=-\infty}^{\infty} \sum_{m=-\infty}^{\infty} [c_1(n) - c_2(n)] [c_1^*(m) - c_2^*(m)],$$

where $c_1^*(m)$ and $c_2^*(m)$ are the complex conjugates of $c_1(m)$ and $c_2(m)$, respectively.

(b) Given that $c(n) \to 0$ when $n \to \infty$, show that the distance between two spectra is approximately equal to the Euclidean distance between the two corresponding cepstral vectors; that is, show that

$$d(S_1, S_2) \approx \|\mathbf{c}_1 - \mathbf{c}_2\|^2,$$

where $\mathbf{c}_i = [c_1 \; c_2 \; \cdots \; c_L]^T$ for some finite integer L.

(c) Hence, discuss the advantages of using cepstral vectors for speech/speaker recognition.

20. Explain why the frame size for the spectral analysis of speech is usually set to 20ms to 30ms.

21. Compare the Hamming window against the rectangular window in the context of their magnitude spectra. Explain the possible advantage(s) of using the Hamming window in spectral analysis of speech.

Chapter 3

EXPECTATION-MAXIMIZATION THEORY

3.1 Introduction

Learning networks are commonly categorized in terms of supervised and unsupervised networks. In unsupervised learning, the training set consists of input training patterns only. In contrast, in supervised learning networks, the training data consist of many pairs of input/output patterns. Therefore, the learning process can benefit greatly from the teacher's assistance. In fact, the amount of adjustment of the updating coefficients often depends on the difference between the desired teacher value and the actual response. As demonstrated in Chapter 5, many supervised learning models have been found to be promising for biometric authentication; their implementation often hinges on an effective data-clustering scheme, which is perhaps the most critical component in unsupervised learning methods. This chapter addresses a data-clustering algorithm, called the expectation-maximization (EM) algorithm, when complete or partial information of observed data is made available.

3.1.1 K-Means and VQ algorithms

An effective data-clustering algorithm is known as K-means [85], which is very similar to another clustering scheme known as the vector quantization (VQ) algorithm [118]. Both methods classify data patterns based on the *nearest-neighbor* criterion.

Verbally, the problem is to cluster a given data set $X = \{x_t; t = 1, \ldots, T\}$ into K groups, each represented by its centroid denoted by $\mu^{(j)}, j = 1, \ldots, K$. The task is (1) to determine the K centroids $\{\mu^{(1)}, \mu^{(2)}, \ldots, \mu^{(K)}\}$ and (2) to assign each pattern x_t to one of the centroids. The nearest-neighbor rule assigns a pattern x to the class associated with its nearest centroid, say $\mu^{(i)}$.

Mathematically speaking, one denotes the centroid associated with x_t as μ_t, where $\mu_t \in \{\mu^{(1)}, \mu^{(2)}, \ldots, \mu^{(K)}\}$. Then the objective of the K-means algorithm is to minimize the following sum of squared errors:

$$E(X) = \sum_t \|x_t - \mu_t\|^2, \tag{3.1.1}$$

where $\|\cdot\|$ is the Euclidean norm.

Let X_k denote the set of data patterns associated with the k-th cluster with the centroid $\mu^{(k)}$ and N_k denotes the number of patterns in the cluster X_k, where $k = 1, \ldots, K$. The learning rule of the K-means algorithm consists of the following two basic steps.

1. *Determine the membership of a data pattern:*

$$\mathbf{x} \in X_k \quad \text{if} \quad \|\mathbf{x} - \mu_k\| < \|\mathbf{x} - \mu_j\| \quad \forall j \neq k. \tag{3.1.2}$$

2. *Updating the representation of the cluster:* In a clustering process, the inclusion (or removal) of a new pattern in a cluster (or from a cluster) affects the representation (e.g., the centroid or variance) of the cluster. Therefore, the centroid should be updated based on the new membership:

$$\mu_j = \frac{1}{N_j} \sum_{\mathbf{x} \in X_j} \mathbf{x}. \tag{3.1.3}$$

Sometimes, the variance of the data cluster is also of great interest (e.g., in Gaussian mixture models). In this case, the variance can be computed as

$$\Sigma_j = \frac{1}{N_j} \sum_{\mathbf{x} \in X_j} (\mathbf{x} - \mu_j)(\mathbf{x} - \mu_j)^T. \tag{3.1.4}$$

3.1.2 Gaussian Mixture Model

The EM scheme can be seen as a generalized version of K-means clustering. The main difference hinges on the notion of a hard-versus-soft membership. A hard membership is adopted in the K-means algorithm, (i.e., a data pattern is assigned to one cluster only). This is not the case with the EM algorithm, where a soft membership is adopted, (i.e., the membership of each data pattern can be distributed over multiple clusters).

The necessity of using a distributed (i.e., soft) membership is the most conspicuous for a Gaussian mixture model (GMM). Given a set of N-independent and identically distributed patterns $X^{(i)} = \{\mathbf{x}_t; t = 1, 2, \ldots, N\}$ associated with class ω_i, the likelihood function $p(\mathbf{x}_t|\omega_i)$ for class ω_i is a mixture of Gaussian distributions; that is,

$$p(\mathbf{x}_t|\omega_i) = \sum_{r=1}^{R} P(\Theta_{r|i}|\omega_i) p(\mathbf{x}_t|\omega_i, \Theta_{r|i}), \tag{3.1.5}$$

where $\Theta_{r|i}$ represents the parameters of the r-th mixture component; R is the total number of mixture components; $p(\mathbf{x}_t|\omega_i, \Theta_{r|i}) \equiv \mathcal{N}(\mathbf{x}; \mu_{r|i}, \Sigma_{r|i})$ is the probability density function of the r-th component; and $P(\Theta_{r|i}|\omega_i)$ is the prior probability of the r-th component. Typically, $\mathcal{N}(\mathbf{x}; \mu_{r|i}, \Sigma_{r|i})$ is a Gaussian distribution with mean $\mu_{r|i}$ and covariance $\Sigma_{r|i}$.

In short, the output of a GMM is the weighted sum of R-component densities. The training of GMMs can be formulated as a maximum likelihood problem, where the mean vectors $\{\mu_{r|i}\}$, covariance matrices $\{\Sigma_{r|i}\}$, and mixture coefficients $\{P(\Theta_{r|i}|\omega_i)\}$ are often estimated by the iterative EM algorithm—the main topic of the current chapter.

3.1.3   Expectation-Maximization Algorithm

The expectation-maximization (EM) algorithm is an ideal candidate for solving parameter estimation problems for the GMM or other neural networks. In particular, EM is applicable to problems, where the observable data provide only partial information or where some data are "missing"—see Figure 3.1(a). Another important class of parameter estimation that can be addressed by EM involves a mixture-of-experts—see Figure 3.1(b). In this class of problems, there are two categories of unknown parameters: one pertaining to the membership function of an expert (or cluster) and the other consisting of the unknown parameters defining individual experts. Let's use a Gaussian mixture model shown in Figure 3.1(b) as an example, where $\pi^{(j)}$ denotes the prior probability of expert j and where $\phi^{(j)} = \{\mu^{(j)}, \Sigma^{(j)}\}$ denotes the parameters (mean and variance) of the expert. This chapter explains why the EM method can serve as a powerful tool for estimating these parameters. It also demonstrates how the EM algorithm can be applied to data clustering.

The EM algorithm is a very general parameter estimation method in that it is applicable to many statistical models, for example, mixture-of-experts (MOE), Gaussian mixture models (GMMs), and vector quantization (VQ). Figure 3.2 depicts the relationship among EM, MOE, GMM, and VQ. In particular, the figure highlights the fact that VQ is a special case of GMM, which in turn is a special case of the more general mixture-of-experts. More important, EM is applicable to all of these models.

The classic EM algorithm can be dated back to Dempster, Laird, and Rubin's paper in 1977 [74]. It is a special kind of quasi-Newton algorithm with a searching direction having a positive projection on the gradient of log-likelihood. Each EM iteration consists of two steps—Estimation (E) and Maximization (M). The M-step maximizes a likelihood function that is refined in each iteration by the E-step. Interested readers can refer to the references [74, 168, 297, 350]

One important feature of the EM algorithm is that it can be applied to problems in which observed data provide "partial" information only or when artificially introducing some information (referred to as "hidden"-state information hereafter) can greatly simplify the parameter estimation process. Figure 3.3 illustrates the concept of hidden and partial data. In Figure 3.3(a), all data (x_1 to x_7) are known.

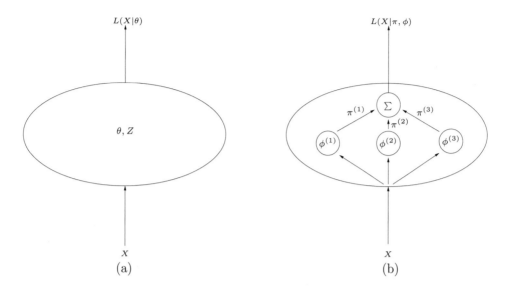

Figure 3.1.   Parameter estimation by EM. (a) EM for general missing data problems, where θ is the nonstructural parameters to be estimated and Z is the set of missing data. (b) EM for hidden-state problems in which the parameter θ can be divided into two groups: $\{\pi^{(j)}\}_{j=1}^{3}$ and $\{\phi^{(j)}\}_{j=1}^{3}$, where $\pi^{(j)}$ represents the prior probability of the j-th expert and $\phi^{(j)}$ defines the density function associated with the j-th expert.

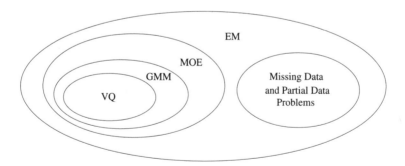

Figure 3.2.   Diagram depicting the relationship among EM, MOE, GMM, VQ and the class of problems known as missing- and partial-data problems.

Let's assume that there are two clusters in the observed data. Although all data constituting the two clusters are observable, one does not know exactly to which cluster each of these data belongs. Lacking this hidden membership information results in a complicated parameter estimation procedure. The estimation procedure, however, can be greatly simplified if this membership information is assumed to be known. For example, if the cluster identities of x_1 to x_7 in Figure 3.3(a) were known, finding the cluster means is reduced to computing the mean of individual clusters separately. Figure 3.3(b) illustrates the idea of partial data. Unlike Figure 3.3(a), the partial-data problem in Figure 3.3(b) contains uncertain data y because y can be equal to 5.0 or 6.0. As a result, the true value of y is unobservable whereas those of x_1 to x_4 are observable. The EM algorithm can solve this partial-data problem effectively by computing the expected value of y. Figure 3.3(c) illustrates the case in which cluster membership information is hidden and only partial information is available. The problem can be viewed as a generalization of the problems in Figure 3.3(a) and Figure 3.3(b). A new type of EM called doubly-stochastic EM is derived in Section 3.4 to address this kind of general problem. Numerical solutions for the problems in Figure 3.3 are provided in later sections.

The concepts of hidden and partial data have been applied to many scientific and engineering applications. For instance, in digital communication, the receiver receives a sequence consisting of $+1$'s and -1's without knowing which bit in the sequence is a $+1$ and which bit is a -1. In such cases, the state of each bit constitutes the missing information. In biometric applications, a MOE is typically applied to model the features of an individual. Each expert is designed to model some of the user-specific features. In such cases, the contribution of individual experts constitutes the hidden information.

EM has been shown to have favorable convergence properties, automatical satisfaction of constraints, and fast convergence. The next section explains the traditional approach to deriving the EM algorithm and proving its convergence property. Section 3.3 covers the interpretion the EM algorithm as the maximization of two quantities: the entropy and the expectation of complete-data likelihood. Then, the K-means algorithm and the EM algorithm are compared. The conditions under which the EM algorithm is reduced to the K-means are also explained. The discussion in Section 3.4 generalizes the EM algorithm described in Sections 3.2 and 3.3 to problems with partial-data and hidden-state. We refer to this new type of EM as the doubly stochastic EM. Finally, the chapter is concluded in Section 3.5.

3.2   Traditional Derivation of EM

Each EM iteration is composed of two steps—Estimation (E) and Maximization (M). The M-step maximizes a likelihood function that is further refined in each iteration by the E-step. This section derives the traditional EM and establishes its convergence property.

3.2.1   General Analysis

The following notations are adopted.

- $X = \{x_t \in \Re^D; t = 1, \ldots, T\}$ is the observation sequence, where T is the number of observations and D is the dimensionality of x_t.

Example 1: Hidden-State Problem

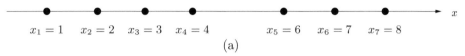

$$x_1 = 1 \quad x_2 = 2 \quad x_3 = 3 \quad x_4 = 4 \qquad x_5 = 6 \quad x_6 = 7 \quad x_7 = 8$$

(a)

Example 2: Partial-Data Problem

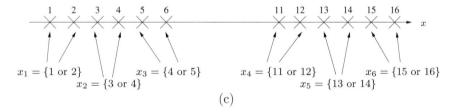

$$x_1 = 1 \qquad x_2 = 2 \qquad x_3 = 3 \qquad x_4 = 4$$

$$y = 5.0 \text{ or } 6.0$$

(b)

Example 3: Doubly-Stochastic
(Partial-Data and Hidden-State)

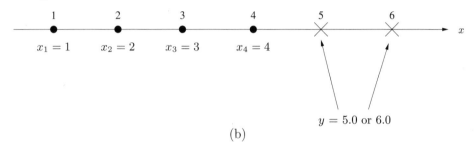

$$x_1 = \{1 \text{ or } 2\} \qquad x_3 = \{4 \text{ or } 5\} \qquad x_4 = \{11 \text{ or } 12\} \qquad x_6 = \{15 \text{ or } 16\}$$
$$x_2 = \{3 \text{ or } 4\} \qquad\qquad x_5 = \{13 \text{ or } 14\}$$

(c)

Figure 3.3. One-dimensional example illustrating the concept of (a) hidden-state, (b) partial-data, and (c) combined partial-data and hidden-state. In (a) the information regarding the cluster membership of x_t is hidden; in (b) y is partial in that its exact value is unknown; and in (c) data x_t provide partial information only because none of their exact values are known. The cluster membership information is also hidden.

- $\mathcal{C} = \{\mathcal{C}^{(1)}, \dots, \mathcal{C}^{(J)}\}$ is the set of cluster mixture labels, where J is the number of mixture components.

- $Z = \{z_t \in \mathcal{C}; t = 1, \dots, T\}$ is the set of missing data (specifying the hidden-state information).

- $\theta = \{\theta^{(j)}; j = 1, \dots, J\}$ is the set of unknown parameters that define the density function for approximating the true probability density of X.

- $\theta^{(j)} = \{\pi^{(j)}, \phi^{(j)}\}$, where $\pi^{(j)}$ denotes the prior probability of the j-th component density and $\phi^{(j)}$ defines the j-th component density.

Note that the combination of observations X and the "hidden-states" Z constitute the complete-data. The likelihood of the complete-data is instrumental in accordance with the EM formulation.

To facilitate the derivation, define

$$L(X|\theta_n) \equiv \log p(X|\theta_n) \tag{3.2.1}$$

as the log-likelihood of the incomplete-data given the current estimate θ_n, where n represents the iteration index; also, define $p(Z, X|\theta_n)$ as the completed data likelihood. According to probability theory,[1] $p(X|\theta_n)$ can be expressed as

$$p(X|\theta_n) = \frac{p(Z, X|\theta_n)}{P(Z|X, \theta_n)}. \tag{3.2.2}$$

Using Eq. 3.2.1 and Eq. 3.2.2, one can write the incomplete-data log-likelihood as follows:

$$
\begin{aligned}
L(X|\theta_n) &\equiv \log p(X|\theta_n) \\
&= [\log p(X|\theta_n)] \sum_Z P(Z|X, \theta_n) && \left(\text{since } \sum_Z P(Z|X, \theta_n) = 1\right) \\
&= \sum_Z P(Z|X, \theta_n) \log p(X|\theta_n) \\
&= \sum_Z P(Z|X, \theta_n) \log \frac{p(Z, X|\theta_n)}{P(Z|X, \theta_n)} && \text{(as a result of Eq. 3.2.2)} \\
&= \sum_Z P(Z|X, \theta_n) \log p(Z, X|\theta_n) - \sum_Z P(Z|X, \theta_n) \log P(Z|X, \theta_n) \\
&= E_Z\{\log p(Z, X|\theta_n)|X, \theta_n\} \\
&\quad - E_Z\{\log P(Z|X, \theta_n)|X, \theta_n\} && \text{(by definition of expectation)} \\
&= Q(\theta_n|\theta_n) + R(\theta_n|\theta_n), && (3.2.3)
\end{aligned}
$$

where $E_Z\{\}$ denotes expectation with respect to Z. Thus, denote

$$Q(\theta|\theta_n) \equiv E_Z\{\log p(Z, X|\theta)|X, \theta_n\} \tag{3.2.4}$$

---

[1] Hereafter, a capital P denotes probabilities and a lowercase p denotes density functions.

and

$$R(\theta|\theta_n) \equiv -E_Z\{\log P(Z|X,\theta)|X,\theta_n\}, \tag{3.2.5}$$

where $R(\theta|\theta_n)$ is an entropy term representing the difference between the incomplete-data likelihood and the expectation of the completed-data likelihood. Interpretation of $R(\theta|\theta_n)$ and its role in the EM algorithm is discussed further in Section 3.3.

3.2.2   Convergence Property of EM

The following demonstrates why the EM algorithm has a general convergence property. The basic idea is via Jensen's inequality. More precisely, it can be shown that if the Q-function in Eq. 3.2.4 is improved in each iteration (in the M-step), then so will be the likelihood function L.

The proof of convergence begins with the observation of the following relationship:

$$L(X|\theta) = \log p(X|\theta) = \log\left\{\sum_Z p(Z,X|\theta)\right\} = \log\left\{\sum_Z P(Z|X,\theta_n)\frac{p(Z,X|\theta)}{P(Z|X,\theta_n)}\right\}. \tag{3.2.6}$$

Using Eq. 3.2.6 and Jensen's inequality, this is obtained:

$$
\begin{aligned}
L(X|\theta) &= \log p(X|\theta) \\
&= \log\left\{\sum_Z P(Z|X,\theta_n)\frac{p(Z,X|\theta)}{P(Z|X,\theta_n)}\right\} \\
&= \log\left\{E_Z\left[\frac{p(Z,X|\theta)}{P(Z|X,\theta_n)}\Big|X,\theta_n\right]\right\} && \text{(by definition of expectation)} \\
&\geq E_Z\left\{\log\left[\frac{p(Z,X|\theta)}{P(Z|X,\theta_n)}\right]\Big|X,\theta_n\right\} && \text{(by Jensen's inequality)} \\
&= \sum_Z P(Z|X,\theta_n)\log\left[\frac{p(Z,X|\theta)}{P(Z|X,\theta_n)}\right] && \text{(by definition of expectation)} \\
&= \sum_Z P(Z|X,\theta_n)\log p(Z,X|\theta) - \sum_Z P(Z|X,\theta_n)\log P(Z|X,\theta_n) \\
&= E_Z\{\log p(Z,X|\theta)|X,\theta_n\} - E_Z\{\log P(Z|X,\theta_n)|X,\theta_n\} \\
&= Q(\theta|\theta_n) + R(\theta_n|\theta_n). \tag{3.2.7}
\end{aligned}
$$

In the M-step of the n-th iteration, θ^* is selected according to

$$\theta^* = \arg\max_\theta Q(\theta|\theta_n). \tag{3.2.8}$$

This means one can always choose a θ^* at iteration n such that

$$Q(\theta^*|\theta_n) \geq Q(\theta_n|\theta_n). \tag{3.2.9}$$

Note that this equation constitutes a sufficient condition to ensure the convergence property of the EM algorithm because, according to Eqs. 3.2.3, 3.2.7, and 3.2.9

$$
\begin{aligned}
L(X|\theta^*) &\geq Q(\theta^*|\theta_n) + R(\theta_n|\theta_n) \\
&\geq Q(\theta_n|\theta_n) + R(\theta_n|\theta_n) \\
&= L(X|\theta_n).
\end{aligned}
$$

Instead of directly maximizing $L(X|\theta)$, the EM algorithm divides the optimization problem into two subproblems: **E**xpectation and **M**aximization.

In each EM iteration, the E-step computes $Q(\theta|\theta_n)$ using a set of presumed model parameters θ_n. The M-step determines the value of θ (say θ^*) that maximizes $Q(\theta|\theta_n)$; that is,

$$
\theta^* = \max_{\theta} \sum_{Z} P(Z|X,\theta_n) \log p(Z,X|\theta). \tag{3.2.10}
$$

This results in (see Problem 8)

$$
p(Z,X|\theta^*) = \frac{P(Z|X,\theta_n)}{\sum\limits_{Z} P(Z|X,\theta_n)}. \tag{3.2.11}
$$

Dividing the optimization into two interdependent steps is most useful if optimizing $Q(\theta|\theta_n)$ is simpler than that of $L(X|\theta_n)$. Figure 3.4 illustrates how the E- and M-steps interplay to obtain a maximum-likelihood solution. The next section explains how to compute $Q(\theta|\theta_n)$ in the E-step and how to maximize $Q(\theta|\theta_n)$ in the M-step.

Generalized EM

In case θ^* in Eq. 3.2.8 is difficult to attain, the EM approach is still applicable if one can *improve* $Q(\theta|\theta_n)$ in each M-step (e.g., by gradient ascent). The algorithm is known as generalized EM. Although convergence of generalized EM is slower than that of the standard EM, it offers a more general and flexible framework for dividing the optimization process into the EM steps.

3.2.3 Complete-Data Likelihood

EM begins with an optimization of a likelihood function, which may be considerably simplified if a set of "missing" or "hidden" data is assumed to be known. The following demonstrates that computing the expectation of the complete-data likelihood in the E-step can be accomplished by finding the expectation of the missing or hidden data.

If $X = \{x_t; t = 1,\ldots,T\}$ contains T statistically independent vectors and $Z = \{z_t \in \mathcal{C}; t = 1,\ldots,T\}$, where $z_t = \mathcal{C}^{(j)}$ means that the j-th mixture generates x_t, then one can write $p(Z,X|\theta)$ as

$$
p(Z,X|\theta) = \prod_{t=1}^{T} p(z_t, x_t|\theta).
$$

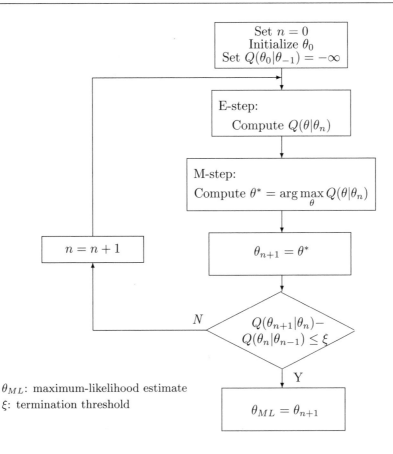

Figure 3.4. The flow of the EM algorithm.

Now, a set of indicator variables is introduced to indicate the status of the hidden-states:[2]

$$\Delta = \{\delta_t^{(j)}; j = 1, \ldots, J \text{ and } t = 1, \ldots, T\}$$

where

$$\delta_t^{(j)} \equiv \delta(z_t, \mathcal{C}^{(j)}) = \begin{cases} 1 & \text{if } x_t \text{ is generated by mixture } \mathcal{C}^{(j)}, \\ 0 & \text{otherwise.} \end{cases}$$

Since for each t only one of the terms in $\{\delta_t^{(j)}; j = 1, \ldots, J\}$ is equal to one and all of the others are equal to 0, one can express $p(Z, X|\theta)$ as follows:

$$p(Z, X|\theta) = \prod_{t=1}^{T} \sum_{j=1}^{J} \delta_t^{(j)} p(x_t, z_t|\theta)$$

---

[2]For illustration simplicity, assume that the missing data is in discrete form or the hidden data is the cluster membership.

$$= \prod_{t=1}^{T} \sum_{j=1}^{J} \delta_t^{(j)} p(x_t, z_t = \mathcal{C}^{(j)}|\theta)$$

$$= \prod_{t=1}^{T} \sum_{j=1}^{J} \delta_t^{(j)} p(x_t, \delta_t^{(j)} = 1|\theta).$$

Hence, the completed-data likelihood is given by

$$\log p(Z, X|\theta) = \sum_{t=1}^{T} \log \left\{ \sum_{j=1}^{J} \delta_t^{(j)} p(x_t, \delta_t^{(j)} = 1|\theta) \right\}$$

$$= \sum_{t=1}^{T} \log \left\{ \sum_{j=1}^{J} \delta_t^{(j)} p(x_t|\delta_t^{(j)} = 1, \theta) P(\delta_t^{(j)} = 1|\theta) \right\}$$

$$= \sum_{t=1}^{T} \log \left\{ \sum_{j=1}^{J} \delta_t^{(j)} p(x_t|\delta_t^{(j)} = 1, \phi^{(j)}) P(\delta_t^{(j)} = 1) \right\}$$

$$= \sum_{t=1}^{T} \sum_{j=1}^{J} \delta_t^{(j)} \log \left[p(x_t|\delta_t^{(j)} = 1, \phi^{(j)}) \pi^{(j)} \right]$$

$$= \sum_{t=1}^{T} \sum_{j=1}^{J} \delta_t^{(j)} \log \left[p(x_t|z_t = \mathcal{C}^{(j)}, \phi^{(j)}) \pi^{(j)} \right], \qquad (3.2.12)$$

where $\pi^{(j)}$ is the mixing coefficient of the j-th mixture. Eq. 3.2.12 uses the fact that $p(x_t|\delta_t^{(j)} = 1, \theta) = p(x_t|\delta_t^{(j)} = 1, \phi^{(j)})$ and $P(\delta_t^{(j)} = 1|\theta) = \pi^{(j)}$. Moreover, because there is only one non-zero term inside the summation $\sum_{j=1}^{J}$, one can extract $\delta_t^{(j)}$ from the log function without affecting the result.

E-Step.   Taking the expectations of Eq. 3.2.12 and using the defintions in Eq. 3.2.4, one obtains

$$Q(\theta|\theta_n) = E_Z\{\log p(Z, X|\theta)|X, \theta_n\}$$

$$= \sum_{t=1}^{T} \sum_{j=1}^{J} E\{\delta_t^{(j)}|x_t, \theta_n\} \log \left[p(x_t|\delta_t^{(j)} = 1, \phi^{(j)}) \pi^{(j)} \right]. \quad (3.2.13)$$

Then, define

$$h_n^{(j)}(x_t) \equiv E\{\delta_t^{(j)}|x_t, \theta_n\} = P(\delta_t^{(j)} = 1|x_t, \theta_n)$$

and denote $\pi_n^{(j)}$ as the j-th mixture coefficient at iteration n. Using the Bayes theorem, one can express $h_n^{(j)}(x_t)$ as

$$h_n^{(j)}(x_t) = P(\delta_t^{(j)} = 1|x_t, \theta_n)$$

$$= \frac{p(x_t|\delta_t^{(j)} = 1, \theta_n)P(\delta_t^{(j)} = 1|\theta_n)}{p(x_t|\theta_n)}$$

$$= \frac{p(x_t|\delta_t^{(j)} = 1, \phi_n^{(j)})P(\delta_t^{(j)} = 1|\theta_n)}{p(x_t|\theta_n)}$$

$$= \frac{p(x_t|\delta_t^{(j)} = 1, \phi_n^{(j)})\pi_n^{(j)}}{\sum_{k=1}^{J} p(x_t|\delta_t^{(k)} = 1, \phi_n^{(k)})\pi_n^{(k)}}. \qquad (3.2.14)$$

The E-step determines the best guess of the membership function $h_n^{(j)}(x_t)$. Once the probability $h_n^{(j)}(x_t)$ are computed for each t and j, $Q(\theta|\theta_n)$ can be considered as a function of θ. In the M-step of each iteration, this function is maximized to obtain the best value of θ (denoted as θ^*). In most cases, the M-step is substantially simplified if $h_n^{(j)}(x_t)$ are known. Therefore, the E-step can be viewed as a preparation step for the M-step.

3.2.4 EM for GMMs

To better illustrate the EM steps, a simple example applying EM to Gaussian mixture models (GMMs) is presented next. The most common forms for the mixture density are the radial basis functions (RBFs) or the more general elliptical basis functions (EBFs). In the latter case, the component density $p(x_t|\delta_t^{(j)} = 1, \phi^{(j)})$ is a Gaussian distribution, with the model parameter of the j-th cluster $\phi^{(j)} = \{\mu^{(j)}, \Sigma^{(j)}\}$ consisting of a mean vector and a full-rank covariance matrix.

Assume a Gaussian mixture model:

$$\theta = \{\pi^{(j)}, \mu^{(j)}, \Sigma^{(j)}; j = 1, \ldots, J\},$$

where $\pi^{(j)}$, $\mu^{(j)}$, and $\Sigma^{(j)}$ denote, respectively, the mixture coefficient, mean vector, covariance matrix of the j-th component density. The GMM's output is given by

$$p(x_t|\theta) = \sum_{j=1}^{J} \pi^{(j)} p(x_t|\delta_t^{(j)} = 1, \phi^{(j)}), \qquad (3.2.15)$$

where

$$p(x_t|\delta_t^{(j)} = 1, \phi^{(j)}) = (2\pi)^{-\frac{D}{2}} |\Sigma^{(j)}|^{-\frac{1}{2}} \exp\left\{-\frac{1}{2}(x_t - \mu^{(j)})^T (\Sigma^{(j)})^{-1} (x_t - \mu^{(j)})\right\} \qquad (3.2.16)$$

is the j-th Gaussian density of the GMM. A closer look at Eqs. 3.2.15 and 3.2.16 reveals that the GMM parameters θ can be divided into two groups: one containing $\pi^{(j)}$s and another containing $\mu^{(j)}$s and $\Sigma^{(j)}$s. The former indicates the importance of individual mixture densities via the prior probabilities $\pi^{(j)}$s, whereas the latter is commonly regarded as the kernel parameter defining the form of the mixture density. Unlike other optimization techniques (e.g., gradient descent) in which unknown parameters can be arranged in any order, the EM approach effectively makes use of the structural relationship among the unknown parameters to simplify the optimization process.

After the initialization of θ_0, the EM iteration is as follows:

- *E-step.* In the n-th iteration, compute $h_n^{(j)}(x_t)$ for each j and t using Eqs. 3.2.14 and 3.2.16. This is followed by the M-step described next.

- *M-step.* Maximize $Q(\theta|\theta_n)$ with respect to θ to find θ^*. Replace θ_n by θ^*. Then, increment n by 1 and repeat the E-step until convergence.

To determine $\mu^{(k)*}$, set $\frac{\partial Q(\theta|\theta_n)}{\partial \mu^{(k)}} = 0$, which gives

$$\mu^{(k)*} = \frac{\sum_{t=1}^{T} h_n^{(k)}(x_t)x_t}{\sum_{t=1}^{T} h_n^{(k)}(x_t)}. \tag{3.2.17}$$

To determine $\Sigma^{(k)*}$, set $\frac{\partial Q(\theta|\theta_n)}{\partial \Sigma^{(k)}} = 0$, which gives

$$\Sigma^{(k)*} = \frac{\sum_{t=1}^{T} h_n^{(k)}(x_t)(x_t - \mu^{(k)*})(x_t - \mu^{(k)*})^T}{\sum_{t=1}^{T} h_n^{(k)}(x_t)}. \tag{3.2.18}$$

To determine $\pi^{(k)*}$, maximize $Q(\theta|\theta_n)$ with respect to $\pi^{(k)}$ subject to the constraint $\sum_{j=1}^{J} \pi^{(j)} = 1$, which gives

$$\pi^{(k)*} = \frac{1}{T} \sum_{t=1}^{T} h_n^{(k)}(x_t). \tag{3.2.19}$$

The detailed derivations of Eq. 3.2.17 to Eq. 3.2.19 are as follows:

$$\frac{\partial Q(\theta|\theta_n)}{\partial \mu^{(k)}} = \sum_{t=1}^{T} \sum_{j=1}^{J} h_n^{(j)}(x_t) \frac{\partial}{\partial \mu^{(k)}} \log \left\{ p(x_t|\delta_t^{(j)} = 1, \phi^{(j)}) \right\}$$

$$= \sum_{t=1}^{T} \sum_{j=1}^{J} h_n^{(j)}(x_t) \frac{1}{p(x_t|\delta_t^{(j)} = 1, \phi^{(j)})} \frac{\partial}{\partial \mu^{(k)}} p(x_t|\delta_t^{(j)} = 1, \phi^{(j)})$$

$$= -\frac{1}{2} \sum_{t=1}^{T} \sum_{j=1}^{J} h_n^{(j)}(x_t) \cdot \frac{\partial}{\partial \mu^{(k)}} \left\{ x_t^T (\Sigma^{(j)})^{-1} x_t - (\mu^{(j)})^T (\Sigma^{(j)})^{-1} x_t \right.$$

$$\left. + (\mu^{(j)})^T (\Sigma^{(j)})^{-1} \mu^{(j)} - x_t^T (\Sigma^{(j)})^{-1} \mu^{(j)} \right\}$$

$$= -\frac{1}{2} \sum_{t=1}^{T} h_n^{(k)}(x_t) \left\{ (0 - (\Sigma^{(k)})^{-1} x_t + (\Sigma^{(k)})^{-1} \mu^{(k)} + ((\Sigma^{(k)})^{-1})^T \mu^{(k)} \right.$$

$$\left. - ((\Sigma^{(k)})^{-1})^T x_t) \right\}$$

$$= \sum_{t=1}^{T} h_n^{(k)}(x_t)(\Sigma^{(k)})^{-1}(\mu^{(k)} - x_t) = 0$$

$$\implies \mu^{(k)*} = \frac{\sum_{t=1}^{T} h_n^{(k)}(x_t)x_t}{\sum_{t=1}^{T} h_n^{(k)}(x_t)}. \tag{3.2.20}$$

To determine $\Sigma^{(k)*}$, $k = 1, \ldots, J$, let $\Lambda^{(k)} = (\Sigma^{(k)})^{-1}$ and set $\frac{\partial Q(\theta|\theta_n)}{\partial \Lambda^{(k)}} = 0$, that is,

$$\frac{\partial Q(\theta|\theta_n)}{\partial \Lambda^{(k)}} = \sum_{t=1}^{T} \sum_{j=1}^{J} h_n^{(j)}(x_t) \frac{\partial}{\partial \Lambda^{(k)}} \log \left\{ p(x_t|\delta_t^{(j)} = 1, \phi^{(j)}) \right\}$$

$$= \sum_{t=1}^{T} \sum_{j=1}^{J} h_n^{(j)}(x_t) \frac{\partial}{\partial \Lambda^{(k)}}$$

$$\left\{ -\frac{D}{2}\log(2\pi) + \frac{1}{2}\log|\Lambda^{(j)}| - \frac{1}{2}(x_t - \mu^{(j)})^T \Lambda^{(j)}(x_t - \mu^{(j)}) \right\}$$

$$= \sum_{t=1}^{T} \sum_{j=1}^{J} h_n^{(j)}(x_t)$$

$$\left\{ \frac{1}{2}\frac{\partial}{\partial \Lambda^{(k)}}\log|\Lambda^{(j)}| - \frac{1}{2}\frac{\partial}{\partial \Lambda^{(k)}}(x_t - \mu^{(j)})^T \Lambda^{(j)}(x_t - \mu^{(j)}) \right\}$$

$$= \frac{1}{2}\sum_{t=1}^{T} h_n^{(k)}(x_t) \left\{ \frac{\partial}{\partial \Lambda^{(k)}}\log|\Lambda^{(k)}| - \frac{\partial}{\partial \Lambda^{(k)}}(x_t - \mu^{(k)})^T \Lambda^{(k)}(x_t - \mu^{(k)}) \right\}$$

$$= \frac{1}{2}\sum_{t=1}^{T} h_n^{(k)}(x_t) \left\{ \frac{1}{|\Lambda^{(k)}|}\frac{\partial|\Lambda^{(k)}|}{\partial \Lambda^{(k)}} - (x_t - \mu^{(k)})(x_t - \mu^{(k)})^T \right\}$$

$$= \frac{1}{2}\sum_{t=1}^{T} h_n^{(k)}(x_t) \left\{ \frac{1}{|\Lambda^{(k)}|}|\Lambda^{(k)}|(\Lambda^{(k)})^{-1} - (x_t - \mu^{(k)})(x_t - \mu^{(k)})^T \right\}$$

$$= \frac{1}{2}\sum_{t=1}^{T} h_n^{(k)}(x_t) \left\{ (\Lambda^{(k)})^{-1} - (x_t - \mu^{(k)})(x_t - \mu^{(k)})^T \right\} = 0 \tag{3.2.21}$$

$$\implies (\Lambda^{(k)*})^{-1} = \Sigma^{(k)*} = \frac{\sum_{t=1}^{T} h_n^{(k)}(x_t)(x_t - \mu^{(k)})(x_t - \mu^{(k)})^T}{\sum_{t=1}^{T} h_n^{(k)}(x_t)} \tag{3.2.22}$$

Note that Eq. 3.2.21 makes use of the identity $\frac{\partial|A|}{\partial A} = |A|A^{-1}$, where A is a symmetric matrix. Note also that one can replace $\mu^{(k)}$ by $\mu^{(k)*}$ in Eq. 3.2.20 to obtain Eq. 3.2.18.

To determine $\pi^{(r)}, r = 1, \ldots, J$, maximize $Q(\theta|\theta_n)$ with respect to $\pi^{(r)}$ subject to the constraint $\sum_{j=1}^{J} \pi^{(j)} = 1$. More specifically, maximize the function $f(\lambda, \pi^{(j)}) = Q(\theta|\theta_n) + \lambda(\sum_{j=1}^{J} \pi^{(j)} - 1)$ where λ is the Lagrange multiplier. Setting $\frac{\partial f(\lambda, \pi^{(j)})}{\partial \pi^{(r)}} = 0$ results in

$$\frac{\partial Q(\theta|\theta_n)}{\partial \pi^{(r)}} + \lambda = 0 \tag{3.2.23}$$

$$\Longrightarrow \lambda = -\sum_{t=1}^{T} h_n^{(r)}(x_t) \frac{\partial}{\partial \pi^{(r)}} \log \pi^{(r)}$$

$$\Longrightarrow \lambda \pi^{(r)*} = -\sum_{t=1}^{T} h_n^{(r)}(x_t). \tag{3.2.24}$$

Summing both size of Eq. 3.2.24 from $r = 1$ to J, one has

$$\lambda \sum_{r=1}^{J} \pi^{(r)*} = -\sum_{t=1}^{T} \sum_{r=1}^{J} h_n^{(r)}(x_t) \tag{3.2.25}$$

$$\Longrightarrow \lambda = -\sum_{t=1}^{T} \sum_{r=1}^{J} h_n^{(r)}(x_t) = -\sum_{t=1}^{T} 1 = -T. \tag{3.2.26}$$

Substituting Eq. 3.2.26 into Eq. 3.2.24 results in

$$\pi^{(k)*} = \frac{1}{T} \sum_{t=1}^{T} h_n^{(k)}(x_t). \tag{3.2.27}$$

Complexity of EM. Let T denote the number of patterns, J the number of mixtures, and D the feature dimension, then the following is a rough estimation of the computation complexity of using EM to train a GMM:

- *E-step.* $\mathcal{O}(TJD + TJ)$ for each epoch.

- *M-step.* $\mathcal{O}(2TJD)$ for each epoch.

Numerical Example 1. This example uses the data in Figure 3.3(a) as the observed data. Assume that when EM begins, $n = 0$ and

$$\theta_0 = \left\{ \pi_0^{(1)}, \{\mu_0^{(1)}, \sigma_0^{(1)}\}, \pi_0^{(2)}, \{\mu_0^{(2)}, \sigma_0^{(2)}\} \right\}$$
$$= \{0.5, \{0, 1\}, 0.5, \{9, 1\}\}.$$

Therefore, one has

$$h_0^{(1)}(x_t) = \frac{\frac{\pi_0^{(1)}}{\sigma_0^{(1)}} e^{-\frac{1}{2}(x_t - \mu^{(1)})^2/(\sigma_0^{(1)})^2}}{\sum_{k=1}^{2} \frac{\pi_0^{(k)}}{\sigma_0^{(k)}} e^{-\frac{1}{2}(x_t - \mu^{(k)})^2/(\sigma_0^{(k)})^2}}$$

$$= \frac{e^{-\frac{1}{2}x_t^2}}{e^{-\frac{1}{2}x_t^2} + e^{-\frac{1}{2}(x_t - 9)^2}} \tag{3.2.28}$$

and

$$h_0^{(2)}(x_t) = \frac{e^{-\frac{1}{2}(x_t - 9)^2}}{e^{-\frac{1}{2}x_t^2} + e^{-\frac{1}{2}(x_t - 9)^2}}. \tag{3.2.29}$$

Table 3.1. Values of $h_0^{(j)}(x_t)$ in Example 1

Pattern Index (t)	Pattern (x_t)	$h_0^{(1)}(x_t)$	$h_0^{(2)}(x_t)$
1	1	1	0
2	2	1	0
3	3	1	0
4	4	1	0
5	6	0	1
6	7	0	1
7	8	0	1

Table 3.2. Values of $Q(\theta|\theta_n)$, $\mu^{(j)}$ and $(\sigma^{(j)})^2$ in the course of EM iterations. Data shown in Figure 3.3(a) were used as the observed data.

Iteration (n)	$Q(\theta|\theta_n)$	$\mu_n^{(1)}$	$(\sigma_n^{(1)})^2$	$\mu_n^{(2)}$	$(\sigma_n^{(2)})^2$
0	$-\infty$	0	1	9	1
1	-43.71	2.50	1.25	6.99	0.70
2	-25.11	2.51	1.29	7.00	0.68
3	-25.11	2.51	1.30	7.00	0.67
4	-25.10	2.52	1.30	7.00	0.67
5	-25.10	2.52	1.30	7.00	0.67

Substituting $X = \{1, 2, 3, 4, 6, 7, 8\}$ into Eqs. 3.2.28 and 3.2.29, Table 3.1 is obtained. Substituting $h_0^{(j)}(x_t)$ in Table 3.1 into Eqs. 3.2.17 through 3.2.19 results in

$$\theta_1 = \{0.57, \{2.50, 1.12\}, 0.43, \{6.99, 0.83\}\}.$$

Then, continue the algorithm by computing $Q(\theta|\theta_1)$—that is, $h_1^{(j)}(x_t)$—which are then substituted into Eqs. 3.2.17 through 3.2.19 to obtain θ_2. Figure 3.5 depicts the movement of the component density functions specified by $\mu^{(j)}$ and $\sigma^{(j)}$ during the EM iterations, and Table 3.2 lists the numerical values of $Q(\theta|\theta_n)$ and θ_n for the first five iterations. It is obvious that the algorithm converges quickly in this example.

3.3 An Entropy Interpretation

The previous section has shown that the EM algorithm is a powerful tool in estimating the parameters of finite-mixture models. This is achieved by iteratively maximizing the expectation of the model's completed-data likelihood function. The model's parameters, however, can also be obtained by maximizing an incomplete-data likelihood function, leading to an entropy interpretation of the EM algorithm.

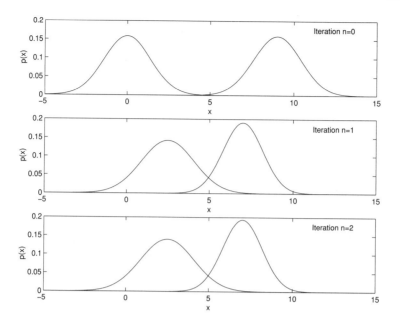

Figure 3.5. Movement of the component density function specified by $\mu^{(j)}$ and $(\sigma^{(j)})^2$ for the first two EM iterations.

3.3.1   Incomplete-Data Likelihood

The optimal estimates are obtained by maximizing

$$
\begin{aligned}
L(X|\theta) &= \sum_{t=1}^{T} \log p(x_t|\theta) \\
&= \sum_{t=1}^{T} \frac{\sum_{j=1}^{J} \pi^{(j)} p(x_t|\delta_t^{(j)} = 1, \phi^{(j)})}{p(x_t|\theta)} \log p(x_t|\theta).
\end{aligned}
\tag{3.3.1}
$$

Define

$$
h^{(j)}(x_t) \equiv \frac{\pi^{(j)} p(x_t|\delta_t^{(j)} = 1, \phi^{(j)})}{p(x_t|\theta)}
$$

such that $\sum_j \pi^{(j)} = 1$ and $p(x_t|\theta) = \sum_j \pi^{(j)} p(x_t|\delta_t^{(j)} = 1, \phi^{(j)})$.[3] Eq. 3.3.1 becomes

$$
L(X|\theta) = \sum_{t=1}^{T} \sum_{j=1}^{J} h^{(j)}(x_t) \log p(x_t|\theta)
$$

---

[3]Note that $h^{(j)}(x_t)$ equals the probability of x_t belonging to the j-th cluster, given x_t and the model—that is, $h^{(j)}(x_t) = Pr(x_t \in j$-th cluster$|x_t, \theta)$; it can be considered a "fuzzy" membership function.

$$
= \sum_{t=1}^{T} \sum_{j=1}^{J} h^{(j)}(x_t) \left\{ \log p(x_t|\theta) - \log \left[\pi^{(j)} p(x_t|\delta_t^{(j)} = 1, \phi^{(j)}) \right] \right.
$$

$$
\left. + \log \left[\pi^{(j)} p(x_t|\delta_t^{(j)} = 1, \phi^{(j)}) \right] \right\}
$$

$$
= \sum_{t=1}^{T} \sum_{j=1}^{J} h^{(j)}(x_t) \left\{ -\log h^{(j)}(x_t) + \log \left[\pi^{(j)} p(x_t|\delta_t^{(j)} = 1, \phi^{(j)}) \right] \right\}
$$

$$
= \sum_{t,j} h^{(j)}(x_t) \log \pi^{(j)} + \sum_{t,j} h^{(j)}(x_t) \log p(x_t|\delta_t^{(j)} = 1, \phi^{(j)})
$$

$$
- \sum_{t,j} h^{(j)}(x_t) \log h^{(j)}(x_t)
$$

$$
\equiv Q + R,
$$

where the first two terms correspond to the Q-term in Eq. 3.2.4 and the second terms corresponds to the R-term in Eq. 3.2.5. This means the maximization of L can be accomplished by maximizing the completed-data likelihood Q, as well as maximizing an entropy term R.

Now, define $s(x_t, \phi^{(j)}) \equiv \log p(x_t|\delta_t^{(j)} = 1, \phi^{(j)})$ so that the likelihood $L(X|\theta)$ can be expressed as:

$$
L(X|\theta) = -\sum_{t,j} h^{(j)}(x_t) \log h^{(j)}(x_t) + \sum_{t,j} h^{(j)}(x_t) \log \pi^{(j)} + \sum_{t,j} h^{(j)}(x_t) s(x_t, \phi^{(j)}).
$$

$$
(3.3.2)
$$

In Eq. 3.3.2, the following three terms have different interpretations:

- The first term can be interpreted as the *entropy* term, which helps induce the membership's fuzziness.

- The second term represents the *prior information*. For each sample x_t, this term grasps the influence (prior probability) of its neighboring clusters; the larger the prior probability, the larger the influence.

- The third term is the *observable-data term*, where $s(x_t, \phi^{(j)})$ represents the influence of the observable data x_t on the total likelihood L.

3.3.2   Simulated Annealing

To control the inference of the entropy terms and the prior information on the total likelihood, one can introduce a temperature parameter σ_T similar to simulated annealing; that is,

$$
L(X|\theta) = -\sigma_T \sum_{t,j} h^{(j)}(x_t) \log h^{(j)}(x_t) \sigma_T \sum_{t,j} h^{(j)}(x_t) \log \pi^{(j)}
$$

$$
+ \sum_{t,j} h^{(j)}(x_t) s^{(j)}(x_t, \phi^{(j)}).
$$

$$
(3.3.3)
$$

Maximization of

$$L(X|\theta) = \sum_t L_t = \sum_t \left[-\sigma_T \sum_j h^{(j)}(x_t) \log h^{(j)}(x_t) + \sigma_T \sum_j h^{(j)}(x_t) \log \pi^{(j)} \right.$$

$$\left. + \sum_j h^{(j)}(x_t) s(x_t, \phi^{(j)}) \right] \tag{3.3.4}$$

can be reformulated as the maximization of L_t under the constraint that

$$\sum_j h^{(j)}(x_t) - 1 = 0.$$

This is achieved by introducing a Lagrange multiplier λ such that

$$\mathcal{L} = L_t + \lambda \left(\sum_j h^{(j)}(x_t) - 1 \right)$$

$$= -\sigma_T \sum_j h^{(j)}(x_t) \log h^{(j)}(x_t) + \sigma_T \sum_j h^{(j)}(x_t) \log \pi^{(j)} + \sum_j h^{(j)}(x_t) s(x_t, \phi^{(j)})$$

$$+ \lambda \left(\sum_j h^{(j)}(x_t) - 1 \right) \tag{3.3.5}$$

is to be maximized. To solve this constrained optimization problem, one needs to apply two different kinds of derivatives, as shown here:

1. $\frac{\partial \mathcal{L}}{\partial h^{(j)}(x_t)} = 0$, which means that

$$-\sigma_T \log h^{(j)}(x_t) - \frac{\sigma_T h^{(j)}(x_t)}{h^{(j)}(x_t)} + \sigma_T \log \pi^{(j)} + s(x_t, \phi^{(j)}) + \lambda = 0$$

that is,

$$h^{(j)}(x_t) = \alpha \pi^{(j)} e^{s(x_t, \phi^{(j)})/\sigma_T}, \tag{3.3.6}$$

where $\alpha = e^{\frac{\lambda}{\sigma_T} - 1}$.

2. $\frac{\partial \mathcal{L}}{\partial \lambda} = 0$, which means that

$$\sum_j h^{(j)}(x_t) - 1 = 0. \tag{3.3.7}$$

Plugging Eq. 3.3.6 into Eq. 3.3.7 results in

$$\alpha = \left(\sum_j \pi^{(j)} e^{s(x_t, \phi^{(j)})/\sigma_T} \right)^{-1}. \tag{3.3.8}$$

Hence, the optimal membership (Eq. 3.3.6) for each data is

$$h^{(j)}(x_t) = \frac{\pi^{(j)} e^{s(x_t, \phi^{(j)})/\sigma_T}}{\sum_k \pi^{(k)} e^{s(x_t, \phi^{(k)})/\sigma_T}}. \tag{3.3.9}$$

It is interesting to note that both Eqs. 3.3.9 and 3.2.14 have the same "marginalized" form. They can be connected by observing that $p(x_t|\delta_t^{(j)} = 1, \phi^{(j)}) \propto e^{s(x_t, \phi^{(j)})/\sigma_T}$ in the case of mixture-of-experts. As an additional bonus, such a connection leads to a claim that the expectation of hidden-states (Eq. 3.2.14) provides an optimal membership estimation.

The role of σ_T can be illustrated by Figure 3.6. For simplicity, only two clusters are considered and both $\pi^{(1)}$ and $\pi^{(2)}$ are initialized to 0.5 before the EM iterations begin. Refer to Figure 3.6(a), where the temperature σ_T is extremely high, there exists a major ambiguity between clusters 1 and 2 (i.e., they have almost equivalent probability). This is because according to Eq. 3.3.9, $h^{(j)}(x_t) \simeq 0.5$ at the first few EM iterations when $\sigma_T \to \infty$. When σ_T decreases during the course of EM iterations, such ambiguity becomes more resolved—cf. Figure 3.6(b). Finally, when σ_T approaches zero, a total "certainty" is reached: the probability that either cluster 1 or 2 will approach 100%—cf. Figure 3.6(c). This can be explained by rewriting Eq. 3.3.9 in the following form (for the case $J = 2$ and $j = 2$):

$$\begin{aligned} h^{(2)}(x_t) &= \frac{\pi^{(2)} e^{s(x_t, \phi^{(2)})/\sigma_T}}{\pi^{(1)} e^{s(x_t, \phi^{(1)}/\sigma_T)} + \pi^{(2)} e^{s(x_t, \phi^{(2)})/\sigma_T}} \\ &= \frac{\frac{\pi^{(2)}}{\pi^{(1)}} e^{s(x_t, \phi^{(2)})/\sigma_T - s(x_t, \phi^{(1)})/\sigma_T}}{1 + \frac{\pi^{(2)}}{\pi^{(1)}} e^{s(x_t, \phi^{(2)})/\sigma_T - s(x_t, \phi^{(1)})/\sigma_T}}. \end{aligned} \tag{3.3.10}$$

In Eq. 3.3.10, when $\sigma_T \to 0$ and $s(x_t, \phi^{(2)}) > s(x_t, \phi^{(1)})$, $h^{(2)}(x_t) \simeq 1.0$, and $h^{(1)}(x_t) \simeq 0.0$. This means that x_t is closer to cluster 2 than to cluster 1. Similarly, $h^{(2)}(x_t) \simeq 0.0$ and $h^{(1)}(x_t) \simeq 1.0$ when $s(x_t, \phi^{(2)}) < s(x_t, \phi^{(1)})$. Therefore, Eq. 3.3.10 suggests that when $\sigma_T \to 0$, there is a hard-decision clustering (i.e., with cluster probabilities equal to either 1 or 0). This demonstrates that σ_T plays the same role as the temperature parameter in the simulated annealing method. It is a common practice to use annealing temperature schedules to force a more certain classification (i.e., starting with a higher σ_T and then gradually decreasing σ_T to a lower value as iterations progress).

3.3.3 EM Iterations

Next, the optimization formulation described in Section 3.2 is slightly modified (but causes no net effect). The EM problem can be expressed as one that maximizes L with respect to both (1) the model parameters $\theta = \{\theta^{(j)}, \forall j\}$ and (2) the membership function $\{h^{(j)}(x_t), \forall t \text{ and } j\}$. The interplay of these two sets of variables can hopefully induce a bootstrapping effect facilitating the convergence process. The list that follows further elaborates on this.

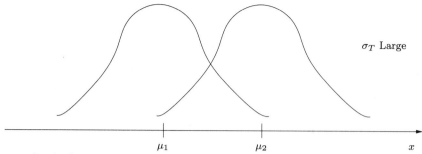

At the beginning of the EM algorithm, σ_T is set to a large value.

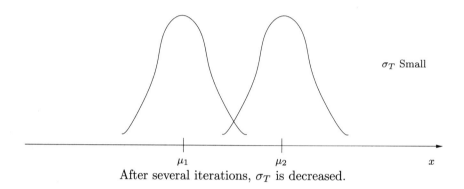

After several iterations, σ_T is decreased.

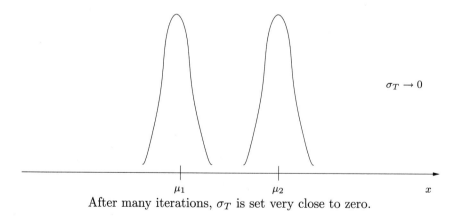

After many iterations, σ_T is set very close to zero.

Figure 3.6. This figure demonstrates how the temperature σ_T can be applied to control the convergence of the EM algorithm.

- In the E-step, while fixing the model parameter $\theta = \{\theta^{(j)}, \forall j\}$, one can find the best cluster probability $h^{(j)}(x_t)$ to optimize L with the constraint $\sum_{j=1}^{J} h^{(j)}(x_t) = 1$, which gave Eq. 3.3.9.

- In the M-step, one searches for the best model parameter $\theta = \{\theta^{(j)}, \forall j\}$ that optimizes L, while fixing the cluster probability $h^{(j)}(x_t), \forall t$ and j.

3.3.4 Special Case: GMM

When θ defines a GMM, $s(x_t, \phi^{(j)})$ becomes

$$s(x_t, \phi^{(j)}) = -\frac{D}{2} \log 2\pi - \frac{1}{2} \log |\Sigma^{(j)}| - \frac{1}{2}(x_t - \mu^{(j)})^T (\Sigma^{(j)})^{-1}(x_t - \mu^{(j)}). \quad (3.3.11)$$

Ignoring terms independent of $h^{(j)}(x_t)$, $\mu^{(j)}$, $\Sigma^{(j)}$, and $\pi^{(j)}$, the likelihood function in Eq. 3.3.2 can be rewritten as:

$$L(X|\theta) = -\sum_{t,j} h^{(j)}(x_t) \log h^{(j)}(x_t) + \sum_{t,j} h^{(j)}(x_t) \log \pi^{(j)} -$$
$$\sum_{t,j} h^{(j)}(x_t) \left\{ \frac{1}{2} \log |\Sigma^{(j)}| + \frac{1}{2}(x_t - \mu^{(j)})^T (\Sigma^{(j)})^{-1}(x_t - \mu^{(j)}) \right\}.$$
$$(3.3.12)$$

Note that the maximization of Eq. 3.3.12 with respect to θ leads to the same maximum likelihood estimtates as shown in Section 3.2.4.

For RBF- and EBF-type likelihood functions, the parameters that maximize $s(x_t, \phi^{(j)})$ can be obtained analytically (see Section 3.2.4), which simplifies the optimization process. On the other hand, if a linear model (e.g. LBF) is chosen to parameterize the likelihood, an iterative method is needed to achieve the optimal solutions in the M-step. In other words, the EM algorithm becomes a double-loop optimization known as the generalized EM. For example, Jordan and Jacobs [168] applied a Fisher scoring method called *iteratively reweighted least squares* (IRLS) to train the LBF mixture-of-experts network.

3.3.5 K-Means versus EM

K-means [85] and VQ [118] are often used interchangeably: They classify input patterns based on the *nearest-neighbor rule*. The task is to cluster a given data set $X = \{x_t; t = 1, \ldots, T\}$ into K groups, each represented by its centroid denoted by $\mu^{(j)}, j = 1, \ldots, K$. The nearest-neighbor rule assigns a pattern x to the class associated with its nearest centroid, say $\mu^{(i)}$. K-means and VQ have simple learning rules and the classification scheme is straightforward. In Eq. 3.3.12, when $h^{(j)}(x_t)$ implements a hard-decision scheme—that is, $h^{(j)}(x_t) = 1$ for the members only, otherwise $h^{(j)}(x_t) = 0$—and $\Sigma^{(j)} = c^2 I \; \forall j$, where c is a constant and I is an

Table 3.3. Learning algorithms as a result of optimizing Eq. 3.3.12 using different kernel types and decision types. RBF and EBF stand for radial basis functions and elliptical basis functions, respectively. Note that EM types of learning occur whenever the decisions in $h^{(j)}(x_t)$ are soft.

Kernel Type	$\Sigma^{(j)}$	$h^{(j)}(x_t)$	Learning Algorithm
RBF	Diagonal	Hard	K-means with Euclidean distance
		Soft	EM with Euclidean distance
EBF	Nondiagonal,	Hard	K-means with Mahalanobis distance
	symmetric	Soft	EM with Mahalanobis distance

identity matrix, the maximization of Eq. 3.3.12 reduces to the minimization of

$$E(h, X) = \sum_t \sum_{j=1}^K h^{(j)}(x_t)\|x_t - \mu^{(j)}\|^2. \qquad (3.3.13)$$

Therefore, the K-means algorithm aims to minimize the sum of squared error with K clusters.

The EM scheme can be seen as a generalized version of K-means clustering. In other words, K-means clustering is a special case of the EM scheme (cf. Figure 3.2). Table 3.3 summarizes the kinds of learning algorithms that the EM formulation Eq. 3.3.12 can produce.

3.4   Doubly-Stochastic EM

This section presents an EM-based algorithm for problems that possesses partial data with multiple clusters. The algorithm is referred to as as a doubly-stochastic EM. To facilitate the derivation, adopt the following notations:

- $X = \{x_t \in \Re^D; t = 1, \ldots, T\}$ is a sequence of partial-data.

- $Z = \{z_t \in \mathcal{C}; t = 1, \ldots, T\}$ is the set of hidden-states.

- $\mathcal{C} = \{\mathcal{C}^{(1)}, \ldots, \mathcal{C}^{(J)}\}$, where J is the number of hidden-states.

- $\Gamma = \{\gamma^{(1)}, \ldots, \gamma^{(K)}\}$ is the set of values that x_t can attain, where K is the number of possible values for x_t.

Also define two sets of indicator variables as:

$$\beta_t^{(k)} = \begin{cases} 1 \text{ if } x_t = \gamma^{(k)} \\ 0 \text{ otherwise} \end{cases}$$

and

$$\delta_t^{(j)} = \left\{ \begin{array}{l} 1 \text{ if } z_t = \mathcal{C}^{(j)} \\ 0 \text{ otherwise.} \end{array} \right.$$

Using these notations and those defined in Section 3.2, $Q(\theta|\theta_n)$ can be written as

$$
\begin{aligned}
Q(\theta|\theta_n) &= E\left\{ \log p(X, Z|\theta, \Gamma)|X, \Gamma, \theta_n \right\} \\
&= E\left\{ \log \prod_{t=1}^{T} p(x_t, z_t|\theta, x_t \in \Gamma) \middle| X, \Gamma, \theta_n \right\} \\
&= E\left\{ \sum_{t=1}^{T}\sum_{j=1}^{J}\sum_{k=1}^{K} \beta_t^{(k)}\delta_t^{(j)} \log p(x_t, z_t|\theta, x_t \in \Gamma) \middle| X, \Gamma, \theta_n \right\} \\
&= \sum_{t=1}^{T}\sum_{j=1}^{J}\sum_{k=1}^{K} E\left\{ \beta_t^{(k)}\delta_t^{(j)} \middle| x_t \in \Gamma, \theta_n \right\} \log p(x_t = \gamma^{(k)}, z_t = \mathcal{C}^{(j)}|x_t \in \Gamma, \theta) \\
&= \sum_{t=1}^{T}\sum_{j=1}^{J}\sum_{k=1}^{K} P(x_t = \gamma^{(k)}, z_t = \mathcal{C}^{(j)}|x_t \in \Gamma, \theta_n) \cdot \\
&\qquad \log p(x_t = \gamma^{(k)}, z_t = \mathcal{C}^{(j)}|x_t \in \Gamma, \theta) \\
&= \sum_{t=1}^{T}\sum_{j=1}^{J}\sum_{k=1}^{K} P(x_t = \gamma^{(k)}|z_t = \mathcal{C}^{(j)}, x_t \in \Gamma, \theta_n) P(z_t = \mathcal{C}^{(j)}|x_t \in \Gamma, \theta_n) \cdot \\
&\qquad \log\left[p(x_t = \gamma^{(k)}|x_t \in \Gamma, z_t = \mathcal{C}^{(j)}, \theta) P(z_t = \mathcal{C}^{(j)}|x_t \in \Gamma, \theta) \right] \\
&= \sum_{t=1}^{T}\sum_{j=1}^{J}\sum_{k=1}^{K} g_n^{(k,j)}(x_t) h_n^{(j)}(x_t) \cdot \\
&\qquad \log\left[p(x_t = \gamma^{(k)}|x_t \in \Gamma, z_t = \mathcal{C}^{(j)}, \phi^{(j)}) \pi^{(j)} \right], \qquad (3.4.1)
\end{aligned}
$$

where

$$g_n^{(k,j)}(x_t) = P(x_t = \gamma^{(k)}|z_t = \mathcal{C}^{(j)}, x_t \in \Gamma, \theta_n) \text{ and}$$

$$h_n^{(j)}(x_t) = P(z_t = \mathcal{C}^{(j)}|x_t \in \Gamma, \theta_n).$$

If θ defines a GMM—that is, $\theta = \{\pi^{(j)}, \mu^{(j)}, \sum^{(j)}\}_{j=1}^{J}$—then

$$
\begin{aligned}
Q(\theta|\theta_n) &= \sum_{t=1}^{T}\sum_{j=1}^{J}\sum_{k=1}^{K} (2\pi)^{-\frac{D}{2}} |\Sigma_n^{(j)}|^{-\frac{1}{2}} \exp\left\{ -\frac{1}{2}(\gamma^{(k)} - \mu_n^{(j)})^T (\Sigma_n^{(j)})^{-1}(\gamma^{(k)} - \mu_n^{(j)}) \right\} \cdot \\
&\qquad h_n^{(j)}(x_t)\left\{ -\frac{D}{2}\log(2\pi) - \frac{1}{2}\log|\Sigma^{(j)}| \right. \\
&\qquad \left. -\frac{1}{2}(\gamma^{(k)} - \mu^{(j)})^T (\Sigma^{(j)})^{-1}(\gamma^{(k)} - \mu^{(j)}) + \log \pi^{(j)} \right\}.
\end{aligned}
$$

3.4.1 Singly-Stochastic Single-Cluster with Partial Data

This section demonstrates how the general formulation in Eq. 3.4.1 can be applied to problems with a single cluster and partially observable data. Referring to Example 2 shown in Figure 3.3(b), let $X = \{x_1, x_2, x_3, x_4, y\} = \{1, 2, 3, 4, \{5 \text{ or } 6\}\}$ be the observed data, where $y = \{5 \text{ or } 6\}$ is the observation with missing information. The information is missing because the exact value of y is unknown. Also let $z \in \Gamma$, where $\Gamma = \{\gamma^{(1)}, \gamma^{(2)}\} = \{5, 6\}$, be the missing information. Since there is one cluster only and x_1 to x_4 are certain, define $\theta \equiv \{\mu, \sigma^2\}$, set $\pi^{(1)} = 1.0$ and write Eq. 3.4.1 as

$$Q(\theta|\theta_n) = \sum_{t=1}^{4} \sum_{j=1}^{1} \sum_{k=1}^{1} g_n^{(k,j)}(x_t) h_n^{(j)}(x_t) \log p(x_t|\theta)$$

$$+ \sum_{j=1}^{1} \sum_{k=1}^{2} g_n^{(k,j)}(y) h_n^{(j)}(y) \log p(y = \gamma^{(k)}|y \in \Gamma, \theta)$$

$$= \sum_{t=1}^{4} \log p(x_t|\theta) + \sum_{k=1}^{2} P(y = \gamma^{(k)}|y \in \Gamma, \theta_n) \log p(y = \gamma^{(k)}|y \in \Gamma, \theta).$$

$$(3.4.2)$$

Note that the discrete density $p(y = \gamma^{(k)}|y \in \Gamma, \theta)$ can be interpreted as the product of density $p(y = \gamma^{(k)}|y \in \Gamma)$ and the functional value of $p(y|\theta)$ at $y = \gamma^{(k)}$, as shown in Figure 3.7.

Assume that at the start of the iterations, $n = 0$ and $\theta_0 = \{\mu_0, \sigma_0^2\} = \{0, 1\}$. Then, Eq. 3.4.2 becomes

$$Q(\theta|\theta_0) = \sum_{t=1}^{4} \log p(x_t|\theta) + P(y = 5|y \in \Gamma, \theta_0) \log p(y = 5|y \in \Gamma, \theta)$$

$$+ P(y = 6|y \in \Gamma, \theta_0) \log p(y = 6|y \in \Gamma, \theta)$$

$$= \text{Const.} - 4 \log \sigma - \sum_{t=1}^{4} \frac{(t - \mu)^2}{2\sigma^2} - $$

$$\frac{\frac{1}{\sqrt{2\pi}\sigma_0} e^{-\frac{(5-\mu_0)^2}{2\sigma_0^2}}}{\frac{1}{\sqrt{2\pi}\sigma_0} e^{-\frac{(5-\mu_0)^2}{2\sigma_0^2}} + \frac{1}{\sqrt{2\pi}\sigma_0} e^{-\frac{(6-\mu_0)^2}{2\sigma_0^2}}} \left[\frac{(5-\mu)^2}{2\sigma^2} + \log \sigma \right] -$$

$$\frac{\frac{1}{\sqrt{2\pi}\sigma_0} e^{-\frac{(6-\mu_0)^2}{2\sigma_0^2}}}{\frac{1}{\sqrt{2\pi}\sigma_0} e^{-\frac{(5-\mu_0)^2}{2\sigma_0^2}} + \frac{1}{\sqrt{2\pi}\sigma_0} e^{-\frac{(6-\mu_0)^2}{2\sigma_0^2}}} \left[\frac{(6-\mu)^2}{2\sigma^2} + \log \sigma \right] -$$

$$= \text{Const.} - 4 \log \sigma - \sum_{t=1}^{4} \frac{(t - \mu)^2}{2\sigma^2} - \left(\frac{e^{-25/2}}{e^{-25/2} + e^{-36/2}} \right) \left[\frac{(5-\mu)^2}{2\sigma^2} + \log \sigma \right]$$

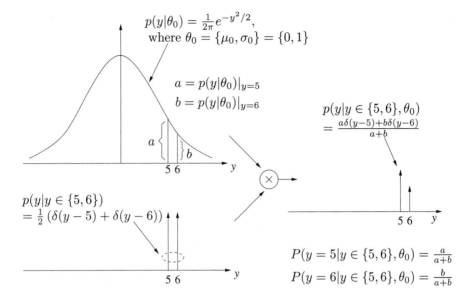

$p(y|\theta_0) = \frac{1}{2\pi}e^{-y^2/2},$
where $\theta_0 = \{\mu_0, \sigma_0\} = \{0, 1\}$

$a = p(y|\theta_0)|_{y=5}$
$b = p(y|\theta_0)|_{y=6}$

$p(y|y \in \{5,6\}, \theta_0)$
$= \frac{a\delta(y-5)+b\delta(y-6)}{a+b}$

$p(y|y \in \{5,6\})$
$= \frac{1}{2}(\delta(y-5)+\delta(y-6))$

$P(y = 5|y \in \{5,6\}, \theta_0) = \frac{a}{a+b}$
$P(y = 6|y \in \{5,6\}, \theta_0) = \frac{b}{a+b}$

Note: Given $y \in \{5, 6\}$, y has non-zero probability only when $y = 5$ or $y = 6$.

Figure 3.7. The relationship between $p(y|\theta_0)$, $p(y|y \in \Gamma)$, $p(y|y \in \Gamma, \theta_0)$, and $P(y = \gamma^{(k)}|y \in \Gamma, \theta_0)$, where $\Gamma = \{5, 6\}$.

$$-\left(\frac{e^{-36/2}}{e^{-25/2} + e^{-36/2}}\right)\left[\frac{(6-\mu)^2}{2\sigma^2} + \log\sigma\right]. \qquad (3.4.3)$$

In the M-step, compute θ_1 according to

$$\theta_1 = \arg\max_\theta Q(\theta|\theta_0).$$

The next iteration replaces θ_0 in Eq. 3.4.3 with θ_1 to compute $Q(\theta|\theta_1)$. The procedure continues until convergence. Table 3.4 shows the value of μ and σ^2 in the course of EM iterations when their initial values are $\mu_0 = 0$ and $\sigma_0^2 = 1$. Figure 3.8 depicts the movement of the Gaussian density function specified by μ and σ^2 during the EM iterations.

3.4.2 Doubly-Stochastic (Partial-Data and Hidden-State) Problem

Here, the single-dimension example shown in Figure 3.9 is used to illustrate the application of Eq. 3.4.1 to problems with partial-data and hidden-states. Review the following definitions:

- $X = \{x_1, x_2, \dots, x_6, y_1, y_2\}$ is the available data with certain $\{x_1, \dots, x_6\}$ and uncertain $\{y_1, y_2\}$ observations.

Table 3.4. Values of μ and σ^2 in the course of EM iterations. Data shown in Figure 3.3(b) were used for the EM iterations.

Iteration (n)	$Q(\theta|\theta_n)$	μ	σ^2
0	$-\infty$	0.00	1.00
1	-29.12	3.00	7.02
2	-4.57	3.08	8.62
3	-4.64	3.09	8.69
4	-4.64	3.09	8.69
5	-4.64	3.09	8.69

- $Z = \{z_1, z_2, \ldots, z_6, z_1', z_2'\}$, where z_t and $z_t' \in \mathcal{C}$ is the set of hidden-states.

- $\Gamma_1 = \{\gamma_1^{(1)}, \gamma_1^{(2)}\} = \{5, 6\}$ and $\Gamma_2 = \{\gamma_2^{(1)}, \gamma_2^{(2)}\} = \{8.9, 9.1\}$ such that $y_1 \in \Gamma_1$ and $y_2 \in \Gamma_2$ are the values attainable by y_1 and y_2, respectively.

- $J = 2$ and $K = 2$.

Using the preceding notations results in

$$
\begin{aligned}
Q(\theta|\theta_n) &= E\left\{\log p(Z, X|\theta, \Gamma_1, \Gamma_2)|X, \theta_n\right\} \\
&= E\left\{\log \prod_{t=1}^{6} p(z_t, x_t|\theta) \prod_{t'=1}^{2} p(z_{t'}, y_{t'}|y_{t'} \in \Gamma_{t'}, \theta)\middle| X, \Gamma_1, \Gamma_2, \theta_n\right\} \\
&= E\left\{\sum_{t=1}^{6} \log p(z_t, x_t|\theta)\middle| X, \Gamma_1, \Gamma_2, \theta_n\right\} \\
&\quad + E\left\{\sum_{t'=1}^{2} \log p(z_{t'}, y_{t'}|y_{t'} \in \Gamma_{t'}, \theta)\middle| X, \Gamma_1, \Gamma_2, \theta_n\right\} \\
&= \sum_{t=1}^{6}\sum_{j=1}^{2} E\left\{\delta_t^{(j)}|x_t, \theta_n\right\} \log p(z_t, x_t|\theta) \\
&\quad + \sum_{t'=1}^{2}\sum_{j=1}^{2}\sum_{k=1}^{2} E\left\{\beta_{t'}^{(k)}\delta_{t'}^{(j)}|y_{t'} \in \Gamma_{t'}, \theta_n\right\} \log p(z_{t'}, y_{t'}|y_{t'} \in \Gamma_{t'}, \theta) \\
&= \sum_{t=1}^{6}\sum_{j=1}^{2} h_n^{(j)}(x_t) \log \left[p(x_t|z_t = \mathcal{C}^{(j)}, \phi^{(j)})\pi^{(j)}\right] \\
&\quad + \sum_{t'=1}^{2}\sum_{j=1}^{2}\sum_{k=1}^{2} P(y_{t'} = \gamma_{t'}^{(k)}|z_{t'} = \mathcal{C}^{(j)}, y_{t'} \in \Gamma_{t'}, \theta_n)
\end{aligned}
$$

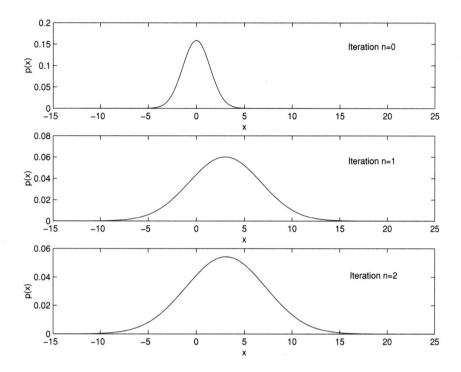

Figure 3.8. Movement of a Gaussian density function during the EM iterations. The density function is to fit the data containing a single cluster with partially observable data.

$$
\begin{aligned}
\cdot P(z_{t'} &= \mathcal{C}^{(j)}|y_{t'} \in \Gamma_{t'}, \theta_n) \\
&\cdot \log\left[p(y_{t'} = \gamma_{t'}^{(k)}|z_{t'} = \mathcal{C}^{(j)}, y_{t'} \in \Gamma_{t'}, \theta)P(z_{t'} = \mathcal{C}^{(j)}|\theta)\right]
\end{aligned}
$$

$$
= \sum_{t=1}^{6}\sum_{j=1}^{2} h_n^{(j)}(x_t)\log\left[p(x_t|z_t = \mathcal{C}^{(j)}, \phi^{(j)})\pi^{(j)}\right]
$$

$$
+ \sum_{t'=1}^{2}\sum_{j=1}^{2}\sum_{k=1}^{2} P(y_{t'} = \gamma_{t'}^{(k)}|z_{t'} = \mathcal{C}^{(j)}, y_{t'} \in \Gamma_{t'}, \theta_n)h_n^{(j)}(y_{t'})
$$

$$
\cdot \log\left[p(y_{t'} = \gamma_{t'}^{(k)}|z_{t'} = \mathcal{C}^{(j)}, y_{t'} \in \Gamma_{t'}, \phi^{(j)})\pi^{(j)}\right]
$$

$$
= \sum_{t=1}^{6}\sum_{j=1}^{2} h_n^{(j)}(x_t)\log\left[p(x_t|z_t = \mathcal{C}^{(j)}, \phi^{(j)})\pi^{(j)}\right]
$$

$$
+ \sum_{t'=1}^{2}\sum_{j=1}^{2}\sum_{k=1}^{2} g_n^{(k,j)}(y_{t'})h_n^{(j)}(y_{t'})
$$

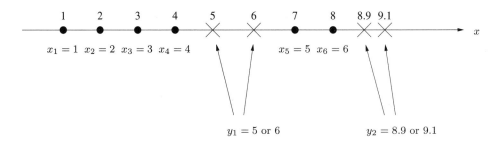

Figure 3.9. Single-dimension example illustrating the idea of hidden-states and partial-data.

$$\cdot \log \left[p(y_{t'} = \gamma_{t'}^{(k)} | z_{t'} = \mathcal{C}^{(j)}, y_{t'} \in \Gamma_{t'}, \phi^{(j)}) \pi^{(j)} \right], \qquad (3.4.4)$$

where $g_n^{(k,j)} = P(y_{t'} = \gamma_{t'}^{(k)} | z_{t'} = \mathcal{C}^{(j)}, y_{t'} \in \Gamma_{t'}, \theta_n)$ is the posterior probability that $y_{t'}$ is equal to $\gamma_{t'}^{(k)}$ given that $y_{t'}$ is generated by cluster $\mathcal{C}^{(j)}$. Note that when the values of y_1 and y_2 are certain (e.g., it is known that $y_1 = 5$, and $\gamma_2^{(1)}$ and $\gamma_2^{(2)}$ become so close that we can consider $y_2 = 9$), then $K = 1$ and $\Gamma_1 = \{\gamma_1\} = \{5\}$ and $\Gamma_2 = \{\gamma_2\} = \{9\}$. In such cases, the second term of Eq. 3.4.4 becomes

$$\sum_{t'=1}^{2} \sum_{j=1}^{2} P(y_{t'} = \gamma_{t'} | z_{t'} = \mathcal{C}^{(j)}, y_{t'} \in \Gamma_{t'}, \theta_n) h_n^{(j)}(y_{t'})$$

$$\log \left[p(y_{t'} = \gamma_{t'} | z_{t'} = \mathcal{C}^{(j)}, y_{t'} \in \Gamma_{t'}, \phi^{(j)}) \pi^{(j)} \right]$$

$$= \sum_{t'=1}^{2} \sum_{j=1}^{2} h_n^{(j)}(y_{t'}) \log \left[p(y_{t'} | z_{t'} = \mathcal{C}^{(j)}, \phi^{(j)}) \pi^{(j)} \right]. \qquad (3.4.5)$$

Replacing the second term of Eq. 3.4.4 by Eq. 3.4.5 and seting $x_7 = y_1$ and $x_8 = y_2$ results in

$$Q(\theta|\theta_n) = \sum_{t=1}^{8} \sum_{j=1}^{2} h_n^{(j)}(x_t) \log \left[p(x_t | z_t = \mathcal{C}^{(j)}, \phi^{(j)}) \pi^{(j)} \right],$$

which is the Q-function of a GMM without partially unknown data with all observable data being certain.

3.5 Concluding Remarks

This chapter has detailed the algorithmic and convergence property of the EM algorithm. The standard EM has also been extended to a more general form called doubly-stochastic EM. A number of numerical examples were given to explain the algorithm's operation. The following summarizes the EM algorithm:

- EM offers an option of "soft" classification.

- EM offers a "soft pruning" mechanism. It is important because features with low probability should not be allowed to unduly influence the training of class parameters.

- EM naturally accommodates model-based clustering formulation.

- EM allows incorporation of prior information.

- EM training algorithm yields probabilistic parameters that are instrumental for media fusion. For linear-media fusion, EM plays a role in training the weights on the fusion layer. This will be elaborated on in subsequent chapters.

Problems

1. Assume that you are given a set of one-dimensional data $X = \{0, 1, 2, 3, 4, 3, 4, 5\}$. Find two cluster centers using

   (a) the K-means algorithm
   (b) the EM algorithm

   You may assume that the initial cluster centers are 0 and 5.

2. Compute the solutions of the single-cluster partial-data problem in Example 2 of Figure 3.3 with the following initial conditions:

$$\theta_0 = \{\mu_0, \sigma_0^2\} = \{-1, 0.5\}$$

3. In each iteration of the EM algorithm, the maximum-likelihood estimates of an M-center GMM are given by

$$\pi_j^{\text{new}} = \frac{\sum_{\mathbf{x} \in X} h_j(\mathbf{x})}{\sum_{\mathbf{x} \in X} 1}, \quad \mu_j^{\text{new}} = \frac{\sum_{\mathbf{x} \in X} h_j(\mathbf{x})\mathbf{x}}{\sum_{\mathbf{x} \in X} h_j(\mathbf{x})}, \text{ and}$$

$$\Sigma_j^{\text{new}} = \frac{\sum_{\mathbf{x} \in X} h_j(\mathbf{x})(\mathbf{x} - \mu_j^{\text{new}})(\mathbf{x} - \mu_j^{\text{new}})^T}{\sum_{\mathbf{x} \in X} h_j(\mathbf{x})},$$

where

$$h_j(\mathbf{x}) = \frac{\pi_j^{\text{old}} p_j(\mathbf{x}|\mu_j^{\text{old}}, \Sigma_j^{\text{old}})}{\sum_{k=1}^{M} \pi_j^{\text{old}} p_k(\mathbf{x}|\mu_j^{\text{old}}, \Sigma_j^{\text{old}})}.$$

X is the set of observed samples. $\{\pi_j^{\text{old}}, \mu_j^{\text{old}}, \Sigma_j^{\text{old}}\}_{j=1}^{M}$ are the maximum likelihood estimates of the last EM iteration, and

$$p_j(\mathbf{x}|\mu_j^{\text{old}}, \Sigma_j^{\text{old}}) = \frac{1}{(2\pi)^{\frac{D}{2}} \left|\Sigma_j^{\text{old}}\right|^{\frac{1}{2}}} \exp\left\{-\frac{1}{2}(\mathbf{x} - \mu_j^{\text{old}})^T (\Sigma_j^{\text{old}})^{-1}(\mathbf{x} - \mu_j^{\text{old}})\right\}.$$

State the condition in which the EM algorithm reduces to the K-means algorithm.

4. You are given a set $X = \{\mathbf{x}_1, \ldots, \mathbf{x}_T\}$ of T unlabeled samples drawn independently from a population whose density function is approximated by a GMM of the form

$$p(\mathbf{x}) \approx p(\mathbf{x}|\theta) = \sum_{j=1}^{M} \pi_j p(\mathbf{x}|\theta_j),$$

where $\theta = \{\theta_j\}_{j=1}^{M} = \{\mu_j, \Sigma_j\}_{j=1}^{M}$ are the means and covariance matrices of the component densities $\{p(\mathbf{x}|\theta_j)\}_{j=1}^{M}$. Assume that π_j are known, θ_j and $\theta_i(i \neq j)$ are functionally independent, and $\mathbf{x}_t$ are statistically independent.

(a) Show that the log-likelihood function is $L(X|\theta) = \sum_{t=1}^{T} \log p(\mathbf{x}_t|\theta)$.

(b) Show that the maximum-likelihood estimate $\hat{\theta}_i$ that maximizes L satisfies the conditions

$$\sum_{t=1}^{T} P(\pi_i|\mathbf{x}_t, \hat{\theta}) \frac{\partial}{\partial \theta_i} \log p(\mathbf{x}_t|\hat{\theta}_i) = 0 \quad i = 1, \ldots, M,$$

where $P(\pi_i|\mathbf{x}_t, \hat{\theta}) = p(\mathbf{x}_t|\hat{\theta}_i)\pi_i/p(\mathbf{x}_t|\hat{\theta})$ is the posterior probability that the i-th cluster generates $\mathbf{x}_t$.

(c) Hence, show that if $\{\Sigma_j\}_{j=1}^{M}$ are known, the maximum-likelihood estimate $\hat{\mu}_i$, $i = 1, \ldots, M$ are given by

$$\hat{\mu}_i = \frac{\sum_{t=1}^{T} P(\pi_i|\mathbf{x}_t, \hat{\theta})\mathbf{x}_t}{\sum_{t=1}^{T} P(\pi_i|\mathbf{x}_t, \hat{\theta})}.$$

(d) Hence, state the conditions where the equation in Problem 4c reduces to the K-means algorithm. State also the condition where the K-means algorithm and the equation in Problem 4c give similar solutions.

5. Based on the normal distribution

$$p(\vec{x}; \vec{\mu}, \Sigma) = \frac{1}{(2\pi)^{D/2} |\Sigma|^{1/2}} \exp\left\{-\frac{1}{2}(\vec{x} - \vec{\mu})^T \Sigma^{-1} (\vec{x} - \vec{\mu})\right\}$$

in D-dimensions, show that the mean vector and covariance matrix that maximize the log-likelihood function

$$L(X; \vec{\mu}, \Sigma) = \log p(X; \vec{\mu}, \Sigma) = \log \prod_{\vec{x} \in X} p(\vec{x}; \vec{\mu}, \Sigma)$$

are, respectively, given by

$$\vec{\hat{\mu}} = \frac{1}{N} \sum_{\vec{x} \in X} \vec{x} \quad \text{and} \quad \hat{\Sigma} = \frac{1}{N} \sum_{\vec{x} \in X} (\vec{x} - \vec{\hat{\mu}})(\vec{x} - \vec{\hat{\mu}})^T,$$

where X is a set of samples drawn from the population with distribution $p(\vec{x}; \vec{\mu}, \Sigma)$, N is the number of samples in X, and T denotes matrix transpose.

Hint: Use the derivatives

$$\frac{\partial}{\partial \vec{x}}\left(\vec{x}^T \vec{y}\right) = \vec{y}, \; \frac{\partial}{\partial \vec{x}}\left(\vec{x}^T A \vec{y}\right) = A\vec{y}, \; \frac{\partial}{\partial \vec{x}}\left(\vec{x}^T A \vec{x}\right) = A\vec{x} + A^T \vec{x},$$
$$\frac{\partial}{\partial A}\left(\vec{x}^T A \vec{x}\right) = \vec{x}\vec{x}^T, \text{ and } \frac{\partial |A|}{\partial A} = |A| A^{-1},$$

where A is a symmetric matrix.

6. Let $p(\mathbf{x}|\Sigma) \equiv \mathcal{N}(\mu, \Sigma)$ be a D-dimensional Gaussian density function with mean vector μ and covariance matrix Σ. Show that if μ is known and Σ is unknown, the maximum-likelihood estimate for Σ is given by

$$\hat{\Sigma} = \frac{1}{N} \sum_{k=1}^{N} (\mathbf{x}_k - \mu)(\mathbf{x}_k - \mu)^T.$$

7. *LBF-Type EM Methods:* The fitness criterion for an RBF-type EM is determined by the closeness of a cluster member to a designated centroid of the cluster. As to its LBF-type counterpart, the fitness criterion hinges on the closeness of a subset of data to a linear plane (more exactly, hyperplane). In an exact fit, an ideal hyperplane is prescribed by a system of linear equations: $A\mathbf{x}_t + \mathbf{b} = \mathbf{0}$, where $\mathbf{x}_t$ is a data point on the plane. When the data are approximated by the hyperplane, then the following fitness function

$$\|A\mathbf{x}_t + \mathbf{b}\|^2$$

should approach zero. Sometimes, the data distribution can be better represented by more than one hyperplane. In a multiplane model, the data can be effectively represented by say N hyperplanes, which may be derived by minimizing the following LBF fitness function:

$$E(h, X) = \sum_t \sum_{j=1}^{K} h^{(j)}(\mathbf{x}_t)\|A\mathbf{x}_t + \mathbf{b}\|^2, \tag{3.5.1}$$

where $h^{(j)}(\mathbf{x}_t)$ is the membership probability satisfying $\sum_j h^{(j)}(\mathbf{x}_t) = 1$. If $h^{(j)}(\mathbf{x}_t)$ implements a hard-decision scheme, then $h^{(j)}(\mathbf{x}_t) = 1$ for the cluster members only, otherwise $h^{(j)}(\mathbf{x}_t) = 0$.

(a) Compare the LBF-based formulation in Eq. 3.5.1 with the RBF-based formulation in Eq. 3.3.13.

(b) Modify an RBF-based EM Matlab code so that it may be applicable to either RBF-based or LBF-based representation.

8. The following is a useful optimization formulation for the derivation of many EM algorithms. Given N known positive values u_n, where $n = 1, \ldots, N$. The problem is to determine the unknown parameters w_n to maximize the criterion function

$$\sum_{n=1}^{N} u_n \log w_n \qquad (3.5.2)$$

under the constraints $w_n > 0$ and $\sum_{n=1}^{N} w_n = 1$.

   (a) Provide a mathematical proof that the optimal parameters have a closed-form solution:
   $$w_n = \frac{u_n}{\sum_{n=1}^{N} u_n} \quad \text{for} \quad n = 1, \ldots, N.$$

   Hints: refer to Eqs. 3.2.23 through 3.2.26.

   (b) As an application example of the EM formulation in Eq. 3.2.10, what are the parameters corresponding to the known positive values u_n and the unknown positive values w_n. Hence, provide a physical meaning of the criterion function in Eq. 3.5.2.

9. As a numerical example of the preceding problem, given $u_1 = 3$, $u_2 = 4$, and $u_3 = 5$, and denote $x = u_1$, $y = u_2$, and $1 - x - y = u_3$, the criterion function can then be expressed as

$$3 \log(x) \; + \; 4 \log(y) + \; 5 \log(1 - x - y).$$

   (a) Write a simple Matlab program to plot the criterion function over the admissible space $0 < x$, $0 < y$, and $x + y < 1$ (verify this range!).

   (b) Show numerically that the maximum value occurs at $x = \frac{1}{4}$ and $x = \frac{1}{3}$.

10. Suppose that someone is going to train a GMM by the EM algorithm. Let T denote the number of patterns, M the number of mixtures, and D the feature dimension. Show that the orders of computational complexity (in terms of multiplications) for each epoch in the E-step is $\mathcal{O}(TMD + TM)$ and that in the M-step is $\mathcal{O}(2TMD)$.

11. Assume that you are given the following observed data distribution:

$$x_1 = 1, x_2 = 2, x_3 = 4, x_4 = 8, \text{ and } x_5 = 9.$$

Assume also that when EM begins, $n = 0$ and

$$\theta_0 = \left\{ \pi_0^{(1)}, \{\mu_0^{(1)}, \sigma_0^{(1)}\}, \pi_0^{(2)}, \{\mu_0^{(2)}, \sigma_0^{(2)}\} \right\}$$
$$= \{0.5, \{1, 1\}, 0.5, \{9, 1\}\}.$$

(a) Derive

$$h_0^{(1)}(x_t) = \frac{\frac{\pi_0^{(1)}}{\sigma_0^{(1)}} e^{-\frac{1}{2}(x_t - \mu^{(1)})^2/(\sigma_0^{(1)})^2}}{\sum_{k=1}^{2} \frac{\pi_0^{(k)}}{\sigma_0^{(k)}} e^{-\frac{1}{2}(x_t - \mu^{(k)})^2/(\sigma_0^{(k)})^2}}.$$

In a similar manner, derive $h_0^{(2)}(x_t)$.

(b) Substituting $X = \{1, 2, 4, 8, 9\}$ into your derivation to obtain the corresponding membership values.

(c) Substitute the derived membership values into Eqs. 3.2.17 through 3.2.19, and compute the values of the new parameters θ_1.

(d) To do more iterations, one can continue the algorithm by computing $Q(\theta|\theta_1)$, which can again be substituted into Eqs. 3.2.17 through 3.2.19 to obtain θ_2. Perform the iterative process until it converges.

12. It is difficult to provide any definitive assurance on the convergence of the EM algorithm to a global optimal solution. This is especially true when the data vector space has a very high dimension. Fortunately, for many inherently offline applications, there is no pressure to produce results in realtime speed. For such applications, the adoption of a user interface to pinpoint a reasonable initial estimate could prove helpful. To facilitate a visualization-based user interface, it is important to project from the original t-space (via a discriminant axis) to a two-dimensional (or three-dimensional) x-space [367, 370] (see Figure 2.7). Create a Matlab program to execute the following steps:

(a) Project the data set onto a reduced-dimensional x-space, via say PCA.

(b) Select initial cluster centers in the x-space by user's pinpoint. Based on the user-pinpointed membership, perform the EM algorithm in the x-space.

(c) Calculate the values of AIC and MDL to select the number of clusters (see the next problem).

(d) Trace the membership information back to the t-space, and use the membership function as the initial condition and further fine-tune the GMM clustering by the EM algorithm in the t-space.

13. One of the most important factors in data clustering is to select the proper number of clusters. Two prominent criteria for such selections are AIC and MDL [5, 314]. From the literature, find out the differences between AIC and MDL criteria. Do you have a preference and why?

14. Given a set of observed data X, develop Matlab code so that the estimated probability density $p(x)$ can be represented in terms of a set of means and variances.

15. Use Matlab to create a three-component Gaussian mixture distribution with different means and variances for each Gaussian component. Ensure that there is some overlap among the distributions.

    (a) Use 2-mean, 3-mean, and K-mean algorithms to cluster the data.

    (b) Compute the likelihood of the true parameters (means and variances that define the three Gaussian components) and the likelihood of your estimates. Which of your estimates is closest to the true distribution in the maximum-likelihood sense?

    (c) Compute the symmetric divergence between the true Gaussian distributions and your estimates. *Hint*: Given two Gaussian distributions Λ_j and Λ_k with mean vectors μ_j and μ_k and covariance matrices Σ_j and Σ_k, their symmetric divergence is

    $$D(\Lambda_j \| \Lambda_k) = \frac{1}{2} tr \left\{ (\Sigma_j)^{-1} \Sigma_k + (\Sigma_k)^{-1} \Sigma_j - 2I \right\}$$
    $$+ \frac{1}{2} (\mu_j - \mu_k)^T \left[(\Sigma_k)^{-1} + (\Sigma_j)^{-1} \right] (\mu_j - \mu_k),$$

    where I is an identity matrix.

    (d) Repeat (a), (b), and (c) with the EM clustering algorithm.

16. Use Matlab to create a mixture density function with three Gaussian component densities. The prior probabilities should be as follows:

    (a) $P(\omega_1) = 0.2$, $P(\omega_2) = 0.3$, and $P(\omega_3) = 0.5$.
    (b) $P(\omega_1) = 0.1$, $P(\omega_2) = 0.1$, and $P(\omega_3) = 0.8$.

    Use 2-mean, 3-mean, and 4-mean VQ algorithms. Compute the likelihood between the data distribution and your estimate. Repeat the problem with the EM clustering algorithm.

Chapter 4

SUPPORT VECTOR MACHINES

4.1 Introduction

The study of linear classifiers with supervised training has a long history. As early as 1936, Fisher's discriminant had already laid the groundwork for statistical pattern recognition. Later, Rosenblatt's perceptron (1956) was the first neural classifier proposed; it shares many similarities with the learning techniques used in the support vector machine developed by Vapnik [358] and in an earlier paper by Boser, Guyon, and Vapnik [33].

Section 4.2 begins by introducing a simple two-class classifier, then goes on to derive the least-squares classifier and the classical Fisher discriminant linear analysis. The decision boundary of the least-squares classifier and the Fisher classifier is dictated (i.e., supported) by *all* of the training data. In contrast, in Section 4.3, the decision boundary of the support vector machine (SVM) hinges on a specially selected set of training data that serve as support vectors. In a linear SVM, some "important" training data are considered as support vectors that are divided into two groups—one group for each class. The objective of the SVM is to maximize the minimal separation margin of these two groups of support vectors. A notion of fuzzy SVMs is proposed to cope with training data that are not linearly separable. The notion of margin of separation is conveniently replaced by a new notion of *fuzzy separation region*. Mathematically, slack variables are introduced so that linearly separable constraints can be relaxed. To further enhance the SVM's classification capability, nonlinear kernels are adopted in Section 4.5. The use of nonlinear kernels leads to more general and flexible nonlinear decision boundaries. For more in-depth treatment of SVMs for pattern classification, readers are referred to these references [40, 55, 257, 337].

4.2 Fisher's Linear Discriminant Analysis

The study of linear classifiers with supervised training has a long history, going back to that of Fisher's linear discriminant analysis (LDA) and perceptron. The devel-

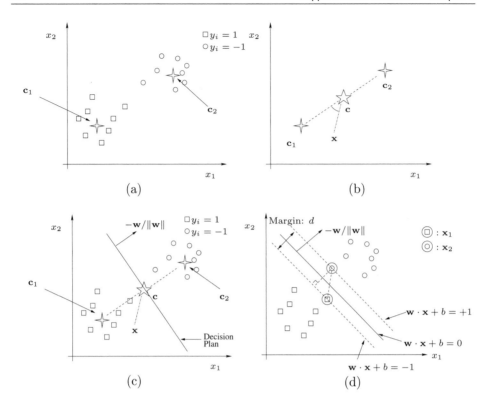

Figure 4.1. Illustration of a two-class classification process, from finding centroids of classes for a basic classifier to the determination of decision and marginal hyperplanes of the SVM classifier. (a) N sets of labeled data $\{\mathbf{x}_i, y_i; i = 1, \ldots, N\} \in X \times \{+1, -1\}$; x_1 and x_2 denote the first two components of $\mathbf{x}$. (b) Classification based on the angle between the unknown vector $\mathbf{x} - \mathbf{c}$ and the vectors formed by the difference between the two class means ($\mathbf{c}_1 - \mathbf{c}_2$). (c) The *decision hyperplane*, or decision boundary, formed by a classifier in which all training data points are considered to be equally important. (d) The decision hyperplane formed by an SVM is shown as the *solid line*; *marginal hyperplanes* are shown as the two *dashed lines*, which are characterized by (the support vectors given in) Eqs. 4.3.5 and 4.3.6. The distance d (as given in Eq. 4.3.8) between the marginal hyperplane (*dashed lines*) and the decision hyperplane (*solid line*) represents the *safety margin* guaranteed by the classifier.

opment of perceptron has led to a vast literature on neural classifiers, especially the class of decision-based neural networks; thus, its discussion is deferred to Chapter 7. This section describes the basic Fisher classifier and highlights its fundamental difference from the support vector machine.

4.2.1   A Two-Class Classifier

Assume that one is given N sets of labeled input/output pairs $\{\mathbf{x}_i, y_i;\ i = 1, \ldots, N\} \in X \times \{+1, -1\}$—shown in Figure 4.1(a)—where X is the set of input data in $\Re^D$ and y_i are the labels. The task is to compute the means of the two classes:

$$\mathbf{c}_1 = \frac{1}{N_1} \sum_{i:y_i=+1} \mathbf{x}_i \quad \text{and} \quad \mathbf{c}_2 = \frac{1}{N_2} \sum_{i:y_i=-1} \mathbf{x}_i, \tag{4.2.1}$$

where N_1 and N_2 are, respectively, the number of labeled pairs for which $y_i = +1$ and $y_i = -1$. During classification, each unknown vector $\mathbf{x}$ is assigned to the class whose mean is closest to $\mathbf{x}$. This can be achieved in two steps:

1. Compute $\mathbf{c} = \frac{1}{2}(\mathbf{c}_1 + \mathbf{c}_2)$.

2. Determine the class of $\mathbf{x}$ by checking whether the vector connecting $\mathbf{x}$ and $\mathbf{c}$ encloses an angle smaller than $\pi/2$ with the vector $\mathbf{w} = \mathbf{c}_1 - \mathbf{c}_2$. More specifically, compute

$$\begin{aligned} y &= \text{sgn}\{(\mathbf{x} - \mathbf{c}) \cdot \mathbf{w}\} \\ &= \text{sgn}\left\{ \left(\mathbf{x} - \frac{(\mathbf{c}_1 + \mathbf{c}_2)}{2} \right) \cdot (\mathbf{c}_1 - \mathbf{c}_2) \right\} \\ &= \text{sgn}\{(\mathbf{x} \cdot \mathbf{c}_1) - (\mathbf{x} \cdot \mathbf{c}_2) + b\}, \end{aligned} \tag{4.2.2}$$

where $b = \frac{1}{2}(\|\mathbf{c}_2\|^2 - \|\mathbf{c}_1\|^2)$.

Figure 4.1(b) illustrates classification procedure.

In the special case where $b = 0$, one has

$$y = \text{sgn}\left\{ \frac{1}{N_1} \sum_{i:y_i=+1} (\mathbf{x} \cdot \mathbf{x}_i) - \frac{1}{N_2} \sum_{i:y_i=-1} (\mathbf{x} \cdot \mathbf{x}_i) \right\}, \tag{4.2.3}$$

which means that *all* data points $\mathbf{x}_i$ are used, each being weighed equally by $1/N_1$ or $1/N_2$, to define the decision plane. Figure 4.1(c) depicts the decision boundary of the two classes specified in Figure 4.1(a).

4.2.2   Fisher's Discriminant Analysis

The earliest and the most well-known example of a linear classifier is Fisher's discriminant [106]. Fisher's idea is based on seeking a direction $\mathbf{w}$ that separates the class means well (when projected onto the direction $\mathbf{w}$) while achieving a small variance around these means.

Given two classes, with means c_1 and c_2, the means of the data projected onto some direction $\mathbf{w}$ can be expressed by

$$\mu_1 = \mathbf{w}^T \mathbf{c}_1 \quad \text{and} \quad \mu_2 = \mathbf{w}^T \mathbf{c}_2. \tag{4.2.4}$$

Similarly, the variances of the projected data can be expressed as

$$\sigma_1^2 = \frac{1}{N_1} \sum_{i:y_i=+1} (\mathbf{w}^T \mathbf{x}_i - \mu_1)^2 \text{ and } \sigma_2^2 = \frac{1}{N_2} \sum_{i:y_i=-1} (\mathbf{w}^T \mathbf{x}_i - \mu_2)^2. \qquad (4.2.5)$$

The goal of Fisher's discriminant is to find the linear projection $\mathbf{w}_{opt}$ that maximizes the contrast criterion—the ratio of intercluster distance to intracluster variance

$$\frac{(\mu_1 - \mu_2)^2}{\sigma_1^2 + \sigma_2^2}. \qquad (4.2.6)$$

Plugging Eqs. 4.2.4 and 4.2.5 into Eq. 4.2.6, the contrast criterion can be expressed as:

$$J(\mathbf{w}) = \frac{\mathbf{w}^T \mathbf{S}_B \mathbf{w}}{\mathbf{w}^T \mathbf{S}_W \mathbf{w}}, \qquad (4.2.7)$$

where the *between-class* and *within-class* scatter matrices are defined as

$$\mathbf{S}_B = (\mathbf{c}_1 - \mathbf{c}_2)(\mathbf{c}_1 - \mathbf{c}_2)^T$$

and

$$\begin{aligned} \mathbf{S}_W &= \mathbf{S}_1 + \mathbf{S}_2 \\ &= \frac{1}{N_1} \sum_{i:y_i=+1} (\mathbf{x}_i - \mathbf{c}_1)(\mathbf{x}_i - \mathbf{c}_1)^T + \frac{1}{N_2} \sum_{i:y_i=-1} (\mathbf{x}_i - \mathbf{c}_2)(\mathbf{x}_i - \mathbf{c}_2)^T, \end{aligned}$$

respectively. The quantity $J(\mathbf{w})$ represents the Fisher discriminant power along the projection direction $\mathbf{w}$.

Fisher's Discriminant Formulated in Eigenvalue Problem

A favorable property of Fisher's discriminant is that the maximization of Eq. 4.2.6 has a global solution, and such a globally optimal solution can be found by solving an eigenvalue problem. More exactly, the *Fisher discriminant power*, defined as the maximum value of $J(\mathbf{w})$, is known to be

$$\max_{\mathbf{w}} J(\mathbf{w}) = \lambda_{\mathbf{S}}, \qquad (4.2.8)$$

where $\lambda_{\mathbf{S}}$ denotes the largest eigenvalue of the matrix $\mathbf{S}_W^{-1}\mathbf{S}_B$. The corresponding optimal projection is therefore along the direction of the principal eigenvector:

$$\mathbf{w}_{opt} = \mathbf{S}_W^{-1}(\mathbf{c}_1 - \mathbf{c}_2). \qquad (4.2.9)$$

In other words, an optimal direction $\mathbf{w}$ can be found by simply inverting the within class scatter matrix $\mathbf{S}_W$.

4.2.3   The Bayes Optimality Property

For convenience of analysis, simple statistical models are often accepted to represent the distribution of data points within the same class. The most popular model is perhaps the single-cluster Gaussian distribution. Such a simple statistical model enjoys the following advantages:

1. The estimation of density parameters (centroid and covariance) for each class is straightforward.

2. There is a closed-form solution for the optimal Bayesian classifier.

More important, assuming that the classes have equal normal conditional distributions, the optimal Bayes classifier is indeed equivalent to Fisher's discriminants [27, 85].[1] It is another promising justification of decision making by using Fisher's classifier.

Note that, for most practical biometric applications, the data are unlikely to fit a single-cluster Gaussian model. Indeed, these applications are often more accurately modelled by a multiple-cluster *Gaussian mixture model* (or GMM). The distribution of the data set within each class is modeled as a mixture of multiple normal distributions, with possibly different means and covariances. Unfortunately, for GMM models, the Bayes optimality property of Fisher's discriminants is no longer valid. (For example, see Problem 3.)

Another concern about using Fisher's criterion is that the criterion aims at maximizing average class distance, projected onto the direction of discrimination, while normalizing intraclass variance. In a sense, its goal is to maximize the average margin (after variance normalization). While it is a reasonable measurement, for certain critical applications, one may want to remove the influence of patterns that are far from the decision boundary, because their influence on decision accuracy can be considered more negligible. This motivates consideration of learning models, such as support vector machines, to be discussed in Section 4.3.

4.2.4   Fisher's Discriminant versus Least-Squares Classifiers

Just like Fisher's criterion, the least-squares criterion makes use of class information to guide its search for a separable projection. However, in the least-squares perspective, it is important to find a linear discriminant function $f(\mathbf{x}) = \mathbf{w}^T\mathbf{x} + b$ such that on the training sample the sum-of-squares error between the outputs $f(\mathbf{x}_i)$ and the known targets y_i is small. Mathematically, one wants to minimize the sum-of-squares error

$$E(\mathbf{w}, b) = \sum_i (f(\mathbf{x}_i) - y_i)^2 = \sum_i (\mathbf{w}^T\mathbf{x}_i + b - y_i)^2.$$

---

[1] This equivalence can be verified by computing the posterior probability using Bayes's theorem, again obtaining the same direction **w**. More exactly, it is equivalent up to a scaling factor to the direction found by Fisher's discriminant. (See Problem 1.)

It can be shown (see Mika [244]) that the solution to the least-squares problem is in the same direction as the solution of Fisher's discriminant (see Problem 2):

$$\mathbf{w} = \beta \mathbf{S}^{-1}(\mathbf{c}_2 - \mathbf{c}_1)$$

and

$$b = \frac{N_2 - N_1 - (N_1 \mathbf{c}_1 + N_2 \mathbf{c}_2)^T \mathbf{w}}{N_1 + N_2},$$

where β is some scaler. This shows that the two classifiers have the same projection direction although they will have a different length and threshold.

4.3 Linear SVMs: Separable Case

The Fisher's criterion aims at maximizing the average class distance, projected onto the direction of discrimination, while having the intraclass variance normalized. While it is a reasonable measurement, for certain critical applications, one may want to remove the influence of patterns that are far from the decision boundary, because their influence on decision accuracy may not be critical. This motivates consideration of learning models (e.g., support vector machines), which, in contrast to Fisher's discriminant, maximize the minimal margin of separation [33, 358]. This notion of separation margin can be further expanded to represent a fuzzy decision region (also called soft-decision region in the literature) under a fuzzy SVM interpretation.

Support vector machines (SVMs) have become popular due to their attractive features and promising performance. Unlike conventional neural networks in which network weights are determined by minimizing the mean-squares error between actual and desired outputs (empirical risk minimization), SVMs optimize their parameters by minimizing classification error (structural risk minimization). Empirical risk minimization relies on the availability of large amounts of training data. In other words, empirical risks will be closed to the true risk when the sample size is large. For problems with a limited amount of training data, even a small empirical risk does not guarantee a low level of true risk. In this circumstance, structural risk minimization, which strikes a balance between empirical risk and complexity of the mapping function (the neural network), is required. Because of this special property of structural risk minimization, SVMs are particularly appropriate for problems in which training data are difficult to obtain. Research has also shown that SVMs possess better generalization capability than conventional neural networks in many application domains, including pattern classification, regression, and density estimation.

To remove the influence of patterns at a distance from the decision boundary, because their influence is usually inconsequential, one might also want to select only a few important data points (known as support vectors) and make the decision exclusively dependent on them. In such cases, an SVM has a decision line as shown in Figure 4.1(d).

For linearly separable problems, such as the one shown in Figure 4.1, support vectors are selected from the training data so that the distance between the two hyperplanes (represented by the dotted lines) passing through the support vectors is at a maximum. Because there is no training data between the two hyperplanes for linearly separable problems, the decision plane, which is midway between the two hyperplanes, should minimize classification error.

4.3.1   Margin of Separation

The SVM approach aims at achieving the largest possible margin of separation. Assume that the decision hyperplane in Figure 4.1(d) can be expressed as

$$\mathbf{x} \cdot \mathbf{w} + b = 0. \tag{4.3.1}$$

Assume also that all training data satisfy the constraints:

$$\mathbf{x}_i \cdot \mathbf{w} + b \geq +1 \quad \text{for} \quad y_i = +1 \quad i = 1, \ldots, N. \tag{4.3.2}$$

$$\mathbf{x}_i \cdot \mathbf{w} + b \leq -1 \quad \text{for} \quad y_i = -1 \quad i = 1, \ldots, N. \tag{4.3.3}$$

If the equality holds for a data point $\mathbf{x}$, then it is said to be right on the *marginal hyperplanes*. Mathematically, the marginal hyperplanes are characterized by

$$\mathbf{x} \cdot \mathbf{w} + b = \pm 1. \tag{4.3.4}$$

Note that data points, say $\mathbf{x}_1$ and $\mathbf{x}_2$, that satisfy

$$\mathbf{x}_1 \cdot \mathbf{w} + b = +1 \quad \text{and} \tag{4.3.5}$$

$$\mathbf{x}_2 \cdot \mathbf{w} + b = -1 \tag{4.3.6}$$

will, respectively, fall on the two hyperplanes that are parallel to the decision plane and orthogonal to $\mathbf{w}$. These two points, denoted as $\mathbf{x}_1$ and $\mathbf{x}_2$, are shown in Figure 4.1(d). Subtracting Eq. 4.3.6 from Eq. 4.3.5 results in

$$\mathbf{w} \cdot (\mathbf{x}_1 - \mathbf{x}_2) = 2$$

$$\Rightarrow \left(\frac{\mathbf{w}}{\|\mathbf{w}\|} \right) \cdot (\mathbf{x}_1 - \mathbf{x}_2) = \frac{2}{\|\mathbf{w}\|}. \tag{4.3.7}$$

Therefore, the distance between the two hyperplanes is

$$2d = \frac{2}{\|\mathbf{w}\|}, \tag{4.3.8}$$

where $2d$ can be considered as the width of separation, which provides an objective measure on how separable the two classes of training data are. The distance d can be considered as the *safety margin* of the classifier. In other words, it can be guaranteed that a pattern can still be correctly classified as long as its deviation from the original training pattern is smaller than d. Equivalently, it means that the pattern retains proximity to the corresponding marginal hyperplane within a distance less than d, thus ensuring a correct decision.

4.3.2 Wolfe Dual Optimization

Now, combine the two constraints in Eqs. 4.3.2 and 4.3.3 into a single constraint to get:

$$y_i(\mathbf{x}_i \cdot \mathbf{w} + b) \geq 1 \quad \forall i = 1, \ldots, N. \tag{4.3.9}$$

During the training phase, the objective is to find the support vectors (e.g., $\mathbf{x}_1$ and $\mathbf{x}_2$) that maximize the *margin of separation d*. Alternatively, the same goal can be achieved by minimizing $\|\mathbf{w}\|^2$. Therefore, the objective is to minimize $\|\mathbf{w}\|^2$ subject to the constraint in Eq. 4.3.9. This is typically solved by introducing Lagrange multipliers $\alpha_i \geq 0$ and a Lagrangian

$$L(\mathbf{w}, b, \alpha) = \frac{1}{2}\|\mathbf{w}\|^2 - \sum_{i=1}^{N} \alpha_i(y_i(\mathbf{x}_i \cdot \mathbf{w} + b) - 1), \tag{4.3.10}$$

where $L(\mathbf{w}, b, \alpha)$ is simultaneously minimized with respect to $\mathbf{w}$ and b and maximized with respect to α_i. Setting

$$\frac{\partial}{\partial b}L(\mathbf{w}, b, \alpha) = 0 \quad \text{and} \quad \frac{\partial}{\partial \mathbf{w}}L(\mathbf{w}, b, \alpha) = 0, \tag{4.3.11}$$

subject to the constraint $\alpha_i \geq 0$, results in

$$\sum_{i=1}^{N} \alpha_i y_i = 0 \quad \text{and} \tag{4.3.12}$$

$$\mathbf{w} = \sum_{i=1}^{N} \alpha_i y_i \mathbf{x}_i. \tag{4.3.13}$$

Substituting Eqs. 4.3.12 and 4.3.13 into Eq. 4.3.10, produces

$$\begin{aligned} L(\mathbf{w}, b, \alpha) &= \frac{1}{2}(\mathbf{w} \cdot \mathbf{w}) - \sum_{i=1}^{N} \alpha_i y_i(\mathbf{x}_i \cdot \mathbf{w}) - \sum_{i=1}^{N} \alpha_i y_i b + \sum_{i=1}^{N} \alpha_i \\ &= \frac{1}{2}\sum_{i=1}^{N} \alpha_i y_i \mathbf{x}_i \cdot \sum_{j=1}^{N} \alpha_j y_j \mathbf{x}_j - \sum_{i=1}^{N} \alpha_i y_i \mathbf{x}_i \cdot \sum_{j=1}^{N} \alpha_j y_j \mathbf{x}_j + \sum_{i=1}^{N} \alpha_i \\ &= \sum_{i=1}^{N} \alpha_i - \frac{1}{2}\sum_{i=1}^{N}\sum_{j=1}^{N} \alpha_i \alpha_j y_i y_j(\mathbf{x}_i \cdot \mathbf{x}_j). \end{aligned} \tag{4.3.14}$$

This results in the following *Wolfe dual optimization* formulation:

$$\max_{\alpha} \sum_{i=1}^{N} \alpha_i - \frac{1}{2}\sum_{i=1}^{N}\sum_{j=1}^{N} \alpha_i \alpha_j y_i y_j(\mathbf{x}_i \cdot \mathbf{x}_j) \tag{4.3.15}$$

subject to

1. $\sum_{i=1}^{N} \alpha_i y_i = 0$ and

2. $\alpha_i \geq 0, \ i = 1, \ldots, N.$

4.3.3   Support Vectors

The solution of the dual optimization problem contains two kinds of Lagrange multipliers α_k: (1) one kind with $\alpha_k = 0$ and (2) the other with non-zero α_k. Data for which the multipliers $\alpha_k = 0$ are considered to be irrelevant to the classification problem. The data points corresponding to non-zero multipliers (i.e., $\alpha_k > 0$) are deemed to be more critical for the classification problem. These critical data points are called *support vectors*.

According to the Karush-Kuhn-Tucker conditions of optimization theory [25, 99], the following equality

$$\alpha_i \left\{ y_i(\mathbf{x}_i \cdot \mathbf{w} + b) - 1 \right\} = 0, \quad \text{for } i = 1, 2, \ldots, N \qquad (4.3.16)$$

must be satisfied at all of the saddle points. Therefore, only those data points with $y_i(\mathbf{x}_i \cdot \mathbf{w} + b) - 1 = 0$ can be support vectors, since they are the only data points that can have a non-zero value of multipliers α_i. These vectors lie on the margin of the separable region as shown by the two dashed lines in Figure 4.1(d). In summary, the support vectors satisfy

$$y_k(\mathbf{x}_k \cdot \mathbf{w} + b) - 1 = 0 \quad \forall k \in \mathcal{S}, \qquad (4.3.17)$$

where $\mathcal{S}$ is a set containing the indexes to the support vectors.

4.3.4   Decision Boundary

Once the values of α_i have been found from Eq. 4.3.15, Eq. 4.3.13 can then be used to determine $\mathbf{w}$:

$$\mathbf{w} = \sum_{i=1}^{N} \alpha_i y_i \mathbf{x}_i.$$

Once $\mathbf{w}$ is found, it can in turn be plugged into Eq. 4.3.17 to determine the value of threshold b:

$$b = 1 - \mathbf{x}_k \cdot \mathbf{w}, \text{ where } y_k = 1 \text{ and } \mathbf{x}_k \in \mathcal{S}.$$

Ultimately, the decision boundary can be derived as follow:

$$f(\mathbf{x}) = \mathbf{w} \cdot \mathbf{x} + b = \sum_{i=1}^{N} y_i \alpha_i (\mathbf{x} \cdot \mathbf{x}_i) + b = 0. \qquad (4.3.18)$$

Analytical Examples

Example 4.1: *3-point problem*

As shown in Figure 4.2(a), there is one data point $(0.0, 0.0)$ for one class and two points in the other: $(0.0, 1.0)$ and $(1.0, 0.0)$. This example presents an analytical

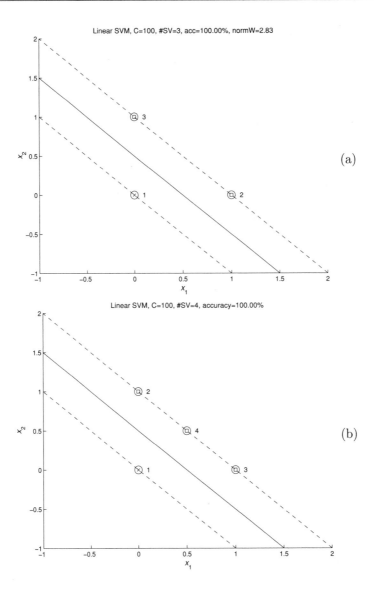

Figure 4.2. Linearly separable problems that can be solved by a linear SVM. (a) A three-point problem in which all data points are SVs. See Example 4.1 in the text for the procedure to find the analytical solutions. (b) A four-point linearly separable problem for Example 4.2; note that this figure only illustrates the case where all four data points are SVs. (c) A four-point linearly separable problem for Example 4.3 in which only three data points are SVs. This figure and others in this chapter were created using the SVM software by Anton Schwaighofer, available from *http://www.igi.tugraz.at/aschwaig*. Used with permission.

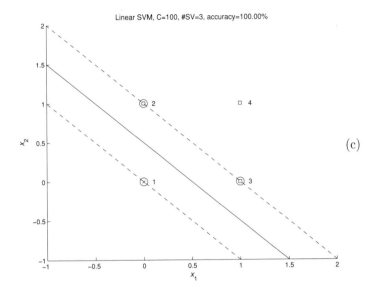

Figure 4.2. (continued)

method to derive the optimal boundary and find the support vectors for the training data points.[2]

Let the training data points be denoted as

$$\mathbf{x}_1 = [0.0 \quad 0.0]^T \qquad y_1 = -1$$
$$\mathbf{x}_2 = [1.0 \quad 0.0]^T \qquad y_2 = +1$$
$$\mathbf{x}_3 = [0.0 \quad 1.0]^T \qquad y_3 = +1.$$

The objective function of interest is now:

$$L(\alpha) = \sum_{i=1}^{3} \alpha_i - \frac{1}{2} \sum_{i=1}^{3} \sum_{j=1}^{3} \alpha_i \alpha_j y_i y_j (\mathbf{x}_i^T \mathbf{x}_j) \qquad (4.3.19)$$

subject to $\alpha_i \geq 0$ and $\sum_{i=1}^{3} \alpha_i y_i = 0$. To solve this constraint optimization problem, another Lagrange multiplier, namely λ, is introduced so that the resulting

---

[2]Traditionally, the solution to the SVM optimization problem is found by quadratic programming techniques. For illustration purpose, a short-cut approach in this example is used. Readers should bear in mind that when not all training vectors are also support vectors (as shown in Example 4.3), this short-cut method does not have a solution.

Lagrangian to be maximized is:

$$F(\alpha, \lambda) = L(\alpha) - \lambda \sum_{i=1}^{3} \alpha_i y_i$$

$$= \alpha_1 + \alpha_2 + \alpha_3 - \frac{1}{2}\alpha_2^2 - \frac{1}{2}\alpha_3^2 - \lambda(-\alpha_1 + \alpha_2 + \alpha_3).$$

Differentiating $F(\alpha, \lambda)$ with respect to the α_i and λ yields the following set of simultaneous equations:

$$1 + \lambda = 0$$
$$1 - \alpha_2 - \lambda = 0$$
$$1 - \alpha_3 - \lambda = 0$$
$$-\alpha_1 + \alpha_2 + \alpha_3 = 0,$$

which give $\alpha_1 = 4$, $\alpha_2 = 2$, $\alpha_3 = 2$, and $\lambda = -1$. Using these Lagrange multipliers, one can compute the weight vector $\mathbf{w}$ and the bias b as follows:

$$\mathbf{w} = \sum_{i=1}^{3} \alpha_i y_i \mathbf{x}_i = -4 \begin{bmatrix} 0 \\ 0 \end{bmatrix} + 2 \begin{bmatrix} 1 \\ 0 \end{bmatrix} + 2 \begin{bmatrix} 0 \\ 1 \end{bmatrix} = \begin{bmatrix} 2 \\ 2 \end{bmatrix}$$

$$b = 1 - \mathbf{w}^T \mathbf{x_2} = 1 - 2 = -1.$$

Therefore, the decision plane is defined by the equation:

$$\mathbf{w}^T \mathbf{x} + b = 0 \Rightarrow \begin{bmatrix} 2 & 2 \end{bmatrix} \begin{bmatrix} x_1 \\ x_2 \end{bmatrix} - 1 = 0,$$

which is the line $x_1 + x_2 - 0.5 = 0$ shown in Figure 4.2(a). The margin of separation is given by $2/\|\mathbf{w}\| = 0.707$.

Example 4.2: *4-point problem with 4 support vectors*

In this example, there are three points in one class and one point in the other, as illustrated in Figure 4.2(b). Applying the method shown in Example 4.1 results in the following linear equations:

$$\alpha_2 + 0.5\alpha_4 = 2$$
$$\alpha_3 + 0.5\alpha_4 = 2$$
$$\alpha_2 + \alpha_3 + \alpha_4 = 4$$
$$-\alpha_1 + \alpha_2 + \alpha_3 + \alpha_4 = 0.$$

This set of equations have two solutions for α_i (i.e., the solution for the Lagrange multiplier is not unique). One set of solutions can have all of the four points as support vectors:

$$\alpha_1 = 4, \alpha_2 = 4/3, \alpha_3 = 4/3, \text{ and } \alpha_4 = 4/3.$$

In another set of solutions, only three points will be support vectors (cf. Problem 5):

$$\alpha_1 = 4, \alpha_2 = 2, \alpha_3 = 2, \text{ and } \alpha_4 = 0.$$

Note: both solutions yield the maximum Langrangian (=4) and an identical decision plane ($\mathbf{w} = [2\ \ 2]^T$ and $b = -1$).

Example 4.3: *4-point problem with 3 support vectors*

In this example, there are three points in one class and one point in the other, as illustrated in Figure 4.2(c). Unlike the four-point problem in Example 4.2, the fourth point in this example is $[1\ \ 1]^T$. Applying the sample procedure described in the preceding two examples, this is obtained:

$$\alpha_2 + \alpha_4 = 2 \qquad (4.3.20)$$
$$\alpha_3 + \alpha_4 = 2 \qquad (4.3.21)$$
$$\alpha_2 + \alpha_3 + 2\alpha_4 = 2 \qquad (4.3.22)$$
$$-\alpha_1 + \alpha_2 + \alpha_3 + \alpha_4 = 0. \qquad (4.3.23)$$

Obviously, Eqs. 4.3.20 and 4.3.21 contradict Eq. 4.3.22. Therefore, there is no solution unless Eq. 4.3.22 is removed from the set of equations. Readers should note that Eq. 4.3.22 is obtained by setting the derivative of the Lagrangian with respective to α_4 to 0. However, if the data point corresponding to α_4 is not selected as a support vector, the derivative does not have to be 0. Therefore, Eq. 4.3.22 should have been removed from the set of equations. After removing this equation, the solution will be identical to those in Example 4.1. Note that the analytical method requires manual identification of support vectors; this is reasonable because otherwise the more computation intensive quadratic programming technique will not be required for software implementation of SVMs.

4.4 Linear SVMs: Fuzzy Separation

Sometimes the training data points are not clearly separable, and most often, they are best characterized as "fuzzily" separable. It is therefore convenient to introduce the notion of a fuzzy (or soft) decision region to cope with such situations. For the nonseparable cases, it is not possible to construct a linear hyperplane decision boundary without incurring classification errors. A fuzzy SVM is a model that allows a more relaxed separation, which in turn would facilitate a more robust decision.

For fuzzy SVM classifiers, it is no longer suitable to talk about the *separation width*. Instead, it is more appropriate to adopt a notion of *fuzzy region* or, more exactly, *fuzzy separation region*. Accordingly, the distance $2d = \frac{2}{\|\mathbf{w}\|}$ takes a new meaning as the *width of fuzzy separation*.

4.4.1 Why Fuzzy Separation?

This section extends discussion of the basic SVM to a fuzzy SVM. From the biometric authentication perspective, the fuzzy SVM serves a practical purpose because it can provide the user with a probabilistic recommendation as opposed to a black-and-white decision. This can be useful for two scenarios:

1. In some applications, there will be combined sensors (e.g., visual and audio) for high-security access control. The decisions made by an individual sensor would best be soft, so that they can be properly weighted when all information is fused to reach a final decision of the joint-sensor classifier.

2. Another scenario is that, even if only a single sensor is used, the information is processed by a hierarchical structure (see Chapter 6); then, information from different local experts or classes will have to be combined. It is often observed that final decisions made by combining local information expressed in terms of soft decisions tends to yield superior performance, when compared with final decisions based on hard local decisions.

If the data patterns are not separable by a linear hyperplane, a set of slack variables $\{\xi = \xi_1, \ldots, \xi_N\}$ is introduced with $\xi_i \geq 0$ such that Eq. 4.3.9 becomes

$$y_i(\mathbf{x}_i \cdot \mathbf{w} + b) \geq 1 - \xi_i \quad \forall i = 1, \ldots, N. \tag{4.4.1}$$

The *slack variables* $\{\xi_i\}_{i=1}^N$ allow some data to violate the constraints noted in Eq. 4.3.9, which defines the minimum safety margin required for training data in the clearly separable case. (Recall that the safety margin is $d = \frac{1}{\|w\|}$.) The slack variables measure the deviation of data points from the marginal hyperplane. Equivalently, they measure how severely the safety margin is violated. In other words, in the fuzzy SVM, the minimum safety margin is relaxed by the amount specified by $\{\xi_i\}$. (In the clearly separable case, no such violation is allowed.) In this sense, the slack variables also indicate the degree of fuzziness embedded in the fuzzy classifier.

4.4.2 Wolfe Dual Optimization

The greater ξ_i is the less confidence one can in the decision, thus consider $\sum_i \xi_i$ to be the upperbound of the training error. The new objective function to be minimized becomes

$$\frac{1}{2}\|\mathbf{w}\|^2 + C\sum_i \xi_i, \qquad \text{subject to} \qquad y_i(\mathbf{x}_i \cdot \mathbf{w} + b) \geq 1 - \xi_i, \tag{4.4.2}$$

where C is a user-defined penalty parameter to penalize any violation of the safety margin for all training data. (Recall that the safety margin is defined to be $d = \frac{1}{\|w\|}$.) Note also that a larger C means a heavier penalty will be imposed for the same level of violation.

The new Lagrangian is

$$L(\mathbf{w}, b, \alpha) = \frac{1}{2}\|\mathbf{w}\|^2 + C\sum_i \xi_i - \sum_{i=1}^N \alpha_i(y_i(\mathbf{x}_i \cdot \mathbf{w} + b) - 1 + \xi_i) - \sum_{i=1}^N \beta_i \xi_i, \tag{4.4.3}$$

where $\alpha_i \geq 0$ and $\beta_i \geq 0$ are, respectively, the Lagrange multipliers to ensure that $y_i(\mathbf{x}_i \cdot \mathbf{w} + b) \geq 1 - \xi_i$ and that $\xi_i > 0$. With some mathematical manipulation (see Problem 4), the Wolfe dual is obtained:

$$\max_{\alpha} \sum_{i=1}^{N} \alpha_i - \frac{1}{2} \sum_{i=1}^{N} \sum_{j=1}^{N} \alpha_i \alpha_j y_i y_j (\mathbf{x}_i \cdot \mathbf{x}_j) \qquad (4.4.4)$$

subject to $0 \leq \alpha_i \leq C$, $i = 1, \ldots, N$, $\sum_{i=1}^{N} \alpha_i y_i = 0$.

Similar to the basic SVM, the output weight vector is

$$\mathbf{w} = \sum_{i=1}^{N} \alpha_i y_i \mathbf{x}_i$$

and the bias term is found by

$$b = 1 - \mathbf{x}_k \cdot \mathbf{w}.$$

where $y_k = 1$ and $\mathbf{x}_k$ is a support vector that lies on the plane $\mathbf{w}^T \mathbf{x} + b = 1$.

Margin of Fuzzy Separation

The distance $2d = \frac{2}{\|\mathbf{w}\|}$ represents the *margin of fuzzy separation*. The larger the margin, the fuzzier is the classification.

1. Note that a smaller C leads to more SVs while a larger C leads to fewer SVs.

2. Also note that a smaller C leads to a wider width of fuzzy separation, while a larger C has a narrower fuzzy separation region.

Analytical Examples

Example 4.4: *5-point linearly nonseparable problem*

This example investigates the effect of varying C on the number of support vectors and the margin width in a linear SVM. Assume that the data points and their class labels are as follows:

$$\begin{aligned} \mathbf{x}_1 &= [0.5 \ \ 0.5]^T & y_1 &= -1 \\ \mathbf{x}_2 &= [0.0 \ \ 0.0]^T & y_2 &= -1 \\ \mathbf{x}_3 &= [0.0 \ \ 1.0]^T & y_3 &= +1 \\ \mathbf{x}_4 &= [1.0 \ \ 0.0]^T & y_4 &= +1 \\ \mathbf{x}_5 &= [1.0 \ \ 1.0]^T & y_5 &= +1. \end{aligned}$$

The decision boundaries created by a linear SVM for this five-point linearly nonseparable problem for $C = 1$ and $C = 100$ are shown in Figure 4.3. Note that for small C, support vectors (Points 1, 3, and 4) fall between the margin because the cost of making a mistake is smaller for small C. On the other hand, when C is large, there is no SV that falls in the fuzzy region because the cost of making an error is higher for large C.

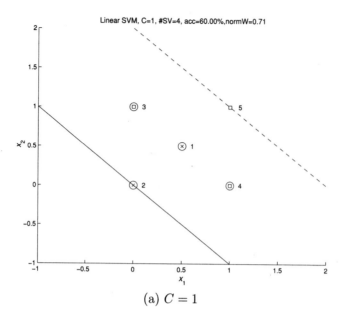

(a) $C = 1$

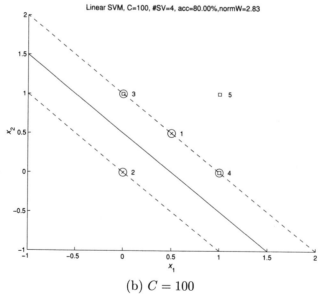

(b) $C = 100$

Figure 4.3. Five-point problem: (a) $C = 1$ and (b) $C = 100$. (See Example 4.4 for details.)

Example 4.5: *Analytical solutions for a 4-point XOR problem*

This example demonstrates how the Wolfe dual formulation can be applied to solve the classic XOR problem shown in Figure 4.4 using linear SVMs. Assume that the training data are defined as

$$\mathbf{x}_1 = [0.0 \quad 0.0]^T \quad y_1 = -1$$
$$\mathbf{x}_2 = [1.0 \quad 0.0]^T \quad y_2 = +1$$
$$\mathbf{x}_3 = [0.0 \quad 1.0]^T \quad y_3 = +1$$
$$\mathbf{x}_4 = [1.0 \quad 1.0]^T \quad y_3 = -1.$$

Using the procedure described in Example 4.1, the following Lagrangian for a linear SVM is obtained:

$$F(\alpha, \lambda) = \alpha_1 + \alpha_2 + \alpha_3 + \alpha_4 - \frac{1}{2}\alpha_2^2 + \alpha_2\alpha_4 - \frac{1}{2}\alpha_3^2 + \alpha_3\alpha_4$$
$$-\alpha_4^2 - \lambda(-\alpha_1 + \alpha_2 + \alpha_3 - \alpha_4). \tag{4.4.5}$$

Differentiating Eq. 4.4.5 with respect to λ and α_i, results in

$$\lambda = -1$$
$$\alpha_2 - \alpha_4 = 2$$
$$\alpha_3 - \alpha_4 = 2$$
$$\alpha_2 + \alpha_3 - 2\alpha_4 = 0$$
$$-\alpha_1 + \alpha_2 + \alpha_3 - \alpha_4 = 0.$$

Since $\mathbf{x}_4$ is not a support vector, the derivative of $F(\alpha, \lambda)$ with respect to α_4 needs not be 0. Therefore, the fourth equation can be ignored, which results in

$$\alpha_1 = 4, \ \alpha_2 = 2, \ \alpha_3 = 2, \ \alpha_4 = 0, \ \mathbf{w} = [2 \ 2]^T \text{ and } b = -1.$$

Therefore, the decision boundary will be identical to that of the three-point problem shown in Example 4.1 and Figure 4.2.

Readers' attention is drawn to the fact that while the analytical method can find a solution for this four-point XOR problem (because it is possible to manually discard the non-SV), the quadratic programming approach cannot find a decision boundary. In particular, the quadratic programming approach leads to the solution: $\alpha_i = C \ \forall \ i$, $\mathbf{w} = [0 \ 0]^T$, and $b = 0$. Therefore, the decision boundary found by the quadratic programming approach reduces to a point in the input space.

4.4.3 Support Vectors in the Fuzzy Region

Because no slack variables were introduced in Section 4.3.2, there is only one kind of support vector in the basic SVM. More exactly, all support vectors lie on the margin of separation. However, with slack variables incorporated, more varieties of support vectors exist, which is elaborated on in this section.

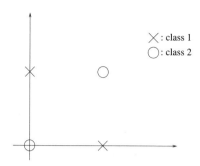

Figure 4.4. The XOR problem. Clearly, the XOR problem is not linearly separable, but it can be separated by a nonlinear decision boundary.

The support vectors of the clearly separable and unclearly separable cases can be selected in almost the same manner. The only difference is that the Karush-Kuhn-Tucker conditions need to adopt a modified expression:

$$\alpha_i \left\{ y_i(\mathbf{x}_i \cdot \mathbf{w} + b) - 1 + \xi_i \right\} = \alpha_i \left\{ y_i f(\mathbf{x}_i) - 1 + \xi_i \right\} = 0 \quad \forall i = 1, \ldots, N. \quad (4.4.6)$$

This equality states the necessary (and sometimes sufficient) conditions for which a set of variables will be optimal under the Wolfe optimization formulation. For example, the condition for a data point not being selected as a support vector is particularly simple, because

$$\alpha_i = 0 \quad \Rightarrow \quad y_i f(\mathbf{x}_i) \geq 1 \text{ and } \xi_i = 0.$$

This relationship suggests that all non-SVs are generically outside the margin area (e.g., data points 4 and 15 in Figure 4.5(b)). A rare exception exists in a degenerate situation in which a data point satisfies

$$\alpha_i = 0 \quad \Rightarrow \quad y_i f(\mathbf{x}_i) = 1 \text{ and } \xi_i = 0.$$

Then the data point will appear on the margin of the separation, even though it is theoretically possible that it is not selected to be a support vector. (See Problem 5.)

On the other hand, for a data point to be selected in the set of support vectors, it must precisely satisfy Eq. 4.4.1:

$$y_i(\mathbf{x}_i \cdot \mathbf{w} + b) \geq 1 - \xi_i.$$

Now, examine this privileged set more closely. There are three types of SVs:

1. "On the margin" type: $\xi_i = 0$

$$0 < \alpha_i < C \quad \Rightarrow \quad y_i f(\mathbf{x}_i) = 1 \text{ and } \xi_i = 0;$$

for example, Points 1, 2, 9, and 11 in Figure 4.5(b) belong to this type of support vector.

2. "Strictly inside the fuzzy region" type: $0 < \xi_i < 2$

$$\alpha_i = C \quad \Rightarrow \quad y_i f(\mathbf{x}_i) \le 1 \ \text{ and } \ 0 < \xi_i < 2;$$

for example, Points 10 and 19 in Figure 4.5(b) belong to this type of support vector.

3. "Outliers" type: $\xi_i \ge 2$

$$\alpha_i = C \quad \Rightarrow \quad y_i f(\mathbf{x}_i) \le 1 \ \text{ and } \ \xi_i \ge 2;$$

for example, Point 20 in Figure 4.5(b) belongs to this type of support vector. The existence of such outliers symbolizes an inconsistent message between training data and future generalization results. Some postprocessing treatment may be necessary to achieve a more agreeable accord. Practically, there are two ways to handle such outliers:

(a) Further widen the fuzzy region so that the outliers can be relocated into the fuzzy region.

(b) Weed the outliers out of the training set in the first place so as to preempt undue influence from the (erroneous) outliers.

Example 4.6: *20-point nonlinearly separable problem*

In this example, a linear SVM is applied to classify 20 data points that are not linearly separable. Figure 4.5 depicts the decision boundary (*solid line*), the two lines (*dashed*) defining the margin, and the support vectors for $C = 0.1$ and $C = 10$. Observe that the margin is wider (i.e., decisions are more fuzzy) and that there are more support vectors when C is small.

Sparsity of Support Vectors

The SVM optimization methods are based on second-order optimality conditions (i.e., Karush-Kuhn-Tucker conditions), which reveal that the solution is sparse. This sparsity is one of the most important properties of SVMs because it makes SVM learning practical for large data sets.

4.4.4   Decision Boundary

Once the values of α_i have been found from Eq. 4.4.4, the vector $\mathbf{w}$ can be determined just as before:

$$\mathbf{w} = \sum_{i=1}^{N} \alpha_i y_i \mathbf{x}_i.$$

Once $\mathbf{w}$ is found, the value of threshold b can be computed exactly like the linearly separable case. This is made possible by exploiting the fact that

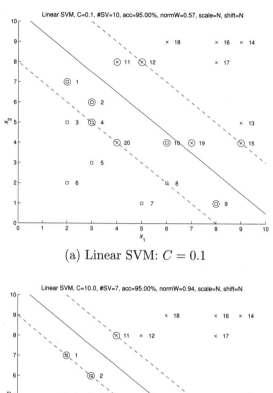

(a) Linear SVM: $C = 0.1$

(b) Linear SVM: $C = 10$

Figure 4.5. Decision boundaries (*solid lines*) and marginal boundaries (*dashed lines*) created by linear and nonlinear SVMs for 20-point data. (a) A linear SVM using $C = 0.1$, (b) a linear SVM with $C = 10$, (c) RBF SVM with $C = 100$, and (d) RBF SVM with $C = 1000$. (In both nonlinear cases (c) and (d), σ^2 in Eq. 4.5.4 was set to 4.) The decision boundaries are almost the same for the two linear cases (a) and (b). The region between the two dashed lines are considered the "fuzzy region." Small C means a larger fuzzy region. This is consistent with the RBF kernel results. *Note*: The data set is obtained from Schwaighofer [149].

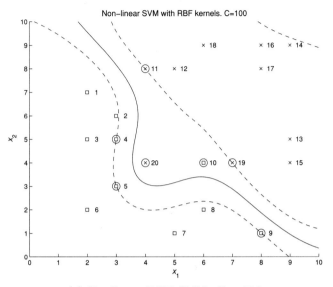

(c) Nonlinear RBF SVM: $C = 100$

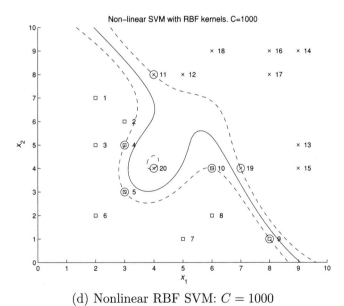

(d) Nonlinear RBF SVM: $C = 1000$

Figure 4.5. (continued)

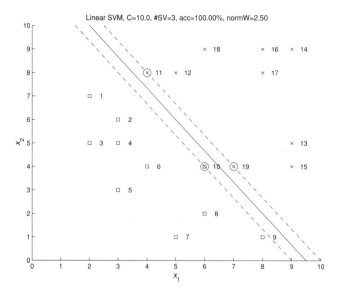

Figure 4.6. The two-class problem in Figure 4.5 but without Point 20. Without this data point, the problem becomes a linearly separable one. Note the change in the decision boundary as compared to Figure 4.5(b).

1. All the marginal support vectors can be readily identified because of the prior knowledge that $0 < \alpha_i < C$ if $\mathbf{x}_i \in$ margin-set.

2. It is also known that $\xi_i = 0$ for all marginal support vectors: $\mathbf{x}_i \in$ margin-set.

Thus, the value of threshold b can be derived by solving the linear equation:

$$\mathbf{x}_i \cdot \mathbf{w} + b - 1 = 0 \quad \text{for all} \quad \xi_i \in \text{margin-set}$$

with $\mathbf{w}$ plugged in. Ultimately, the decision boundary can be derived as follows:

$$f(\mathbf{x}) = \mathbf{w} \cdot \mathbf{x} + b = \sum_{i=1}^{N} y_i \alpha_i (\mathbf{x} \cdot \mathbf{x}_i) + b \;=\; 0. \tag{4.4.7}$$

The decision boundary is dictated by the support vectors. This fact can be demonstrated by removing a support vector in the 20-point problem shown in Figure 4.5. Figure 4.6 illustrates the change in the decision boundary when Point 20 (a support vector in Figure 4.5) is removed from the training data set. Because the problem becomes a linearly separable one after the removal of this point, none of the support vectors falls in the fuzzy region.

4.4.5   Probabilistic Function for Decision Confidence

As shown in Figure 4.5, the region between the two dashed lines can be considered the fuzzy decision region. In this fuzzy region, assigning a probability (confidence) to the classification decision is highly recommended. There are two approaches to assigning probabilities:

1. *Linearly proportional probability function.* Two situations exist in this case: (1) If a data point is interior to the fuzzy region, assign $0.5(\mathbf{w}^T\mathbf{x} + b + 1)$ as the probability of a data point belonging to the "positive-class" (i.e., $y_i = 1$). Consequently, its "negative-class" probability is $1 - 0.5(\mathbf{w}^T\mathbf{x} + b + 1) = 0.5(1 - \mathbf{w}^T\mathbf{x} - b)$. Note that a pattern $\mathbf{x}$ is more likely to belong to the positive-class if $f(\mathbf{x}) > 0$ and to the negative-class if $f(\mathbf{x}) < 0$. (2) If a data point is exterior to the fuzzy region, it falls in the hard-decision region and its class probability will be either 1.0 or 0.0.

2. *Sigmoidal probability function.* Via a sigmoidal function, it is possible to use one unified formula to express the probability for both the hard-decision and soft-decision regions. More exactly, the positive-class probability can be expressed as

$$\frac{1}{1 + e^{-\frac{\mathbf{w}^T\mathbf{x}+b}{\sigma}}} \tag{4.4.8}$$

It is easy to show that the sigmoid value has a range of 0.0 to 1.0 for the entire vector space. Suppose the marginal support vectors are set for a confidence level of 1% or 99%, then

$$\frac{1}{1 + e^{-\frac{1}{\sigma}}} = .99 \quad \Rightarrow \quad \sigma = -\frac{1}{\log 0.010101} \approx 0.22.$$

Figure 4.7 depicts the equal-probability contours of a linear SVM for the 20-point problem using $\sigma = 0.22$ and Eq. 4.4.8. Because the 20 data points shown in Figure 4.7 are not linearly separable, assigning a soft probability as the classification confidence for data in the fuzzy separation region is highly recommended.

4.4.6   Invariance Properties of Linear SVMs

It can be shown that linear SVMs are invariant to both scaling and translation in the input domain. It is easier to establish invariance using the primal formulation (see Eq. 4.4.3). For proving scale invariance, $\mathbf{w}$ in the third term of Eq. 4.4.3 is scaled down by β. For proving translation invariance, b in the third term is replaced by $b - \mathbf{w}^T\mathbf{t}$, where $\mathbf{t}$ is a translation vector. The proofs based on the Wolfe dual formulation, however, will be useful for investigating the invariance properties of nonlinear SVMs because there is no primal formulation for nonlinear SVMs.

Note that the invariance property of linear SVMs can also be proved by using the dual formulation Eq. 4.4.4. (See Problem 7.)

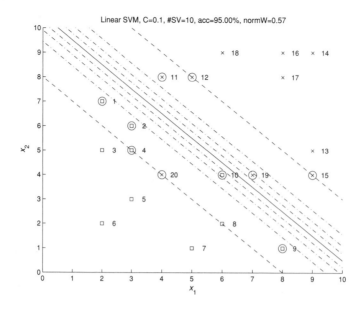

Figure 4.7. Contour plot showing the probabilistic output of a linear SVM for the two-class problem shown in Figure 4.5. The margins (*dashed lines*) correspond to probabilities 0.0 or 1.0. The contour lines between the margins (*dashed-dotted lines*) represent probabilities 0.875, 0.750, and 0.625. Points on the decision boundary (*solid line*) have a probability of 0.5.

4.5 Nonlinear SVMs

It is well known that, by inserting a well-designed nonlinear hidden-layer between the input and output layers, a two-layer network can provide an adequate flexibility in the classification of fuzzily separable data. The original linearly nonseparable data points can be mapped to a new feature space, represented by hidden nodes such that the mapped patterns become linearly separable. This is illustrated in Figure 4.8.

If the hidden nodes can be expressed by a nonlinear function $\{\phi_i(\mathbf{x})\}$, $i = 1, \ldots, K$, a generalized decision function has the following form:

$$f(\mathbf{x}) = \sum_{i=1}^{K} w_i \phi_i(\mathbf{x}) + b = \mathbf{w} \cdot \overrightarrow{\phi}(\mathbf{x}) + b,$$

where

$$\overrightarrow{\phi}(\mathbf{x}) \equiv [\phi_1(\mathbf{x}) \ \phi_2(\mathbf{x}) \ \cdots \ \phi_K(\mathbf{x})]^T.$$

The nonlinear functions $\{\phi_i(x)\}_{i=1}^{K}$ for the hidden nodes can be adaptively trained by various supervised learning algorithms. For this purpose, many different neural

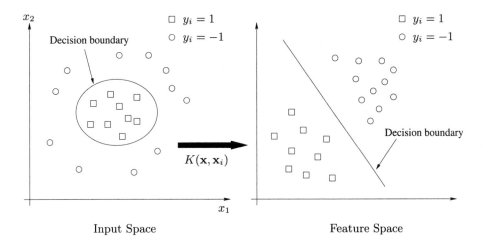

Figure 4.8. Use of a kernel function to map nonlinearly separable data in the input space to a high-dimensional feature space, where the data become linearly separable.

networks are discussed in Chapter 5. This section looks into one particular category of learning algorithms, including SVM, which exploits the vital fact that such a decision function retains a linear expression in terms of the parameters $\mathbf{w}$ and b. As a result, the optimization can be effectively carried out by simple linear techniques such as least-squares error, Fisher's discriminant, or SVMs.

4.5.1   Two-Layer SVM Network Architectures

Recall that a linear decision function for the SVM derived previously has the following form:

$$f(\mathbf{x}) = \sum_{i=1}^{N} y_i \alpha_i (\mathbf{x} \cdot \mathbf{x}_i) + b.$$

To obtain a nonlinear decision boundary, thus enhancing the discrimination power, replace the inner-product $(\mathbf{x} \cdot \mathbf{x}_i)$ with a nonlinear kernel $K(\mathbf{x}, \mathbf{x}_i)$ and obtain

$$f(\mathbf{x}) = \sum_{i=1}^{N} y_i \alpha_i K(\mathbf{x}, \mathbf{x}_i) + b. \tag{4.5.1}$$

Nonlinear Kernels

The decision function in Eq. 4.5.1 can be implemented by a two-layer architecture depicted in Figure 4.9, where it is shown that the original input space is mapped to a new feature space, manifested by the middle hidden-layer in the network.

The basic idea behind the nonlinear SVM is to use a kernel function $K(\mathbf{x}, \mathbf{x}_i)$ to map the data $\mathbf{x}$ from the input space to the feature space on which the mapped data become linearly separable. The hidden units on the hidden-layer are represented by the kernel function adopted by the SVM. There are various forms of nonlinear kernels. Three typical examples are:

$$\text{Polynomial Kernel}: K(\mathbf{x}, \mathbf{x}_i) = \left(1 + \frac{\mathbf{x} \cdot \mathbf{x}_i}{\sigma^2}\right)^p, \quad p > 0 \qquad (4.5.2)$$

$$\text{RBF Kernel}: K(\mathbf{x}, \mathbf{x}_i) = \exp\left\{-\frac{\|\mathbf{x} - \mathbf{x}_i\|^2}{2\sigma^2}\right\} \qquad (4.5.3)$$

$$\text{Sigmoidal Kernel}: K(\mathbf{x}, \mathbf{x}_i) = \frac{1}{1 + e^{-\frac{\mathbf{x} \cdot \mathbf{x}_i + b}{\sigma^2}}} \qquad (4.5.4)$$

Therefore, the SVM can be reformulated to imitate several prominent two-layer neural models (see Chapter 5), including polynomial learning machines, radial basis

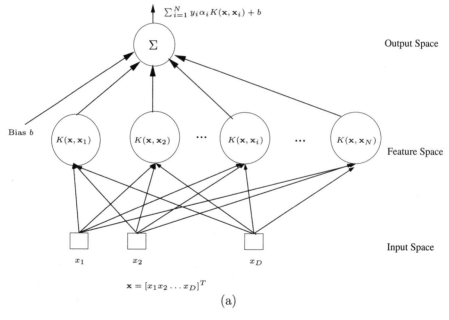

(a)

Figure 4.9. The two-layer architecture of an SVM. Employing the nonlinearity inherent in the hidden-layer in a two-layer network, the original linearly nonseparable input space can be converted into a linearly separable feature space. (a) The two-layer architecture of an SVM with all possible kernels displayed; and (b) simplified SVM architecture: the only support vectors that need to be retained are shown as hidden nodes. Here, the support vectors are indexed by the set $\mathcal{S} = \{s_1, s_2, \ldots, s_M\}$, where $s_m \in \{1, 2, \ldots, N\}$.

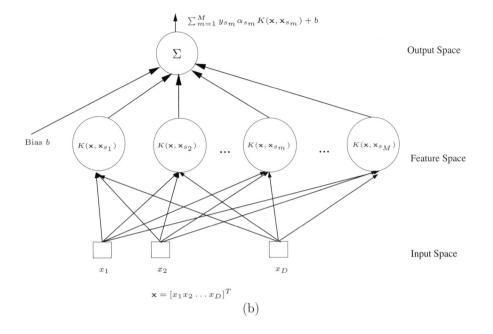

$$\sum_{m=1}^{M} y_{s_m} \alpha_{s_m} K(\mathbf{x}, \mathbf{x}_{s_m}) + b$$

Output Space

Feature Space

Bias b

Input Space

$$\mathbf{x} = [x_1 x_2 \ldots x_D]^T$$

(b)

Figure 4.9. (continued)

function networks, and linear-basis multi-layer perceptrons (with a single sigmoidal hidden-layer).

Lagrangian Optimization in Dual Domain

Recall the Lagrangian in the linear case:

$$L(\alpha) = \sum_{i=1}^{N} \alpha_i - \frac{1}{2} \sum_{i=1}^{N} \sum_{j=1}^{N} \alpha_i \alpha_j y_i y_j \mathbf{x}_i \cdot \mathbf{x}_j.$$

Suppose that the original training patterns X are not linearly separable; to enhance the discrimination ability, the linear inner product $(\mathbf{x}_i \cdot \mathbf{x}_j)$ is replaced by a nonlinear kernel function $K(\mathbf{x}_i, \mathbf{x}_j)$. This leads to a generalized Lagrangian as follows:

$$L(\alpha) = \sum_{i=1}^{N} \alpha_i - \frac{1}{2} \sum_{i=1}^{N} \sum_{j=1}^{N} \alpha_i \alpha_j y_i y_j K(\mathbf{x}_i, \mathbf{x}_j). \tag{4.5.5}$$

1. *Clearly Separable Case.* The Lagrange multipliers $\{\alpha_i\}$ are determined by maximizing $L(\alpha)$ subject to $\sum_{i=1}^{N} \alpha_i y_i = 0$ and $\alpha \geq 0$, $i = 1, \ldots, N$.

2. *Fuzzy Separable Case.* The Lagrange multipliers $\{\alpha_i\}$ are determined by maximizing $L(\alpha)$ subject to $\sum_{i=1}^{N} \alpha_i y_i = 0$ and $0 \leq \alpha \leq C$, $i = 1, \ldots, N$,

where C is a factor used to control the violation of the *safety margin* rule. Of course, a larger C means that a heavier penalty is imposed.[3]

Once the values of the Lagrange multipliers $\{\alpha_i\}$ are obtained, then the decision function can be derived as (cf. Eq. 4.3.18):

$$f(\mathbf{x}) = \sum_{i=1}^{N} y_i \alpha_i K(\mathbf{x}, \mathbf{x}_i) + b, \qquad (4.5.6)$$

which matches the output of the network shown in Figure 4.9. In fact, the output of the network could be one of the following:

1. If a hard decision is preferred, the class of a pattern $\mathbf{x}$ can be determined based on the sign of $f(\mathbf{x})$.

2. The network could also output the probability (or confidence) of a pattern $\mathbf{x}$ belonging to a class if a fuzzy classification is desired. Assuming that the sigmoidal probability function is adopted, the probability of $\mathbf{x}$ belonging to the positive-class is expressed as:

$$\frac{1}{1 + e^{-\frac{f(\mathbf{x})}{\sigma}}}. \qquad (4.5.7)$$

An example of probability contours based on this sigmoidal probability mapping function is shown in Figure 4.10. Note that a pattern $\mathbf{x}$ is more likely to belong to the positive-class (resp. negative-class) if $f(\mathbf{x}) > 0$ (resp. if $f(\mathbf{x}) < 0$).

4.5.2   Hidden Nodes and Retrieving Speed

The two-layer architecture, as shown in Figure 4.9, can be adopted for classification in the retrieving or testing phase. The retrieving speed is proportional to the active hidden nodes required for decision making.

It is important to note that Figure 4.9(a) shows an upper-bound scenario that all training patterns are incorporated into the hidden-layer. In reality, the number of hidden nodes that actively participate in the output function is much smaller— compared to Figure 4.9(b). According to Eq. 4.5.6, only those hidden nodes corresponding to the support vectors (i.e., with $\alpha_i \neq 0$) can influence the decision boundary. It is therefore sufficient to retain only the support vectors in the hidden-layer. Thus, the actual number of hidden nodes could be much smaller.

Therefore, network complexity is closely tied to the number of support vectors selected by the SVM learning process. The number of support vectors (or hidden

---

[3]Although the primal problem formulation does not always exist for nonlinear SVM, the penalty parameter C still inherits the property that a larger (resp. smaller) C more heavily penalizes (resp. tolerate more leniently) the violation of the safety margin.

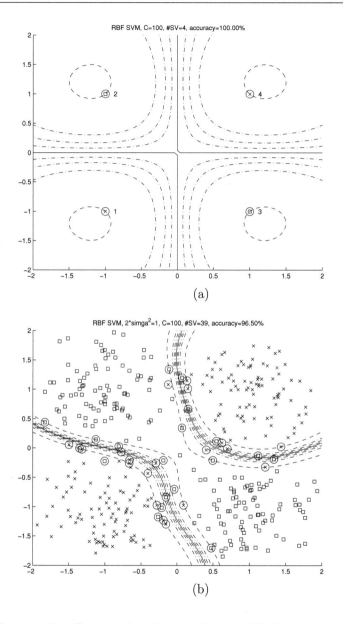

Figure 4.10. Contour plots showing the probabilistic outputs of RBF SVMs in (a) a four-point XOR problem and (b) a noisy XOR problem. The margins (*dashed lines*) correspond to probabilities 0.0 or 1.0. The contour lines between the margins (*dashed-dotted lines*) represent probabilities 0.875, 0.750, and 0.625. Points on the decision boundary (solid line) have a probability of 0.5. The probability mapping function in Eq. 4.5.7 was used to map the SVM output $f(\mathbf{x})$ to probability.

units) is dictated by the so-called Karush-Kuhn-Tucker conditions discussed in Section 4.3.3. For example, if the RBF-kernel SVM is applied to the training data, the number of critical support vectors will be determined automatically, which will in turn dictate the required number of hidden units (i.e., the radial basis functions) in the hidden-layer. The number of SVs can easily get out of control, especially in the case of a large fuzzy region (i.e., small C). A disadvantage of using smaller C lies in the fact that it may select a large number of support vectors, thus slowing down processing speed in the retrieval phase.

Finite-Decomposition of Nonlinear Kernels

Fortunately, the number of hidden nodes can still be substantially reduced if the kernel can be expressed either exactly or by approximation as a finite decomposable representation. A nonlinear kernel is said to be *finite-decomposable* if and only if there exists an integer K such that:

$$K(\mathbf{x}, \mathbf{x}') = \overrightarrow{\phi}(\mathbf{x}) \cdot \overrightarrow{\phi}(\mathbf{x}') = \sum_{k=1}^{K} \phi^{(k)}(\mathbf{x})\phi^{(k)}(\mathbf{x}').$$

In other words, the nonlinear kernel $K(\mathbf{x}, \mathbf{x}')$ can be expressed as an inner product of two vectors induced in a finite-dimensional (i.e., K-dimensional) feature space. Note that not all kernels are finite decomposable. For example, while the second-order polynomial kernel is finite decomposable, the RBF kernels are not.

Note that finite-decomposable kernels are somewhat related to the eigenfunction decomposition of kernels, but they in their own right have different requirements.[4] (See Problem 18.) Note that the development of the SVM neither needs nor depends on the eigenfunction decomposition property, which appears to be unnecessarily restrictive to most practical applications.

Two-Layer Architecture for a Finite-Decomposable Kernel

As in Eq. 4.5.5, the nonlinear kernel Lagrangian can be constructed without having to consider the existence of the feature space itself. However, if a finite-dimensional feature space exists and is easily identified, this fact can be harnessed effectively to simplify the network architecture and improve retrieval performance.

---

[4]If the following orthogonality relationship holds true

$$\int_{b}^{a} \phi^{(k)}(\mathbf{x})\phi^{(i)}(\mathbf{x})dx = 0 \quad \text{if} \quad k \neq j,$$

then the decomposition also happens to qualify as an *eigenfunction decomposition* as formally introduced in the Mercer theorem [63, 241]. The eigenfunction decomposition property can be verified by this equality:

$$\int_{b}^{a} K(\mathbf{x}, \mathbf{x}')\phi^{(k)}(\mathbf{x}')dx' = \lambda_k \phi^{(k)}(\mathbf{x})) \quad \text{for all} \quad k = 1, \ldots, K.$$

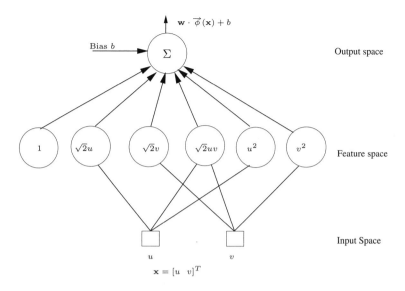

Figure 4.11. For finite-decomposable kernels, the number of hidden nodes can be independent of the size of the training data and SVs. For example, the second-order polynomial kernel SVM can always be implemented with six hidden nodes, with one of the hidden node's output always being equal to 1.0.

Example 4.7: *Finite-decomposable polynomial kernel*

The simplest example is the second-order polynomial kernel on a two-dimensional vector space. Suppose a two-dimensional vector $\mathbf{x} = [u \ \ v]^T$, then the kernel can be expressed as

$$K(\mathbf{x}, \mathbf{x}') = (1 + \mathbf{x} \cdot \mathbf{x}')^2 = \overrightarrow{\phi}(\mathbf{x}) \cdot \overrightarrow{\phi}(\mathbf{x}'),$$

where one denotes

$$\overrightarrow{\phi}(\mathbf{x}) = [1 \ \ \sqrt{2}u \ \ \sqrt{2}v \ \ \sqrt{2}uv \ \ u^2 \ \ v^2]^T. \tag{4.5.8}$$

In conclusion, for any two-dimensional classification problem, including the famous XOR, the feature dimension can be fixed to $K = 6$.

With a finite-decomposable kernel, the primal problem can also be formulated in a fashion similar to its linear counterpart. The primal problem is to minimize:

$$\frac{1}{2}\|\mathbf{w}\|^2 + C\sum_i \xi_i \qquad \text{subject to} \qquad y_i(\overrightarrow{\phi}(\mathbf{x}_i) \cdot \mathbf{w} + b) \geq 1 - \xi_i. \tag{4.5.9}$$

Consequently, the penalty parameter C inherits the property that a larger (resp. smaller) C penalizes more heavily (resp. tolerate more leniently) the violation of the safety margin.

The decision function can be derived as (cf. Eq. 4.3.18):

$$f(\mathbf{x}) = \sum_{i=1}^{N} y_i \alpha_i K(\mathbf{x}, \mathbf{x}_i) + b = \sum_{i=1}^{N} y_i \alpha_i \overrightarrow{\phi}(\mathbf{x}_i) \cdot \overrightarrow{\phi}(\mathbf{x}) + b = \mathbf{w} \cdot \overrightarrow{\phi}(\mathbf{x}) + b = 0, \quad (4.5.10)$$

where

$$\mathbf{w} = \sum_{i=1}^{N} y_i \alpha_i \overrightarrow{\phi}(\mathbf{x}_i). \quad (4.5.11)$$

Thus the number of hidden nodes, in terms of $\phi(\mathbf{x})$-function, can be fixed to K without having to consider how many SVs are selected. The *fixed* network architecture for such a special case is shown in Figure 4.11. It guarantees a constant size of hidden-layer, thereby saving implementation cost. In short, when the number of SVs is potentially large, finite-decomposable kernels, such as the second-order polynomial kernel, become a more attractive alternative because they result in a slimmer architecture and thus a faster retrieving speed.

4.5.3   Training Criteria and Running Time

The retrieving performance depends very much on the network structure itself. Taking a structural perspective, the SVM can be seen as a special case of multi-layer neural models and vice versa. From the training criteria perspective, however, there are fundamental differences [166]. It is important to have a thorough understanding of the issues related to the training criteria as well as the computational cost of the SVM.

SVM versus LSE Learning Criteria

As shown in Section 4.2, LSE and Fisher classifiers produce almost the same decision function. In fact, both the LSE and Fisher training criteria are tuned to be sensitive to *all* of the training data points. For comparison purposes, another prominent classifier is introduced here: RBF neural networks, which are usually trained by the back-propagation (BP) algorithm or a two-stage learning algorithm consisting of K-means, K-nearest neighbors, and least-squares methods (cf. Chapter 5). Just like the LSE classifier, the RBF neural networks are (1) based on a quadratic loss function (i.e., the mean-squared error), and (2) sensitive to the broad base of data distribution.

The key advantage to using a quadratic loss function lies in its elegant mathematical analysis and the existence of a closed-form solution—the latter is vital for lowering computational cost. However, such computational savings often come at the expense of compromised training accuracy.[5] Because a quadratic loss function

---

[5]Note that better training accuracy does not necessarily imply a better generalization performance, which in itself is an important measurement to be discussed in Section 4.5.4.

is proportional to the squared errors, the data points further away from the decision boundary are given higher (and sometimes excessive) weighting to influence the decision boundary.

Via a geometrical criterion, the SVM effectively tackles this concern from two fronts:

1. The SVM criterion is sensitive only to marginal data points, instead of the overall data distribution, ruling out any possible influence from data points far away from the decision boundary.

2. The support vector learning algorithm minimizes a loss function that is linearly proportional to the degree of violation of a set of safety margins. Although such a criterion is not exactly the same as that of minimizing the probability of classification error, it is considered to be much less sensitive than the mean-squared error criterion.[6] In fact, the SVM can be considered an extension of the minimum absolute error criterion of minimax theory because it basically follows the principle of structural risk minimization.

Running Time in Training Phase

Unlike the popular BP algorithm, the support vector learning algorithm can operate only in batch mode. More critically, in terms of running time, SVMs appear to be slower than the back-propagation algorithm for a similar generalization performance [128].

The support vectors are usually a fraction of the total number of examples constituting the training sample. Such sparsity is desirable for the retrieving phase. However, the selection often must endure a long training process. The fundamental difficulty causing the slower processing time is that all training data points are potential candidates for the support vectors. Moreover, there is no control over the number of data points selected to be support vectors by the learning algorithm. In other words, an effective selection of support vectors remains a difficult and challenging issue, particularly for fuzzily separable patterns.

The selection of support vectors, which involves solving a quadratic programming problem, requires a long training time. Unfortunately, the memory size required for running the quadratic programming problem grows with the square of the size of the training data set. Such a lengthy computation time discourages or sometimes prohibits the SVMs from being directly applied to large-scale real-life problems.

Dealing with Large Data Sets

A novel decomposition scheme for cutting the running time of an SVM was proposed by Osuna et al. [267]. The scheme emphasizes the support vector multipliers that

---

[6]For neural networks, which aim to minimize classification error, see the discussion in Chapter 7.

are around $\alpha = 0$ and $\alpha = C$. Reportedly, the proposed algorithm can attain reasonable optimization with satisfactory running time even for large-scale applications with up to 100,000 data points.

In addition, there are several ways to deal with large problem sets. For example, in the decomposition method [33, 267], a small, arbitrary subset of the training data is used to create an initial SVM, which is then tested by the remaining data. The distances between the test data and the margin are sorted in descending order, and the first N of these data together with the N_s support vectors already found are used to retrain the SVM. This process is repeated until all data have been used for training and all of them satisfy the Karush-Kuhn-Tucker conditions. In case the training data are obtained from a number of variable-length sequences (e.g., feature vectors extracted from speech utterances), it is possible to significantly reduce the amount of training data by computing the Fisher scores [160]. In this approach, the whole feature sequence of an utterance is replaced by a Fisher score, which is a vector in a high-dimensional space. SVMs are then trained on the vectors in this space.

Another possibility to reduce the amount of training data is to perform K-means or EM clustering on the large data set to find some representative training vectors prior to SVM training [364]. Two approaches worthy of closer investigation are suggested here:

1. One uses clustering to lower data resolution, resulting in more sparse training data; for example, see Section 4.5.5.

2. The second adopts some sort of prescreening scheme using, for example, Fisher scores as the screening criterion.

4.5.4 Generalization Performance of Nonlinear SVMs

Two kinds of training performances are often used to judge the success/failure of a learning algorithm.

1. *Training Accuracy.* For training accuracy, the test patterns are simply samples drawn from the original training patterns. Therefore, training accuracy is a good indication of the *memorization* capability of the trained model.

2. *Generalization Accuracy.* For generalization accuracy, in contrast, test patterns do not necessarily come from the original training patterns. Instead, they are drawn arbitrarily based on the same distribution function that originally generated the training patterns.

In short, the distinction between training and generalization accuracies lies in the test patterns adopted. Although high training accuracy guarantees good memorization capability, it does not necessarily yield good generalization accuracy. This suggests a fundamental tradeoff between training and generalization accuracies.

Impact of the C Parameter on Generalization Performance

Robustness of nonlinear SVMs is an important design consideration because it exerts a major influence on generalization performance. Just like the previous linear SVM case, the larger the value of C, the fewer critical SVs are selected to determine the final decision boundary. These support vectors are vital for the SVM to attain high training accuracy. Thus, the larger the value of C, the better the training accuracy.

However, from the generalization perspective, these selective SVs are not a good representation of the general statistical distribution of training data. A decision boundary with a focus on a small number of selected SVs is unlikely to deliver a good performance for test data drawn from a general distribution. So, when C is too large, it tends to lead to spurious or more curved decision boundaries. In other words, it suffers a side effect known as *overtraining*.

The robustness and generalization performance of nonlinear SVMs depends heavily on the degree of fuzziness of the support vector machine chosen. Because C is the penalty parameter to curb the training data's violation on the safety margin, thus it plays a key role in controlling the degree of fuzziness. A small, carefully adjusted C tends to relax the penalty on violating the safety margin, and thus enhances generalization performance, although sometimes at the expense of training accuracy.

Simulation Examples

To verify the preceding claims, experiments have been conducted in which RBF- and polynomial-kernel SVMs with different values of C were used for solving the noisy XOR problem. As shown in Figures 4.12(a) through 4.12(d), SVMs with small values of C yield smoother (and more desirable) boundaries as compared to those with large values of C. However, this phenomenon is not obvious in the polynomial-kernel SVMs, as shown in Figures 4.13(a) through (d). In other words, the decision boundaries created by polynomial-kernel SVMs seem to be less sensitive with respect to the change in C. This property can be considered an advantage of the polynomial kernels.

Recall that, in linear SVMs, a larger C delivers a narrower fuzzy separation region. In the nonlinear experiments, it has been observed that a larger C does make the fuzzy separation region a bit narrower. However, it can be observed that a larger C does cause the decision boundary (and marginal boundaries) to become more curved. Conversely, a smaller C usually results in a less curved decision boundary, and thus a more robust classifier. Therefore, choosing smaller C's for the purpose of yielding a more robust classifier and a more predictable generalization performance is recommended.

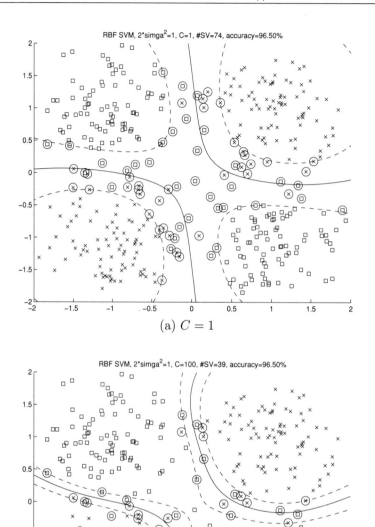

(a) $C = 1$

(b) $C = 100$

Figure 4.12. Experimental results of RBF-kernel SVMs on the noisy XOR problem with various values of C. In all cases, the width parameter σ of the RBF kernel is set to 0.707.

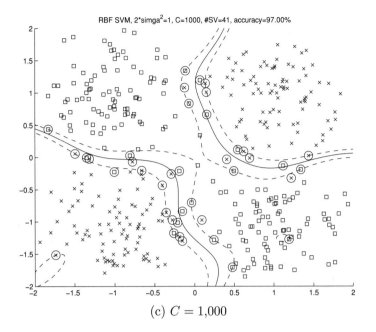

(c) $C = 1,000$

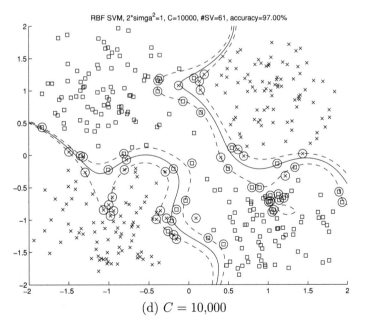

(d) $C = 10,000$

Figure 4.12. (continued)

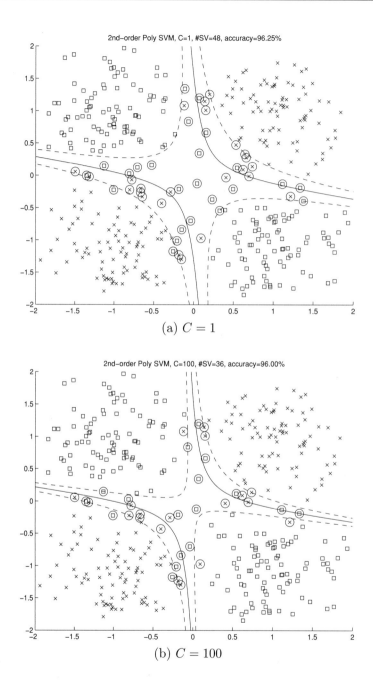

(a) $C = 1$

(b) $C = 100$

Figure 4.13. Results of second-order polynomial-kernel SVMs on the noisy XOR problem with different values of C.

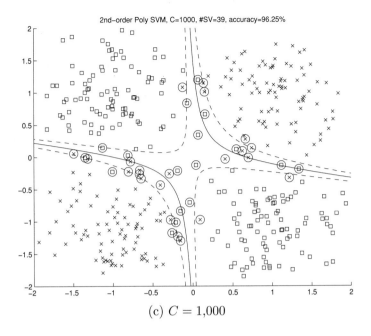

(c) $C = 1,000$

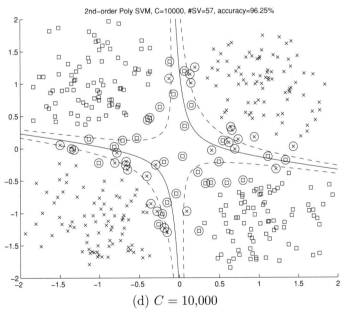

(d) $C = 10,000$

Figure 4.13. (continued)

4.5.5   Effects of Scaling and Shifting

Shift Invariance Properties of Nonlinear SVMs

Shift invariance means that if training data are shifted by a translation vector $\mathbf{t}$

$$\mathbf{x} \leftarrow \mathbf{x} + \mathbf{t},$$

then the decision boundary will also be shifted by the same amount and the shape of the boundary will remain unchanged.

Recall from Section 4.4.6 that linear SVMs are shift invariant. On the other hand, it can be shown that the polynomial SVM is *not* shift invariant. Unlike its RBF counterpart, the polynomial kernel does not warrant a natural formulation to cope with the change in coordinates. Therefore, an optimal selection of coordinate systems can become critical for the polynomial kernel.

Example 4.8: *Effect of coordinate translation on the XOR problem*

For an analytical example, adopt a second-order polynomial kernel for the design of an SVM for the XOR problem. It is well known that a polynomial kernel function $K(\mathbf{x}, \mathbf{x}_i)$ may be expressed in terms of monomials of various orders. More precisely, denote $\mathbf{x} = [u \ v]^T$, then

$$K(\mathbf{x}, \mathbf{x}_i) = 1 + 2u_i u + 2v_i v + 2u_i v_i uv + u_i^2 u^2 + v_i^2 v^2.$$

A. The Original XOR Problem

As shown in Figure 4.14(a), there are two data points, $(0,0)$ and $(1,1)$, for one class and two points for the other, $(1,0)$ and $(0,1)$. Based on the vectors $\mathbf{x}_i$ and class labels y_i defined in Example 4.5, the Lagrangian function of interest is now:

$$
\begin{aligned}
L(\alpha) \ = \ & \alpha_1 + \alpha_2 + \alpha_3 + \alpha_4 + \alpha_1\alpha_2 + \alpha_1\alpha_3 + 4\alpha_2\alpha_4 + 4\alpha_3\alpha_4 \\
& -\alpha_1\alpha_4 - \alpha_2\alpha_3 - .5\alpha_1^2 - 2\alpha_2^2 - 2\alpha_3^2 - 4.5\alpha_4^2
\end{aligned}
$$

subject to $\alpha_i \geq 0$ and $\sum_{i=1}^{4} \alpha_i y_i = 0$. Optimizing $L(\alpha)$ with respect to the Lagrange multipliers yields the following set of simultaneous equations:

$$
\begin{aligned}
\lambda - \alpha_1 + \alpha_2 + \alpha_3 - \alpha_4 \ &= \ -1 \\
-\lambda + \alpha_1 - 4\alpha_2 - \alpha_3 + 4\alpha_4 \ &= \ -1 \\
-\lambda + \alpha_1 - \alpha_2 - 4\alpha_3 + 4\alpha_4 \ &= \ -1 \\
\lambda - \alpha_1 + 4\alpha_2 + 4\alpha_3 - 9\alpha_4 \ &= \ -1 \\
\alpha_1 - \alpha_2 - \alpha_3 + \alpha_4 \ &= \ 0.
\end{aligned}
$$

Then the optimal Lagrange multipliers can be derived as $\alpha_1 = 10/3$, $\alpha_2 = 8/3$, $\alpha_3 = 8/3$, and $\alpha_4 = 2$. (The optimum value of $L(\alpha)$ is 18.) Thus, all of the training vectors are selected as SVs. The threshold is derived to be $b = -1$. Recall that

$$\vec{\phi}(\mathbf{x}) = [1 \ \sqrt{2}u \ \sqrt{2}v \ \sqrt{2}uv \ u^2 \ v^2]^T,$$

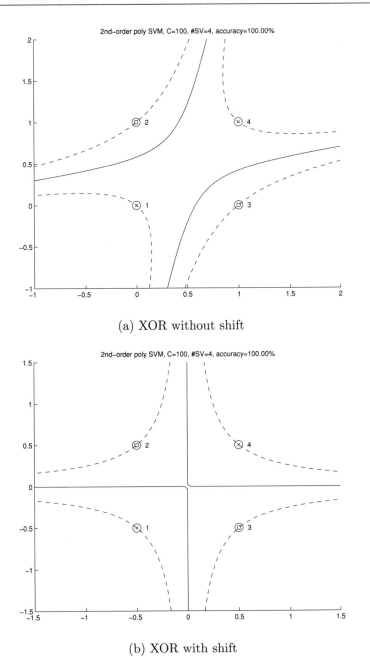

(a) XOR without shift

(b) XOR with shift

Figure 4.14. Effect of the translation of training data on polynomial SVMs based on the XOR problem: (a) Without shift and (b) training data are shifted $(\mathbf{x}_i \leftarrow \mathbf{x}_i - [0.5 \ \ 0.5]^T)$ so that they are symmetric about the origin.

From Eq. 4.5.11, the optimum weight vector is found to be

$$\mathbf{w}_o = -10/3 \times \overrightarrow{\phi}(\mathbf{x}_1) + 8/3\sqrt{2} \times \overrightarrow{\phi}(\mathbf{x}_2) + 8/3 \times \overrightarrow{\phi}(\mathbf{x}_3) - 2 \times \overrightarrow{\phi}(\mathbf{x}_4)$$
$$= [0 \ \ 2\sqrt{2}/3 \ \ 2\sqrt{2}/3 \ \ -2\sqrt{2} \ \ 2/3 \ \ 2/3]^T.$$

which in turn leads to the following optimal decision boundary:

$$\mathbf{w} \cdot \overrightarrow{\phi}(\mathbf{x}) + b = [0 \ \ 2\sqrt{2}/3 \ \ 2\sqrt{2}/3 \ \ -2\sqrt{2} \ \ 2/3 \ \ 2/3]^T \cdot$$
$$[1 \ \ \sqrt{2}u \ \ \sqrt{2}v \ \ \sqrt{2}uv \ \ u^2 \ \ v^2]^T + b$$
$$= \frac{2}{3}u^2 + \frac{2}{3}v^2 - 4uv + \frac{4}{3}u + \frac{4}{3}v - 1 = 0.$$

This decision boundary is illustrated in Figure 4.14(a), which is somewhat less desirable than that in Figure 4.14(b).

B. Coordinate-Shifted XOR Problem

As shown in Figure 4.14(b), there are two data points $(-0.5, -0.5)$ and $(0.5, 0.5)$ for one class and two points for the other, $(-0.5, 0.5)$ and $(0.5, -0.5)$. Now the Lagrangian becomes:

$$L(\alpha) = \alpha_1 + \alpha_2 + \alpha_3 + \alpha_4 + \alpha_1\alpha_2 + \alpha_1\alpha_3 + \alpha_2\alpha_4 + \alpha_3\alpha_4$$
$$-.25\alpha_1\alpha_4 - .25\alpha_2\alpha_3 - 1.125(\alpha_1^2 + \alpha_2^2 + \alpha_3^2 + \alpha_4^2)$$

subject to $\alpha_i \geq 0$ and $\sum_{i=1}^4 \alpha_i y_i = 0$. Solving the following simultaneous saddle-point equations:

$$\lambda - 2.25\alpha_1 + \alpha_2 + \alpha_3 - .25\alpha_4 = -1$$
$$-\lambda + \alpha_1 - 2.25\alpha_2 - .25\alpha_3 + 4\alpha_4 = -1$$
$$-\lambda + \alpha_1 - .25\alpha_2 - 2.25\alpha_3 + 4\alpha_4 = -1$$
$$\lambda - .25\alpha_1 + \alpha_2 + \alpha_3 - 2.25\alpha_4 = -1$$
$$\alpha_1 - \alpha_2 - \alpha_3 + \alpha_4 = 0$$

results in $\alpha_1 = 2$, $\alpha_2 = 2$, $\alpha_3 = 2$, $\alpha_4 = 2$, and $b = 0$. From Eq. 4.5.11, the optimum weight vector is

$$\mathbf{w}_o = [0 \ \ 0 \ \ -2/\sqrt{2} \ \ 0 \ \ 0 \ \ 0]^T.$$

This leads to a simple decision function:

$$\mathbf{w} \cdot \overrightarrow{\phi}(x) = -4uv = 0.$$

which can be easily verified to satisfy the shifted XOR classification problem. As shown in Figures 4.14(a) and 4.14 (b), the two decision boundaries are very different: one for the original XOR and the other for the shifted XOR. The contention is that the latter decision boundary is better than the former because the latter has a centered data distribution, which seems to be a plus for the performance of polynomial-kernel SVM.

Example 4.9: *Effect of shifting or coordinate translation on polynomial SVMs*

Figures 4.15(a) and 4.15(b) illustrate the effect of shifting the training data on polynomial-kernel SVMs in the noisy XOR problem. Evidently, polynomial SVMs are not invariant to coordinate translation.

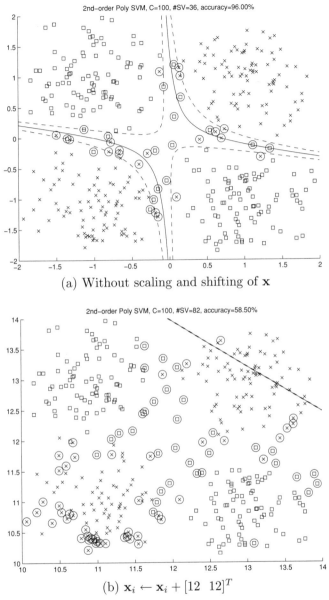

(a) Without scaling and shifting of $\mathbf{x}$

(b) $\mathbf{x}_i \leftarrow \mathbf{x}_i + [12\ \ 12]^T$

Figure 4.15. The effect of scaling and translation of training data on polynomial SVMs. The figures show the decision boundary and SVs for (a) training data $\mathbf{x}_i$ without scaling and shifting, (b) training data shifted by $[12\ \ 12]^T$ (i.e., $\mathbf{x}_i \leftarrow \mathbf{x}_i + [12\ \ 12]^T$), (c) training data scaled by a factor of 5 (i.e., $\mathbf{x}_i \leftarrow 5\mathbf{x}_i$), and (d) training data scaled by a factor of 5 and shifted by $[12\ \ 12]^T$ (i.e., $\mathbf{x}_i \leftarrow 5\mathbf{x}_i + [12\ \ 12]^T$). (Figure continued on next page.)

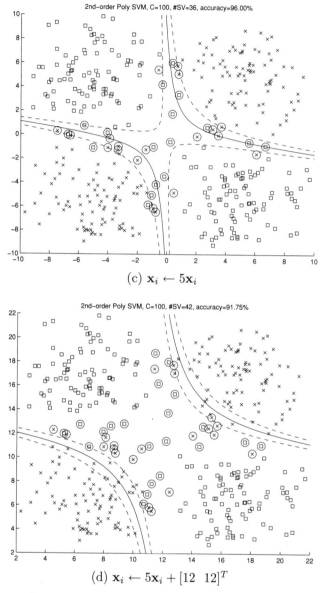

(c) $\mathbf{x}_i \leftarrow 5\mathbf{x}_i$

(d) $\mathbf{x}_i \leftarrow 5\mathbf{x}_i + [12 \quad 12]^T$

Figure 4.15. (continued) Note that for (c) and (d), the kernel function is also scaled: $K(\mathbf{x}, \mathbf{x}_i) = (1 + \mathbf{x} \cdot \mathbf{x}/25)^2$. It can be seen that polynomial SVMs are invariant to scaling after proper adjustment on the kernel function; however, they are not invariant to translation of the training data. *Note*: Because of the translation in (b) and (d), the shape of the decision boundaries may be different in different simulation runs—for some runs, the decision boundaries are outside of the visible range.

Scale Invariance Properties of Nonlinear SVMs

Experiments have been conducted on RBF-kernel and polynomial-kernel SVMs on the noisy XOR problem and the results are shown in Figures 4.12(a) through 4.12(d) and Figures 4.13(a) through 4.13(d).

For all proposed kernel functions, the degree of nonlinearity depends on the parameter σ. The smaller σ, the more nonlinear the kernels become. Moreover, the more nonlinear the kernels, the better the training performance. This does not necessarily mean better generalization performance, however. For polynomial SVMs, the degree of the polynomial also plays an important role in controlling nonlinearity: the larger the degree, the more nonlinear the SVMs become.

Example 4.10: *Effect of scaling on polynomial SVMs*

> Figures 4.15(a) and 4.15(c) illustrate the effect of scaling on polynomial SVMs in the noisy-XOR problem. Evidently, polynomial SVMs are invariant to scaling. Readers, however, should bear in mind that the scale invariance property of polynomial SVMs requires proper selection of the normalization parameter σ. If this parameter is not proportionally scaled by the same amount as the training data, the decision boundary will not be proportionally scaled and the Lagrange multipliers corresponding to the scaled data will also be different from those of the unscaled data.

If necessary, the scale invariance of SVM can be enforced for RBF kernels as well as polynomial kernels. Suppose that the data are now scaled by β:

$$\mathbf{x} \leftarrow \beta \mathbf{x}.$$

To enforce the scale invariance of SVMs, the following adjustment needs to be made:

$$\sigma \leftarrow \beta \sigma.$$

It can be shown that the decision boundary will also be proportionally scaled by the same constant β.

Since the nonlinear version is formulated only in its dual domain (i.e., its formal primal problem is not well defined), the simple proof used in Section 4.4.6 cannot be applied. Here an alternative proof in the dual domain is provided. Assume the data are scaled by a constant factor β (i.e., $\mathbf{x} \leftarrow \beta \mathbf{x}$). Denote the scaled training data as $\mathbf{x}_i'$ (such that $\mathbf{x}_i' = \beta \mathbf{x}$) and the Lagrange multipliers corresponding to the scaled data as α_i'. The Lagrangian for the scaled data is then

$$
\begin{aligned}
L(\alpha') &= \sum_{i=1}^{N} \alpha_i' - \frac{1}{2} \sum_{i=1}^{N} \sum_{j=1}^{N} \alpha_i' \alpha_j' y_i y_j K_1(\mathbf{x}_i', \mathbf{x}_j') \\
&= \sum_{i=1}^{N} \alpha_i' - \frac{1}{2} \sum_{i=1}^{N} \sum_{j=1}^{N} \alpha_i' \alpha_j' y_i y_j \left(1 + \frac{\mathbf{x}_i' \cdot \mathbf{x}_j'}{\sigma'^2}\right)^2 \\
&= \sum_{i=1}^{N} \alpha_i' - \frac{1}{2} \sum_{i=1}^{N} \sum_{j=1}^{N} \alpha_i' \alpha_j' y_i y_j \left(1 + \frac{\beta \mathbf{x}_i \cdot \beta \mathbf{x}_j}{\beta^2 \sigma^2}\right)^2
\end{aligned}
$$

$$= \sum_{i=1}^{N} \alpha'_i - \frac{1}{2} \sum_{i=1}^{N} \sum_{j=1}^{N} \alpha'_i \alpha'_j y_i y_j K(\mathbf{x}_i, \mathbf{x}_j), \tag{4.5.12}$$

where the following has been defined:

$$K_1(\mathbf{x}'_i, \mathbf{x}'_j) \equiv \left(1 + \frac{\mathbf{x}'_i \cdot \mathbf{x}'_j}{\sigma'^2}\right)^2 = \left(1 + \frac{\beta \mathbf{x}_i \cdot \beta \mathbf{x}_j}{\beta^2 \sigma^2}\right)^2 = K(\mathbf{x}_i, \mathbf{x}_j). \tag{4.5.13}$$

Because Eq. 4.5.12 has the same form as $L(\alpha)$, the Lagrange multipliers for the scaled SVM will not change (i.e., $\alpha'_i = \alpha_i \; \forall i$). Note also that the bias b remains unchanged because

$$b' = 1 - \sum_{i=1}^{N} \alpha'_i y_i K_1(\mathbf{x}'_j, \mathbf{x}'_i) = 1 - \sum_{i=1}^{N} \alpha_i y_i K(\mathbf{x}_j, \mathbf{x}_i) = b, \qquad y_j = 1.$$

Since one has $K_1(\mathbf{x}', \mathbf{x}'_i) = K(\mathbf{x}, \mathbf{x}_i)$ and $\alpha'_i = \alpha_i \; \forall i$, it is obvious that

$$\sum_{i=1}^{N} \alpha'_i y_i K_1(\mathbf{x}', \mathbf{x}'_i) + b = \sum_{i=1}^{N} \alpha_i y_i K(\mathbf{x}, \mathbf{x}_i) + b = 0.$$

Therefore, the decision boundary is proportionally scaled.

Theoretically, it is also possible to use an alternative kernel function

$$K_2(\mathbf{x}', \mathbf{x}'_i) = \left(\beta^2 + \frac{\mathbf{x} \cdot \mathbf{x}_i}{\sigma^2}\right)^2 = \beta^4 K(\mathbf{x}, \mathbf{x}_i) \tag{4.5.14}$$

to handle the scaling problem. It can be shown that the Lagrange multipliers for the scaled SVM become α_i / β^4 and the corresponding Lagrangian is $L(\alpha)/\beta^4$. (See Problem 8.) However, Eq. 4.5.14 should be used with caution because its range is significantly larger than that of Eq. 4.5.13, which may cause numerical difficulty in the quadratic programming optimizer [364].

It is obvious that polynomial SVMs are not translation invariant, because

$$K_3(\mathbf{x}'_i, \mathbf{x}'_j) = \left(1 + \frac{(\mathbf{x}_i + \mathbf{t}) \cdot (\mathbf{x}_j + \mathbf{t})}{\sigma^2}\right)^2 \neq K(\mathbf{x}, \mathbf{x}_i),$$

where $\mathbf{x}' = \mathbf{x} + \mathbf{t}$, and $\mathbf{t}$ is a translation vector.

The scale invariance and translation invariance properties of RBF SVMs can be proved in a similar way and are left to be done in an exercise.

Combine EM and RBF-SVM

The RBF kernel, $K(\mathbf{x}, \mathbf{x}_i) = \exp\left\{-\frac{\|\mathbf{x} - \mathbf{x}_i\|^2}{2\sigma^2}\right\}$, can benefit from an optimal selection of the parameter σ.[7] Therefore, the following hybrid *supervised and unsupervised learning* strategy, called EM-SVM, can be proposed.

---

[7] Proper adjustment on the scaling parameter σ is also required to improve the results of polynomial SVMs. In addition, since SVM is *not* translation invariant, the performance of polynomial SVMs could be further improved by using "data-centralized" coordinates.

1. In the *unsupervised learning* phase, the clustering methods (e.g., K-means or EM) are instrumental in revealing the intraclass subcluster structure. The K-means or EM techniques can be adopted to effectively derive the means and variances of intraclass subclusters (cf. Chapter 3).

2. In the *supervised learning phase*, two options can be considered:

   (a) If the size of original data points is manageable, select an appropriate σ according to the estimated variances. Since scaling can have a significant affect, an RBF-SVM can potentially deliver better performance with properly chosen variances.

   (b) More aggressively, if the original data points are too numerous, then it is imperative to force a reduction in training data: All data points in that subcluster will be represented by a single data point located at the cluster centroid. With such drastically reduced training data, running time of the EM-SVM learning can be substantially expedited.

Furthermore, the effectiveness of the training process is accomplished in two ways:

1. As discussed previously, fuzziness is achieved by insisting on smooth boundaries, which can be done by incorporating a restraining penalty factor C into the SVM.

2. Fuzziness is also achieved by controlling the resolution in the data space in terms of the RBF's radius: To avoid sharp and spurious boundaries in local regions, it helps if data are viewed with a somewhat coarser resolution, very much like passing through a smoothing low-pass filter.

Thus, the generalization performance of RBF-SVM is in a sense doubly assured.

4.6   Biometric Authentication Application Examples

SVMs are classifiers, which have demonstrated high generalization capabilities in many different tasks, including the object recognition problem. In Huang et al. [154], SVMs were applied to eye detection, which is often a vital step in face detection. In Popovici and Thiran [281], a method for face class modeling in the eigenfaces space, using a large-margin classifier similar to SVMs, was proposed. The paper also addresses the issue of how to effectively train the SVM to improve generalization. In Heisele et al. [133], a hierarchical SVM classifier is developed by (1) growing image parts by minimizing theoretical bounds on the error probability of an SVM, and (2) then combining component-based face classifiers in a second stage. This represents a new scheme for automatically learning the discriminative components of objects with SVM classifiers. Moreover, experimental results in face classification show considerable robustness against rotations.

In Sadeghi et al. [327], different face identity verification methods on the BANCA database were compared. The SVM classifier was found to be superior when a large enough training set was available. (However, when the training set size is limited, it is outperformed by a *normalized correlation* method in the LDA space.) In Antonini et al. [9], the effect of representation under the face detection/localization context was studied. An ICA representation and an SVM classification were proposed to extract facial features, with the objective of finding a projection subspace offering better robustness against illumination or other environmental variations. The method was successfully tested on the BANCA database. In Deniz et al. [75], two types of experiments were made on face databases, both achieving a very high recognition rate. In one experiment, SVMs and PCAs were combined, while in another, SVMs and ICA (Independent Component Analysis) were combined for face detection and recognition. The experimental results suggest that SVMs are relatively insensitive to the representation space, as the results using the PCA/SVM combination were not far from the ICA/SVM combination. However, since the training time for ICA is much longer than that of PCA, the paper recommends the best practical combination is PCA with SVM.

Several studies suggest that SVMs can be used for speaker modeling [120, 336, 364, 388]. In these studies, SVMs were used as binary classifiers and were trained to discriminate true-speakers' speech from that of impostors'. Due to an overlap between true-speakers' speech and impostors' speech in the feature space, the training of SVM-based speaker models tended to be computationally intensive. A more interesting way of using SVMs for speaker verification is to classify the *dynamic* scores derived from GMMs or generative speaker models by a discriminative model [43, 97, 179, 256]. Unlike the classical GMM-based speaker recognition approach, where the kernels (typically Gaussians) characterize frame-level statistics, the SVM-based approach considers the entire sequence of an utterance as a single entity (i.e., the entire sequence is mapped to a single point in a high-dimensional feature space).[8]

The generative model provides an efficient representation of the data, whereas the discriminative model creates the best decision boundary to separate the speaker and the impostor classes. In Fine et al. [97], Fisher scores [160] were obtained from a generative model (GMM), which was plugged into the discriminative SVMs to create a better decision boundary for separating different classes. In Campbell [43], the Fisher kernel was replaced by a linear discriminant kernel; and in Moreno and Ho [256], it was replaced by the exponential of the symmetric Kullback-Leibler divergence. Wan and Renals [365] compared dynamic kernels against static kernels and found that sequence-level discrimination, using the log-likelihood ratio kernel, achieved the best results.

SVMs have also been applied to learn the decision function that aims to classify scores produced by the client and the impostor models on a two-dimensional score plane [23] or a three-dimensional score space [196]. Given an utterance X, state-of-

---

[8]In the literature, this type of kernel is referred to as an utterance-based dynamic kernel or a sequence kernel.

the-art speaker verification systems typically compute the ratio (or log-difference) between the likelihood of the true speaker $p(X|\text{spk})$ and the likelihood of an impostor $p(X|\text{imp})$. Each frame in X then results in a point on the "$\log p(X|\text{spk})$-$\log p(X|\text{imp})$" plane. The Bayes decision rule amounts to finding the best *straight line* to separate the points on this plane. Instead of using the Bayes decision rule, an SVM with nonlinear kernels can be applied to create a nonlinear decision function to classify data on the plane. Experimental results have demonstrated that decision functions formed by the SVMs are more flexible than Bayes's straight lines, which leads to a lower error rate.

SVMs have also been adopted for the combination of different modalities of identity verification such as face and speakers. It was demonstrated that a combination of modalities did outperform individual modalities; and in particular, the combination of two lowest-performance modules (i.e., Face and VoiceTI) actually outperformed even the best single module (i.e., VoiceTD)—according to the experiments conducted in Ben-Yacoub [21] and Luettin and Ben-Yacoub [218]. In Liu et al. [216], nonstandard SVMs were used to fuse the scores coming from a speaker verifier and a verbal information verifier. It was found that the SVM-based fusion scheme outperforms the conventional combination rule by 50%.

Problems

1. Suppose that in a two-class problem the class-conditional densities for the two classes can be described by two single Gaussian normal distributions with equal variance but different means. Show that the Fisher projection direction **w** in Eq. 4.2.9 can be derived by computing the posterior probability using the Bayes theorem.

2. Show that the Fisher's linear discriminant analysis bears a strong connection to the least-squares error classifier. More precisely,

   (a) show that they have the same projection direction;

   (b) show that they do not necessarily have the same threshold b.

3. This problem will show that the Fisher discriminant analysis is not always optimal. The two data points $[1, 10]$, $[1, 0]$ are the centers of two clusters belonging to the first class and $[-1, 0]$, $[-1, -10]$ are the centers of another two clusters belonging to the second class. If the two-dimensional data are projected onto one-dimensional space with projection angle θ (a line with an angle of inclination θ with respect to the horizontal axis of the two-dimensional space), the projected coordinates of these four centers on the one-dimensional space will be $\cos\theta + 10\sin\theta$, $\cos\theta$, $-\cos\theta$, and $-\cos\theta - 10\sin\theta$, leading to a classification error rate:

$$0.25\left[\text{erfc}\left(\frac{\cos\theta + 10\sin\theta}{\sqrt{2}\sigma}\right) + \text{erfc}\left(\frac{\cos\theta}{\sqrt{2}\sigma}\right)\right],$$

where $\sigma = 3$ denotes the cluster variance.

(a) Show that Fisher's linear projection angle is $53°$ with an error of 21%.

(b) By differentiating the preceding equation with respect to θ, verify that the best projection angle should be $38°$ and that the minimum error rate is 20%.

4. By differentiating Eq. 4.4.3 with respect to $\mathbf{w}$ and b, show that the solution to the Wolfe dual formulation (Eq. 4.4.4) minimizes the objective function Eq. 4.4.2.

5. Consider the linear separable case (without slack variables). Recall that all support vectors must meet the conditions of Eq. 4.3.5 or Eq. 4.3.6. In your opinion, is the reverse not necessarily true? If not true, show an example in which some training data points satisfying Eq. 4.3.5 or Eq. 4.3.6 are not SVs.

6. This problem determines how to set the parameter σ in the sigmoidal probability for fuzzy SVM. If the positive-class confidence is expressed as

$$\frac{1}{1 + e^{-\frac{\mathbf{w} \cdot \mathbf{x} + b}{\sigma}}}$$

and the marginal support vectors are set to have a confidence level of 2% or 98%, determine the value of σ.

7. This exercise guides you through the steps to show the scale invariance property of linear SVMs using the dual formulation shown in Eq. 4.4.4. Denote the original training data as $\mathbf{x}_i$ and the scaled training data as $\mathbf{x}'_i$ such that $\mathbf{x}'_i = \beta \mathbf{x}$ where β is a constant. Also denote the Lagrange multipliers and the Lagrangian corresponding to the SVM in the scaled domain as α'_i and $L(\alpha'_i)$, respectively.

(a) Show that $L(\alpha') = \frac{1}{\beta^2} L(\alpha)$ if $\alpha'_i = \alpha_i / \beta^2$.

(b) Show that the output weights of the linear SVM in the scaled domain is $\mathbf{w}' = \frac{1}{\beta} \mathbf{w}$ and that the bias term is $b' = 1 - \mathbf{w}' \cdot \mathbf{x}'_i = b$.

(c) Hence, show that the distance from the origin to the decision plane in the scaled domain is $b'/\|\mathbf{w}'\| = \beta b/\|\mathbf{w}\|$.

8. Assume that the polynomial kernel of an SVM has the form $K(\mathbf{x}, \mathbf{x}_i) = (1 + \mathbf{x} \cdot \mathbf{x}_i)^2$. When the training data is scaled by a factor β, the kernel function can be changed to $K'(\mathbf{x}', \mathbf{x}'_i) = (\beta^2 + \mathbf{x}' \cdot \mathbf{x}'_i)^2$, where $\mathbf{x}_i = \beta \mathbf{x}_i \; \forall i$.

(a) Show that the Lagrangian corresponding to the scaled data is given by $L(\alpha') = L(\alpha)/\beta^4$, where $L(\alpha)$ is the Lagrangian corresponding to the unscaled data.

(b) Hence, show that the decision boundary of an SVM that uses $(\beta^2 + \mathbf{x}' \cdot \mathbf{x}'_i)^2$ as the kernel function will produce a scaled version of the decision boundary given by the SVM that uses $(1 + \mathbf{x} \cdot \mathbf{x}_i)^2$ as the kernel function.

(c) Discuss the potential problem of using $(\beta^2 + \mathbf{x}' \cdot \mathbf{x}'_i)^2$ as the kernel function, especially when $\beta \gg 1$.

(d) Suggest a better polynomial kernel function to address the scaling issue.

9. Show that SVMs with an RBF-kernel function of the form

$$K(\mathbf{x}, \mathbf{x}_i) = \exp\left\{ -\frac{\|\mathbf{x} - \mathbf{x}_i\|}{2\sigma^2} \right\}$$

are invariant to both scaling and translation if the value of σ can be proportionally scaled. Give the scaling factor for σ if the input data is scaled by β.

10. Figure 4.16(a) shows a two-class problem in which there are two data points—$(0,0)$ and $(0.45, 0.45)$—in one class and three data points—$(1,0)$, $(0,1)$, and $(1,1)$—in the other. Analytically derive the optimal decision boundary and the corresponding SVs for the following SVMs:

   (a) Linear

   (b) Nonlinear, with second-order polynomial kernels

   (c) Nonlinear, with RBF kernels

11. As shown in Figure 4.16(b), which is a shifted version of Figure 4.16(a), there are two data points—$(-0.5, -0.5)$ and $(-0.05, -0.05)$—for one class and three points for the other—$(0.5, -0.5)$, $(-0.5, 0.5)$, and $(0.5, 0.5)$. Derive the optimal boundary and find the support vectors for the following SVMs:

   (a) Linear

   (b) Nonlinear, with second-order polynomial kernels

   (c) Nonlinear, with RBF kernels

12. There are two points—$(0,0)$ and $(0,2)$—for one class and two points—$(0,-1)$ and $(0,1)$—for the other class. Show analytically whether the data patterns can be separated by

   (a) Nonlinear SVMs, with RBF kernels

   (b) Nonlinear SVMs, with second-order polynomial kernels

   (c) Nonlinear SVMs, with higher-order polynomial kernels

   Compare your results with those of Problem 10.

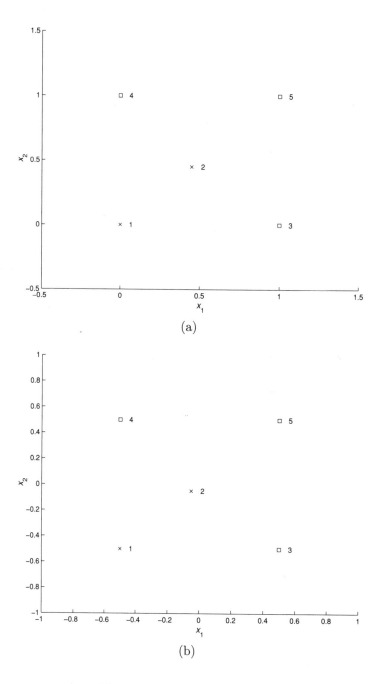

Figure 4.16. A 5-point problem.

13. *Support vectors for clearly separable SVM classifiers.* Consider a clearly separable SVM classifier. Let **A** denote the set of data points with $\alpha_i > 0$ and **B** denote the set of data points for which $y_i(\mathbf{x}_i \cdot \mathbf{w} + b) = 1$.

   (a) Show that $\mathbf{A} \subseteq \mathbf{B}$ according to the Karush-Kuhn-Tucker conditions [25, 99].

   (b) Under what condition will $\mathbf{A} = \mathbf{B}$?

   (c) State the condition for which $\mathbf{A} \neq \mathbf{B}$.

14. *Support vectors for fuzzily separable SVM classifiers.* Consider a fuzzily separable SVM classifier. Let **A** the denote the set of data points with $\alpha_i > 0$ and **C** denote the set of data points such that $y_i(\mathbf{x}_i \cdot \mathbf{w} + b) = 1 - \xi_i$ and $\xi_i > 0$. Are the two sets equivalent? If not equivalent, is one set a subset of the other? Explain why?

15. For two-dimensional training data, suppose that training data points are generically random and that no more than two data points are colinear. What is the maximum number of *marginal* SVs selected?

16. Verify that the inner-product kernel is a symmetric function of its arguments, that is,

$$K(\mathbf{x}, \mathbf{x}') = K(\mathbf{x}', \mathbf{x}).$$

17. Consider the linear inner-product kernel on a two-dimensional vector space $\mathbf{x} = [u \ v]^T$:

$$K(\mathbf{x}, \mathbf{x}_i') = (\mathbf{x} \cdot \mathbf{x}_i').$$

Assuming the lower limit of the integral is $b = -a$, the eigenfunction decomposition property can be verified by the following equality:

$$\int_{-a}^{a} K(\mathbf{x}, \mathbf{x}')\phi(\mathbf{x}')d\mathbf{x}' = \lambda\phi(\mathbf{x}).$$

   (a) Is $\phi(\mathbf{x}) = u$ an eigenfunction?

   (b) Is $\phi(\mathbf{x}) = u + v$ an eigenfunction?

Hint: Note that, for two-dimensional vector space, a double integral must be used:

$$\int_{-a}^{a} K(\mathbf{x}, \mathbf{x}')\phi(\mathbf{x}')d\mathbf{x}' \equiv \int_{-a}^{a}\int_{-a}^{a} K(\mathbf{x}, \mathbf{x}')\phi(\mathbf{x}')du'dv'.$$

18. Consider a second-order polynomial kernel on a two-dimensional vector space $\mathbf{x} = [u \ v]^T$:

$$K(\mathbf{x}, \mathbf{x}_i) = (1 + \mathbf{x} \cdot \mathbf{x}_i)^2.$$

Again assuming the lower limit of the integral is $b = -a$:

(a) Is $\phi(\mathbf{x}) = u$ an eigenfunction?

(b) Is $\phi(\mathbf{x}) = u + v$ an eigenfunction?

(c) Show that $\phi(\mathbf{x}) = u^2$ is *not* an eigenfunction.

(d) Under what condition on a will the function $\phi(\mathbf{x}) = u^2$ be deemed an eigenfunction?

19. *Robust derivation of the threshold b.* Once $\mathbf{w}$ is fixed, the following methods can be used to determine the threshold b. Which method is more robust? Explain your selection.

(a) Use $b = 1 - \mathbf{w}^T \mathbf{x}_i$, where $\mathbf{x}_i$ is a support vector and $y_i = 1$.

(b) Determine b from the mean value of

$$1 - \mathbf{w}^T \mathbf{x}_i, \quad \mathbf{x}_i \in \mathcal{S}^+$$
$$-1 - \mathbf{w}^T \mathbf{x}_j, \quad \mathbf{x}_j \in \mathcal{S}^-,$$

where $\mathcal{S}^+$ and $\mathcal{S}^-$ contain support vectors for which the corresponding target values (y_i and y_j) are equal to $+1$ and -1, respectively.

Design a more robust method that makes use of all of the training data (SVs and non-SVs) for determining b [40].

Chapter 5

MULTI-LAYER NEURAL NETWORKS

5.1 Introduction

Neural networks have played a major role in many applications (e.g., see [171, 368]). In particular, they have demonstrated convincing performance in the detection and recognition of object classes. This is mainly due to their capability to cope with a variety of cues such as texture, intensity, edge, color and motion. In the context of personal identification, neural networks can facilitate detection or recognition of high-level features extracted from facial images or speakers' voices.

Supervised-learning networks represent the mainstream of development in neural networks for biometric authentication. Some examples of well-known pioneering networks include the perceptron network [321], ADALINE/MADALINE [382], and various multi-layer networks [247, 269, 326, 377]. The two phases involved in a supervised-learning network are: *retrieving phase* and *learning phase*. In the retrieving phase, the objective is to determine to which class a pattern belongs based on the winner of the output values. The *output values* are functions of the input values and network weights, called *discriminant function*. In the learning phase, the weights are trained so that the (learned and/or unlearned) patterns are more likely to be correctly classified.

This chapter presents several important variants of multi-layer networks. First, the popular and well-known back-propagation (BP) algorithms for training multi-layer networks are explored. It is widely recognized that the ultimate performance of BP algorithms are known to be sensitive to initial conditions. Taking convergence behavior into consideration, the text here looks into a two-stage training strategy so that the back-propagation training of multi-layer networks will be more likely to start with a reliable initial setting. To present more comprehensive coverage of the learning algorithms of multi-layer networks, a complementary learning scheme, known as the genetic algorithm, is also presented.

Advanced network structures can be more effectively employed to integrate different training disciplines such as local experts, unsupervised modules, and supervised-learning strategies. Chapter 6 presents the concept of modular net-

139

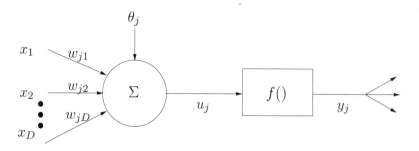

Figure 5.1. McCulloch-Pitts model of a neuron.

works, mixture-of-experts, and hierarchical network structures. Both expert-based and class-based modular networks are introduced. Chapter 7 demonstrates how existing unsupervised modules (e.g., RBF or LBF clustering) and supervised-learning (e.g., reinforced/antireinforced) strategies can be naturally combined to lead to a decision-based neural network.

5.2 Neuron Models

McCulloch and Pitts [237] proposed a binary threshold unit as a computational model for an artificial neuron (see Figure 5.1). Basically, this neuron model can be characterized by the functional descriptions of the *connection network* and the *neuron activation*. Each neuron receives input values x_j's, which are propagated through a network of unidirectional connections from other neurons in the network. Associated with each connection, there is a *synaptic weight* (denoted by w_{ij}), which dictates the effect of the j-th neuron on the i-th neuron. The inputs to the i-th neuron are accumulated, together with the external threshold θ_i, to yield the *net value u_i*. The mapping is mathematically described by a *basis function*. The net value u_i will then be further transformed by a nonlinear *activation function f* to yield an output value y_i.

5.2.1 Basis Functions (Net Functions)

The connection networks are mathematically represented by a basis function $u(\mathbf{w}_j, \mathbf{x})$, where $\mathbf{w}_j$ stands for the weight vector and $\mathbf{x}$ for the input vector. The basis function has two common forms:

1. *Linear basis functions (LBFs).* LBFs are a kind of hyperplane function (i.e., first-order basis functions). Specifically, the net value of the j-th neuron is a linear combination of the inputs:

$$u_j = u(\mathbf{w}_j, \mathbf{x}) = \sum_{i=1}^{D} w_{ji} x_i, \qquad (5.2.1)$$

where w_{ji} is the weight connecting the j-th neuron and its i-th input and D is the number of inputs to the j-th neuron.

2. *Radial basis functions (RBFs).* RBFs are a type of hypersphere function that involves a second-order (nonlinear) basis function. The net value u_j of the j-th neuron is a function of the distance between a reference pattern $\mathbf{w}_j$ and the input $\mathbf{x}$:

$$u_j = u(\mathbf{w}_j, \mathbf{x}) = \sqrt{\sum_{i=1}^{D}(x_i - w_{ji})^2}. \qquad (5.2.2)$$

5.2.2 Activation Functions (Neuron Functions)

The net value as expected by the basis function, $u(\mathbf{w}, \mathbf{x})$, will be immediately transformed by a nonlinear activation function of the neuron. For example, the most common activation functions are *step, ramp, sigmoidal,* and *Gaussian* functions (cf. Figure 5.2). More specifically,

- Sigmoidal function: $\qquad f(u_j) = \dfrac{1}{1 + e^{-(u_j - \theta_j)/\sigma}}, \qquad (5.2.3)$

  where $u_j = \sum_{i=1}^{D} w_{ji} x_i$ and θ_j is a bias term;

- Gaussian function: $\qquad f(u_j) = e^{-u_j^2/\sigma^2}, \qquad (5.2.4)$

  where $u_j = \sqrt{\sum_{i=1}^{D}(x_i - w_{ji})^2}$ and σ represents the spread of $f(u_j)$.

Note that when $\sigma \to 0$, then the sigmoidal function in Figure 5.2(c) becomes the step function shown in Figure 5.2(a) (i.e., the step function is a special case of the sigmoidal function).

5.2.3 Discriminant Functions (Output Functions)

For a neural network classifier, the output values in the output layer are usually expressed as an output vector $\mathbf{y} = [y_1, y_2, \ldots, y_K]^T$, where T denotes vector transpose. The output $\mathbf{y}$ is a function of the input vector $\mathbf{x}$ and the whole network weights

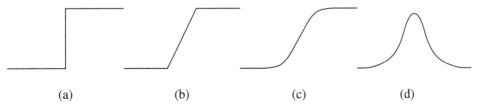

| (a) | (b) | (c) | (d) |

Figure 5.2. Four examples of activation functions: (a) step function, (b) ramp function, (c) sigmoidal function, and (d) Gaussian function.

W—that is, $\mathbf{y} = \phi(\mathbf{x}, W)$. Since the output values of these neurons are used to determine the class to which the input pattern belongs (e.g., $\mathbf{x}$ is classified to class i if $y_i > y_j$, $\forall j \neq i$), the function $\phi(\mathbf{x}, W)$ is called the *discriminant function*.

5.3   Multi-Layer Neural Networks

Multi-layer networks are one of the most popular neural-network models. In a multi-layer network, the basis function of each neuron can be a linear basis function (LBF) with the activation function being either the step function or the sigmoidal function. Alternatively, RBF-type neurons can be adopted, which results in the RBF/EBF neural networks.

Multi-layer networks are typically trained by supervised learning, meaning that teacher information is used to train the network parameters. Depending on the nature of the teacher's information, there can be two approaches to supervised learning: One is based on the correctness of the decision and the other is based on the optimization of a training cost criterion. Of the latter, the (least-squares error) approximation-based formulation represents the most important special case. The decision-based and approximation-based formulations differ in the teacher's information that they contain and the ways in which they use it. A brief comparison follows.

Approximation-Based Formulation

An approximation-based formulation can be viewed as an *approximation/regression* for the trained data set, which is denoted as input/teacher pairs:

$$[\mathcal{X}, \mathcal{T}] = \left\{ [\mathbf{x}^{(1)}, \mathbf{t}^{(1)}], [\mathbf{x}^{(2)}, \mathbf{t}^{(2)}], \ldots, [\mathbf{x}^{(N)}, \mathbf{t}^{(N)}] \right\},$$

where N is the number of training pairs. The desired values at the output nodes corresponding to the input patterns $\mathbf{x}^{(n)}$ are assigned as the teacher's values. *The objective of network training is to find the optimal weights to minimize the error between the teacher value and actual response.* A popular criterion is the minimum-squares error between the teacher's and the actual response. To acquire a more versatile nonlinear approximation capability, multi-layer networks (together with the back-propagation learning rule) are usually adopted.

Decision-Based Formulation

In the decision-based formulation, the teacher only tells the correctness of the classification for each training pattern. The teacher is a set of symbols, $\mathcal{T} = \{\mathbf{t}^{(n)}\}$, labeling the correct class for each input pattern. Unlike the approximation formulation, the teacher's exact values are not required. *The objective of the training is to find a set of weights that yields the correct classification.* In other words, the teacher does not need to know the exact output values of the neurons. This consideration leads to a network structure adopted by the so-called decision-based neural network (DBNN), to be discussed in Chapter 7.

5.3.1   Multi-Layer Neural Models

The most prominent neural models, based on the approximation-based formulation, are the multi-layer networks—also known as the multi-layer perceptrons (MLPs) in the literature. There are basically two types of multi-layer networks: LBF (Eq. 5.2.1) and RBF (Eq. 5.2.2), which are both highly flexible in that they can approximate almost any multivariate distribution function. Generally, a two-layer network should be adequate as a universal approximator of any nonlinear function [138]. Furthermore, it was demonstrated in [213] that, even for extreme situations, a three-layer network suffices to separate any (convex or nonconvex) polyhedral decision region from the background.

Generally, as long as there is an adequate number of nonlinear hidden nodes, a multi-layer network with correctly trained synaptic weights will be able to separate any two classes of data distribution. The synaptic weights can be adaptively trained by the BP learning algorithm. Figure 5.3(b) depicts how the representation space of the hidden-layer can be trained so that the two classes become more linearly separable step by step.

Figure 5.4 depicts the architecture of a three-layer neural network. For training approximation-based networks, the objective is to train the weights $\{w_{ji}(l); j = 1, \ldots, M_l; i = 1, \ldots, M_{l-1}; l = 1, \ldots, L\}$, where M_l is the number of nodes (including the bias) in the l-th layer and L is the number of layers, so as to minimize the least-squares error between the teacher's and actual responses [326]. More specifically, minimize

$$E = \frac{1}{2} \sum_{n=1}^{N} \sum_{k=1}^{K} [t_k^{(n)} - a_k^{(n)}(L)]^2, \tag{5.3.1}$$

where $a_k^{(n)}(L)$ is the output at node k in the output layer, N is the number of training patterns, and $K = M_L$ is the dimension of the output space. The back-propagation algorithm can be applied to any type of energy function, and it offers an effective approach to the computation of the energy gradients.

The input units are represented by $x_i \equiv a_i(0)$ and the output units by $y_i \equiv a_i(L)$, where L is the number of layers. A multi-layer network is characterized by the following feed-forward equations:

$$u_j(l) = u(\mathbf{w}_j(l), \mathbf{a}(l-1)) \tag{5.3.2}$$
$$a_j(l) = f(u_j(l)) \qquad 1 \leq j \leq M_l \text{ and } 1 \leq l \leq L, \tag{5.3.3}$$

where $\mathbf{w}_j(l) = [w_{j1}(l) \ \cdots \ w_{j,M_{l-1}}(l)]$ and $\mathbf{a}(l-1) = [a_1(l-1) \ \cdots \ a_{M_{l-1}}(l-1)]$. The activation function, denoted as $f(\cdot)$, is very often a *sigmoid function* for LBF neurons or a Gaussian activation function for RBF neurons.

5.3.2 The Classic XOR Problem

The most well-known linearly nonseparable example is the exclusive-or (XOR) problem, the input/output mapping of which is

$$
\begin{array}{rcl}
(0,\ 0) & \longrightarrow & 0 \\
(1,\ 1) & \longrightarrow & 0 \\
(1,\ 0) & \longrightarrow & 1 \\
(0,\ 1) & \longrightarrow & 1.
\end{array}
\tag{5.3.4}
$$

The four input data, labeled by "×" and "○", are depicted in Figure 4.4. Note that the XOR data are not linearly separable because there is no straight line that can separate the two classes. This means that no single-layer network could qualify as an XOR classifier. It is easy to show that a two-layer network, shown in Figure 5.3(a), can solve the XOR problem if the lower weights are set as follows:

$$
\begin{array}{rcl}
w_{11} & \longrightarrow & 0.6 \\
w_{12} & \longrightarrow & 1.5 \\
w_{21} & \longrightarrow & 0.2 \\
w_{22} & \longrightarrow & 0.4.
\end{array}
\tag{5.3.5}
$$

Based on the net function in Eq. 5.2.1 and the neuron function in Eq. 5.2.3 (setting $\sigma \to 0$ and the threshold $\theta = 0.5$), the net values and neuron values at the hidden-layer can be derived as follows:

$$
\begin{array}{llll}
\text{Input} & & \text{Net value} & \text{Hidden neurons' output} \\
(0,\ 0) & \Rightarrow & (0,\ 0) & (0,\ 0) \\
(1,\ 1) & \Rightarrow & (2.1,\ 0.6) & (1,\ 1) \\
(1,\ 0) & \Rightarrow & (0.6,\ 0.2) & (1,\ 0) \\
(0,\ 1) & \Rightarrow & (1.5,\ 0.4) & (1,\ 0)
\end{array}
\tag{5.3.6}
$$

Thus, by inspection, the mapped data on the hidden-layer have now become linearly separable. Note also that the choice of synaptic weights is highly nonunique. In general, synaptic weights can be adaptively trained by the BP learning algorithm.

5.4 The Back-Propagation Algorithms

The most prominent algorithm for training multi-layer networks is the so-called back-propagation (BP) algorithm. The algorithm, independently proposed by Werbos [377], Parker [269], and Rumelhart et al. [326], offers an efficient computational speedup for training multi-layer networks. The objective is to train the weights $w_{ji}(l)$ so as to minimize the mean-squared error E in Eq. 5.3.1. The basic gradient-type learning formula is

$$
w_{ji}^{(n+1)}(l) = w_{ji}^{(n)}(l) + \Delta w_{ji}^{(n)}(l) \qquad l = 1, \ldots, L
\tag{5.4.1}
$$

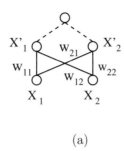

(a)

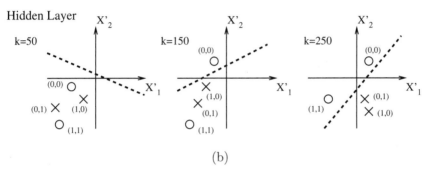

(b)

Figure 5.3. In this simulation of a two-layer BP network for the XOR problem, the hidden-unit representations change with the iterations: (a) A two-layer network and (b) the change of weights (representation) gradually adjusts the coordinates. Eventually, when iteration number $k > 250$, the representation becomes linearly separable by the upper layer.

with the n-th training pattern, $\mathbf{a}^{(n)}(0)$, and its corresponding teacher $\mathbf{t}^{(n)}$, $n = 1, 2, \ldots, N$, presented. The derivation of the BP algorithm follows a chain-rule technique:

$$
\begin{aligned}
\Delta w_{ji}^{(n)}(l) &= -\eta \frac{\partial E}{\partial w_{ji}^{(n)}(l)} \\
&= -\eta \frac{\partial E}{\partial a_j^{(n)}(l)} \cdot \frac{\partial a_j^{(n)}(l)}{\partial w_{ji}^{(n)}(l)} \\
&= \eta \, \delta_j^{(n)}(l) f'(u_j^{(n)}(l)) \frac{\partial u_j^{(n)}(l)}{\partial w_{ji}^{(n)}(l)},
\end{aligned}
\tag{5.4.2}
$$

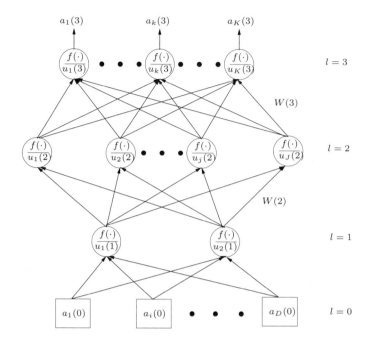

Figure 5.4. The architecture of multi-layer neural networks with $L = 3$.

where the error signal $\delta_j^{(n)}(l)$ is defined as

$$\delta_j^{(n)}(l) \equiv -\frac{\partial E}{\partial a_j^{(n)}(l)}.$$

The error signal $\delta_j^{(n)}(l)$ can be obtained recursively by back-propagation:

- *Initial (Top) Layer.* For the recursion, the initial value (of the top layer), $\delta_j^{(n)}(L)$, can be obtained easily as follows:

$$\begin{aligned} \delta_j^{(n)}(L) &\equiv -\frac{\partial E}{\partial a_j^{(n)}(L)} \\ &= t_j^{(n)} - a_j^{(n)}(L). \end{aligned} \tag{5.4.3}$$

For an energy function other than the LSE, the initial condition can be similarly derived.

- *Recursive Formula.* The general BP recursive formula for the error signal, $\delta_i^{(n)}(l)$, can be derived as follows:

$$\delta_j^{(n)}(l) \equiv -\frac{\partial E}{\partial a_j^{(n)}(l)}$$

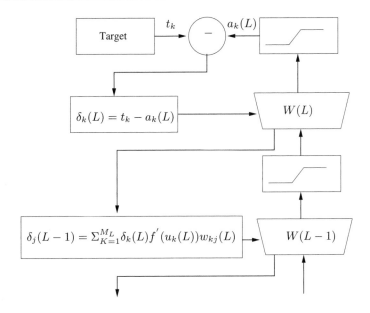

Figure 5.5. The schematic diagram for the BP process.

$$= -\sum_{k=1}^{M_{l+1}} \frac{\partial E}{\partial u_k^{(n)}(l+1)} \frac{\partial u_k^{(n)}(l+1)}{\partial a_j^{(n)}(l)} \tag{5.4.4}$$

in the sequence of $l = L - 1, \ldots, 1$. From Eq. 5.3.3, it follows that

$$\delta_j^{(n)}(l) = \sum_{k=1}^{M_{l+1}} \delta_k^{(n)}(l+1) \; f'(u_k^{(n)}(l+1)) \; \frac{\partial u_k^{(n)}(l+1)}{\partial a_j^{(n)}(l)}. \tag{5.4.5}$$

The recursive formula is the key to back-propagation learning. It allows the error signal of a lower layer $\delta_j^{(n)}(l)$ to be computed as a linear combination of the error signals of the upper layer $\delta_k^{(n)}(l+1)$. In this manner, the error signals $\delta_j^{(n)}(\cdot)$ are back propagated through all of the layers from the top down. This also implies that the influences from an upper to a lower layer (and vice versa) can only be effected via the error signals of the intermediate layers.

The general training algorithm described before is applicable to two types of prominent multi-layer networks: multi-layer LBF networks and multi-layer RBF networks.

5.4.1 BP Algorithm for Multi-Layer LBF Networks

A multi-layer linear basis function (LBF) network is characterized by the following equations:

$$u_j(l) = \sum_{i=1}^{M_{l-1}} w_{ji}(l)a_i(l-1) + \theta(l) = \sum_{i=0}^{M_{l-1}} w_{ji}(l)a_i(l-1) \tag{5.4.6}$$

$$a_j(l) = f(u_j(l)) \qquad 1 \le j \le M_l \text{ and } 1 \le l \le L, \tag{5.4.7}$$

where the input units are represented by $x_i \equiv a_i(0)$, the output units by $y_i \equiv a_i(L)$, and where L is the number of layers and $f(\cdot)$ is the sigmoidal activation function. Note that the bias term $\theta(l)$ has been absorbed into the summation so that $w_{j0}(l) = \theta(l)$ and $a_0(l) = 1$ $\forall l = 0, \ldots, L-1$. Substituting Eq. 5.4.6 into Eq. 5.4.2 and Eq. 5.4.5, the following back-propagation rule is obtained:

$$w_{ji}^{(n+1)}(l) = w_{ji}^{(n)}(l) + \eta \delta_j^{(n)}(l) f'(u_j^{(n)}(l)) a_i^{(n)}(l-1) \quad l = 1, \ldots, L, \tag{5.4.8}$$

where the error signal $\delta_j^{(n)}(l)$ is defined as

$$\delta_j^{(n)}(l) = \sum_{k=1}^{M_{l+1}} \delta_k^{(n)}(l+1) f'(u_k^{(n)}(l+1)) w_{kj}^{(n)}(l+1) \quad l = L-1, \ldots, 1 \tag{5.4.9}$$

with the initial condition given by Eq. 5.4.3. Figure 5.5 illustrates the BP process.

5.4.2 BP Algorithm for Multi-Layer RBF Networks

In the 1980s, the multi-layer LBF networks trained by the BP algorithm [326] attracted considerable attention from the neural network community. While these networks have shown promise in many application domains, during the last two decades, statisticians and the neural network community found that kernel-based neural networks (or equivalent multi-layer RBF networks) possess a number of advantages over multi-layer LBF networks. These advantages include rapid training speed; better generalization; and more important, the ability to divide a complex task into a number of subtasks (i.e., modularization—see Section 6.3). These positive properties have been demonstrated in multiple theoretical studies and can be found in many real-world applications [186].

In a multi-layer RBF network, each neuron in the hidden-layer is composed of a radial basis function, which also serves as an activation function. The weighting parameters in an RBF network are the centers and widths of these neurons. The output functions are the linear combination of these radial basis functions. A more general form of the RBF networks is the elliptical basis function (EBF) networks where the hidden neurons compute the Mahalanobis distance between the centers and the input vectors. It has been shown that RBF networks have the same asymptotic approximation power as multi-layer LBF networks [279].

Figure 5.6 depicts the architecture of an RBF/EBF network with D inputs, M-basis functions (hidden nodes), and K outputs (see p. 152). The input layer distributes the D-dimensional input patterns, $\mathbf{x}^{(n)}$, to the hidden-layer. Each hidden unit is a Gaussian-basis function of the form

$$\phi_j(\mathbf{x}^{(n)}) = \exp\left\{-\frac{1}{2}\left(\mathbf{x}^{(n)} - \mu_j\right)^T \Sigma_j^{-1}\left(\mathbf{x}^{(n)} - \mu_j\right)\right\} \quad j = 1,\ldots,M, \quad (5.4.10)$$

where μ_j and Σ_j are the mean vector and covariance matrix of the j-th basis function, respectively. The k-th output is a linear weighted sum of the basis functions' outputs; that is,

$$y_k(\mathbf{x}^{(n)}) = w_{k0} + \sum_{j=1}^{M} w_{kj}\phi_j(\mathbf{x}^{(n)})$$

$$= \sum_{j=0}^{M} w_{kj}\phi_j(\mathbf{x}^{(n)}) \quad n = 1,\ldots,N \text{ and } k = 1,\ldots,K, \quad (5.4.11)$$

where $\mathbf{x}^{(n)}$ is the n-th input vector, w_{k0} is a bias term, and $\phi_0(\cdot) \equiv 1$. To apply RBF/EBF networks for pattern classification, each class is assigned a group of hidden units as depicted in Figure 5.6, and each group is trained to represent the patterns belonging to its corresponding class.

In addition to multi-layer LBF networks, the BP algorithm described before can also be applied to multi-layer RBF networks. Denote the mean vector and covariance matrix of the j-th kernel of a multi-layer RBF network as $\mu_j = [\mu_{j1},\ldots,\mu_{jD}]^T$ and $\Sigma_j = \text{diag}\{\sigma_{j1}^2,\ldots,\sigma_{jD}^2\}$, respectively, where D is the number of inputs (feature dimension). For a multi-layer RBF network with K outputs, further denote

$$e_k(\mathbf{x}^{(n)}) = t_k^{(n)} - y_k(\mathbf{x}^{(n)}), \quad k = 1,\ldots,K \text{ and } n = 1,\ldots,N \quad (5.4.12)$$

as the error between the k-th target output and the k-th actual output, where $y_k(\mathbf{x}^{(n)})$ is evaluated using Eqs. 5.4.10 and 5.4.11. The total squared error for all training patterns becomes

$$E = \frac{1}{2}\sum_{n=1}^{N}\sum_{k=1}^{K}\left[t_k^{(n)} - y_k(\mathbf{x}^{(n)})\right]^2 = \frac{1}{2}\sum_{n=1}^{N}\sum_{k=1}^{K}\left[e_k(\mathbf{x}^{(n)})\right]^2. \quad (5.4.13)$$

The instantaneous update formulae[1] for the parameters of the network are defined as:

$$w_{kj}^{(n+1)} = w_{kj}^{(n)} + \Delta w_{kj}^{(n)} \quad k = 1,\ldots,K \text{ and } j = 0,\ldots,M \quad (5.4.14)$$

$$\mu_{ji}^{(n+1)} = \mu_{ji}^{(n)} + \Delta\mu_{ji}^{(n)} \quad j = 0,\ldots,M \text{ and } i = 1,\ldots,D \quad (5.4.15)$$

$$\sigma_{ji}^{(n+1)} = \sigma_{ji}^{(n)} + \Delta\sigma_{ji}^{(n)} \quad j = 0,\ldots,M \text{ and } i = 1,\ldots,D, \quad (5.4.16)$$

---

[1]They are equivalent to the sequential update mode of the BP algorithm.

where n denotes the presentation of the n-th pattern in a particular epoch. Note that n also serves as the iteration index within an epoch. Referring to the symbols defined in the beginning of Section 5.4, observe that $y_k(\mathbf{x}^{(n)}) = a_k^{(n)}(L) = f(u_k^{(n)}(L)) = u_k^{(n)}(L)$, where $L = 2$ and $f(u_k(L)) = u_k(L)$. Therefore, based on Eq. 5.4.2, the gradient of the output weights is as follows:

$$
\begin{aligned}
\Delta w_{kj}^{(n)} &= \eta_w \delta_k^{(n)}(L) f'(u_k^{(n)}(L)) \frac{\partial u_k^{(n)}(L)}{\partial w_{kj}^{(n)}(L)} \\
&= \eta_w e_k(\mathbf{x}^{(n)}) \cdot 1 \cdot \phi_j(\mathbf{x}^{(n)}) \\
&= \eta_w e_k(\mathbf{x}^{(n)}) \phi_j(\mathbf{x}^{(n)}).
\end{aligned}
\tag{5.4.17}
$$

For the hidden nodes' weights $\mu_{ji}^{(n)}$ and $\sigma_{ji}^{(n)}$, observe also that $\phi_j(\mathbf{x}^{(n)}) = a_j^{(n)}(1) = f(u_j^{(n)}(1))$, where $f(u) = \exp(u)$ (see Eq. 5.2.4) and $u_j^{(n)}(1)$ has the form

$$
u_j^{(n)}(1) = -\frac{1}{2} \sum_{i=1}^{D} \left(\frac{x_i^{(n)} - \mu_{ji}^{(n)}}{\sigma_{ji}^{(n)}} \right)^2 \quad j = 1, \ldots, M.
$$

Therefore, one can apply Eqs. 5.4.2 and 5.4.5 to obtain $\Delta \mu_{ji}^{(n)}$, as follows:

$$
\begin{aligned}
\Delta \mu_{ji}^{(n)} &= \eta_\mu \delta_j^{(n)}(1) f'\left(u_j^{(n)}(1) \right) \frac{\partial u_j^{(n)}(1)}{\partial \mu_{ji}^{(n)}} \\
&= \eta_\mu \sum_{k=1}^{K} \delta_k^{(n)}(2) f'\left(u_k^{(n)}(2) \right) \frac{\partial u_k^{(n)}(2)}{\partial a_j^{(n)}(1)} f'\left(u_j^{(n)}(1) \right) \frac{\partial u_j^{(n)}(1)}{\partial \mu_{ji}^{(n)}} \\
&= \eta_\mu \sum_{k=1}^{K} e_k(\mathbf{x}^{(n)}) \cdot 1 \cdot w_{kj}^{(n)} \phi_j(\mathbf{x}^{(n)}) \frac{x_i^{(n)} - \mu_{ji}^{(n)}}{(\sigma_{ji}^{(n)})^2},
\end{aligned}
\tag{5.4.18}
$$

where one has made use of the fact that $\delta_k^{(n)}(2) = e_k(\mathbf{x}^{(n)})$, $f'(u_k^{(n)}(2)) = 1$, $f'(u_j^{(n)}(1)) = f(u_j^{(n)}(1)) = \phi_j(\mathbf{x}^{(n)})$, and $a_j^{(n)}(1) = \phi_j(\mathbf{x}^{(n)})$. Similarly, $\Delta \sigma_{ji}^{(n)}$ is computed as

$$
\begin{aligned}
\Delta \sigma_{ji}^{(n)} &= \eta_\sigma \delta_j^{(n)}(1) f'\left(u_j^{(n)}(1) \right) \frac{\partial u_j^{(n)}(1)}{\partial \sigma_{ji}^{(n)}} \\
&= \eta_\sigma \sum_{k=1}^{K} \delta_k^{(n)}(2) f'\left(u_k^{(n)}(2) \right) \frac{\partial u_k^{(n)}(2)}{\partial a_j^{(n)}(1)} f'\left(u_j^{(n)}(1) \right) \frac{\partial u_j^{(n)}(1)}{\partial \sigma_{ji}^{(n)}} \\
&= \eta_\sigma \sum_{k=1}^{K} e_k(\mathbf{x}^{(n)}) \cdot 1 \cdot w_{kj}^{(n)} \phi_j(\mathbf{x}^{(n)}) \frac{(x_i^{(n)} - \mu_{ji}^{(n)})^2}{(\sigma_{ji}^{(n)})^3}
\end{aligned}
\tag{5.4.19}
$$

Alternatively, the following gradient descent rules can be directly applied to obtain the same set of update formulae (see Problem 13):

$$w_{kj}^{(n+1)} = w_{kj}^{(n)} - \eta_w \frac{\partial E}{\partial w_{kj}^{(n)}} \tag{5.4.20}$$

$$\mu_{ji}^{(n+1)} = \mu_{ji}^{(n)} - \eta_\mu \frac{\partial E}{\partial \mu_{ji}^{(n)}} \tag{5.4.21}$$

$$\sigma_{ji}^{(n+1)} = \sigma_{ji}^{(n)} - \eta_\sigma \frac{\partial E}{\partial \sigma_{ji}^{(n)}}. \tag{5.4.22}$$

To ensure positivity of σ_{ji}, define $\sigma_{ji}^2 = \exp\{\alpha_{ji}\}$ and replace σ_{ji} in Eq. 5.4.22 with α_{ji}; that is,

$$\alpha_{ji}^{(n+1)} = \alpha_{ji}^{(n)} - \eta_\alpha \frac{\partial E}{\partial \alpha_{ji}^{(n)}}.$$

Note that

$$2\sigma_{ji} \frac{\partial \sigma_{ji}}{\partial \alpha_{ji}} = \exp(\alpha_{ji}) = \sigma_{ji}^2 \quad \text{i.e.} \quad \frac{\partial \sigma_{ji}}{\partial \alpha_{ji}} = \frac{\sigma_{ji}}{2}$$

results in

$$\frac{\partial E}{\partial \alpha_{ji}^{(n)}} = -\sum_{k=1}^{K} e_k(\mathbf{x}^{(n)}) w_{kj} \phi_j(\mathbf{x}^{(n)}) (x_i^{(n)} - \mu_{ji}^{(n)})^2 \frac{(\sigma_{ji}^{(n)})^{-2}}{2}.$$

5.5   Two-Stage Training Algorithms

The convergence behavior of the BP algorithm for multi-layer networks has always been an issue of great concern. This is because the algorithm tends to be sensitive to initial conditions. Therefore, the convergence problem can be greatly alleviated if proper initial conditions can be estimated. This motivates the two-stage training strategy discussed next.

5.5.1   Two-Stage Training for RBF Networks

The two-stage training of RBF networks starts with the estimation of the kernel means using the K-means algorithm

$$\mu_j = \frac{1}{N_j} \sum_{\mathbf{x} \in X_j} \mathbf{x} \tag{5.5.1}$$

$$\Sigma_j = \frac{1}{N_j} \sum_{\mathbf{x} \in X_j} (\mathbf{x} - \mu_j)(\mathbf{x} - \mu_j)^T, \tag{5.5.2}$$

where

$$\mathbf{x} \in X_j \quad \text{if} \quad \|\mathbf{x} - \mu_j\| < \|\mathbf{x} - \mu_k\| \quad \forall j \neq k, \tag{5.5.3}$$

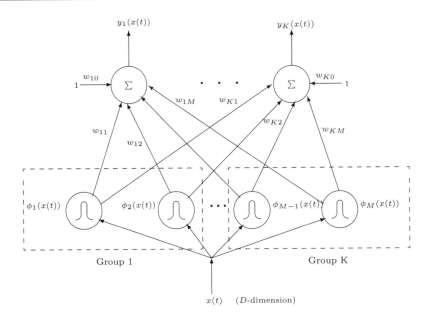

Figure 5.6. Architecture of a K-output EBF network.

N_j is the number of patterns in the cluster X_j, and $\|\cdot\|$ is the Euclidean norm. After the kernel parameters are determined, the kernel outputs $\{\phi_j(\mathbf{x}^{(n)}); j = 0, \ldots, M\}$ are computed using the input/teacher training pairs $\{\mathbf{x}^{(n)}, \mathbf{d}^{(n)}\}$ to form the matrix equation

$$\mathbf{\Phi W} = \mathbf{D}, \tag{5.5.4}$$

where

$$\mathbf{\Phi} = \begin{bmatrix} \phi_0(\mathbf{x}^{(1)}) & \phi_1(\mathbf{x}^{(1)}) & \cdots & \phi_M(\mathbf{x}^{(1)}) \\ \phi_0(\mathbf{x}^{(2)}) & \phi_1(\mathbf{x}^{(2)}) & \cdots & \phi_M(\mathbf{x}^{(2)}) \\ \vdots & \vdots & \ddots & \vdots \\ \phi_0(\mathbf{x}^{(N)}) & \phi_1(\mathbf{x}^{(N)}) & \cdots & \phi_M(\mathbf{x}^{(N)}) \end{bmatrix},$$

$$\mathbf{W} = \begin{bmatrix} w_{10} & w_{20} & \cdots & w_{K0} \\ w_{11} & w_{21} & \cdots & w_{K1} \\ \vdots & \vdots & \ddots & \vdots \\ w_{1M} & w_{2M} & \cdots & w_{KM} \end{bmatrix}, \text{ and } \mathbf{D} = \begin{bmatrix} d_1^{(1)} & d_2^{(1)} & \cdots & d_K^{(1)} \\ d_1^{(2)} & d_2^{(2)} & \cdots & d_K^{(2)} \\ \vdots & \vdots & \ddots & \vdots \\ d_1^{(N)} & d_2^{(N)} & \cdots & d_K^{(N)} \end{bmatrix}.$$

Therefore, the training of the output weights amounts to finding the inverse of $\mathbf{\Phi}$. However, in most cases, $\mathbf{\Phi}$ is not a square matrix. For a small amount of training data, a reliable way to solve Eq. 5.5.4 is to use the technique of singular value decomposition [285]. In this approach, the matrix $\mathbf{\Phi}$ is decomposed into the product $\mathbf{U\Lambda V}^T$, where $\mathbf{U}$ is an $N \times (M + 1)$ column-orthogonal matrix, $\mathbf{\Lambda}$ is an

$(M + 1) \times (M + 1)$ diagonal matrix containing the singular values, and $\mathbf{V}$ is an $(M + 1) \times (M + 1)$ orthogonal matrix. The weight vectors $\{\mathbf{w}_k\}_{k=1}^K$ are given by

$$\mathbf{w}_k = \mathbf{V}\mathbf{\Lambda}^{-1}\mathbf{U}^T\mathbf{d}_k, \tag{5.5.5}$$

where $\mathbf{d}_k$ is the k-th column of $\mathbf{D}$ and $\mathbf{\Lambda} = \text{diag}\{\lambda_0, \ldots, \lambda_M\}$ contains the singular values of $\mathbf{\Phi}$. To ensure numerical stability, set $1/\lambda_i$ to zero for those λ_i less than a machine-dependent precision threshold. This approach effectively removes the effect of the corresponding dependent columns in $\mathbf{\Phi}$. For an overdetermined system, singular value decomposition gives a solution that is the best approximation in the least squares sense. On the other hand, when the amount of training data are sufficiently large to ensure that $\mathbf{\Phi}$ is of full rank, the squared error between the actual output and the desired output can be minimized:

$$E(\mathbf{w}_k) = (\mathbf{\Phi}\mathbf{w}_k - \mathbf{d}_k)^T(\mathbf{\Phi}\mathbf{w}_k - \mathbf{d}_k) \qquad k = 1, \ldots, K$$

to give the normal equations of the form

$$\mathbf{w}_k = (\mathbf{\Phi}^T\mathbf{\Phi})^{-1}\mathbf{\Phi}^T\mathbf{d}_k. \tag{5.5.6}$$

Compared to the gradient-based training scheme for multi-layer RBF networks described before, the two-stage training approach has two advantages:

1. Unlike the gradient-based approach, it is guaranteed that the basis functions are localized.

2. The training time of the two-stage approach is much shorter than the gradient-based approach because the training of the output weights becomes a least-squares problem, and there are efficient algorithms to determine the weights of the hidden-layer.

3. The error function $E(\mathbf{w}_k)$ is quadratic with respect to the output weights. As a result, no local minimums of $E(\mathbf{w}_k)$ exist.

Probabilistic Interpretation

Research has found that RBF networks are capable of estimating Bayesian posterior probabilities [311]. In fact, the link of RBF networks with statistical inference methods is more obvious than that of the LBF networks.

As noted in Section 5.2, each hidden neuron in an RBF network feeds the inputs into the radial basis function (Eq. 5.2.2) and generates the output using the Gaussian activation function (Eq. 5.2.4). For example, the output of the j-th neuron is given by

$$f_j(\mathbf{w}_j, \mathbf{x}) = c \cdot e^{-\sum_{i=1}^{D}(x_i - w_{ji})^2/\sigma^2}, \tag{5.5.7}$$

where $\mathbf{x}$ is a D-dimensional input vector. Usually the vector $\mathbf{w}_j$ is a randomly selected input pattern, but it would be more efficient to use the mean vector to represent multiple patterns of the same cluster. Also, the spread parameter σ can be extended to a covariance matrix Σ so that the neuron has more flexibility.

The network output k is a linear combination of the hidden neurons' outputs:

$$\phi_k(\mathbf{w}, \mathbf{x}) = \sum_{j=1}^{M} \pi_{kj} f_j(\mathbf{w}_j, \mathbf{x}), \qquad (5.5.8)$$

where $\mathbf{w} = \{\mathbf{w}_1, \mathbf{w}_2, \ldots, \mathbf{w}_M\}$ are the parameters (means and covariances) of the M-hidden neurons, and π_{kj} specifies the connection weight between output neuron k and hidden neuron j. Note that RBF networks are closely related to the Parzen windows approach [270] in that the estimated class-conditional densities are a linear superposition of window functions—one for each training pattern. Moreover, if the weighting parameters π_js follow the constraints $\sum_j \pi_{kj} = 1$ and $\pi_{kj} \geq 0 \; \forall j$ and k, RBF networks are similar to the *mixture of Gaussian distributions*. Richard and Lippmann [311] have shown that if the output values of an RBF network are limited to [0,1], then with a large enough training set, the network outputs approximate the Bayesian posterior probabilities. Since most RBF learning schemes do not restrict the output values to [0,1], Bridle [35] suggested normalizing the outputs by a softmax-type function of the form:

$$\phi'_k(\mathbf{w}, \mathbf{x}) = \frac{\phi_k(\mathbf{w}, \mathbf{x})}{\sum_{n=1}^{K} \phi_n(\mathbf{w}, \mathbf{x})}. \qquad (5.5.9)$$

According to Eq. 5.5.8, if $f_j(\mathbf{w}_j, \mathbf{x})$ is a density function, $\phi_k(\mathbf{w}, \mathbf{x})$ will also be a density function. As a result, one can write the posterior probability of class ω_k as

$$P(\omega_k|\mathbf{x}) = \frac{P(\omega_k)\phi_k(\mathbf{w}, \mathbf{x})}{\sum_{k=1}^{K} P(\omega_k)\phi_k(\mathbf{w}, \mathbf{x})}. \qquad (5.5.10)$$

Comparing Eq. 5.5.9 with Eq. 5.5.10, one can see that the normalized RBF network outputs approximate the posterior probabilities with the assumption that $P(\omega_i) = P(\omega_j), \forall i, j$.

RBF Networks versus EBF Networks

Notice that when $\Sigma_j = \sigma_j^2 I$ in Eq. 5.4.10, the EBF network reduces to an RBF network. Although RBF networks have fewer parameters than EBF networks; their disadvantage is that they usually require a larger number of function centers to achieve similar performance [220]. The major reason is that, in EBF networks, each basis function has a full covariance matrix or a diagonal covariance matrix with different variances for each component, which forms a better representation of the probability density function; whereas in RBF networks, the distribution is modeled by a collection of diagonal covariance matrices with equal variances. Figure 5.7 shows the decision boundaries and contour lines of the basis functions formed by RBF and EBF networks with various numbers of function centers. The objective

Table 5.1. Mean vectors and covariance matrices of the four clusters in the noisy XOR problem. Each cluster contains 500 samples.

Class 1		Class 2	
Mean vector	Covariance matrix	Mean vector	Covariance matrix
$\begin{pmatrix} -1 \\ 1 \end{pmatrix}$	$\begin{pmatrix} 0.9 & 0.2 \\ 0.2 & 0.6 \end{pmatrix}$	$\begin{pmatrix} -1 \\ -1 \end{pmatrix}$	$\begin{pmatrix} 0.5 & -0.2 \\ -0.2 & 1.0 \end{pmatrix}$
$\begin{pmatrix} 1 \\ -1 \end{pmatrix}$	$\begin{pmatrix} 0.7 & 0.0 \\ 0.0 & 0.7 \end{pmatrix}$	$\begin{pmatrix} 1 \\ 1 \end{pmatrix}$	$\begin{pmatrix} 0.7 & 0.1 \\ 0.1 & 0.5 \end{pmatrix}$

Table 5.2. Recognition accuracy achieved by the RBF and the EBF networks in the noisy XOR problem. All figures are based on five simulation runs with different initial conditions. $\mathcal{N}_{rbf}$ and $\mathcal{N}_{ebf}$ denote the numbers of free parameters in the RBF and EBF networks, respectively.

Number of Centers	RBF Network		EBF Network	
	$\mathcal{N}_{rbf}$	Rec. accuracy	$\mathcal{N}_{ebf}$	Rec. accuracy
4	22	88.47%	30	91.14%
8	42	90.57%	58	90.87%
16	82	89.77%	114	90.86%

is to classify four Gaussian clusters with mean vectors and covariance matrices, as shown in Table 5.1, into two classes. This can be considered as a noisy XOR problem in which the noise components in the input domain are interdependent.

Figures 5.7(a) and 5.7(b) indicate that the boundaries formed by the EBF network with four centers are better for modeling the noise source characteristic than that formed by the RBF networks. Table 5.2 shows the recognition accuracy obtained by the RBF networks and the EBF networks. The accuracy is based on 2,000 test vectors drawn from the same population as the training set. The result demonstrates that the EBF networks can attain higher recognition accuracy than the RBF networks. It also suggests that increasing the size of the RBF networks could slightly increase recognition accuracy, but recognition accuracy is still lower than that of the EBF network with the smallest size (compare 90.57% with 91.14%). Increasing the number of function centers will also make the decision boundaries unnecessarily complicated, as shown in Figures 5.7(e) and 5.7(f).

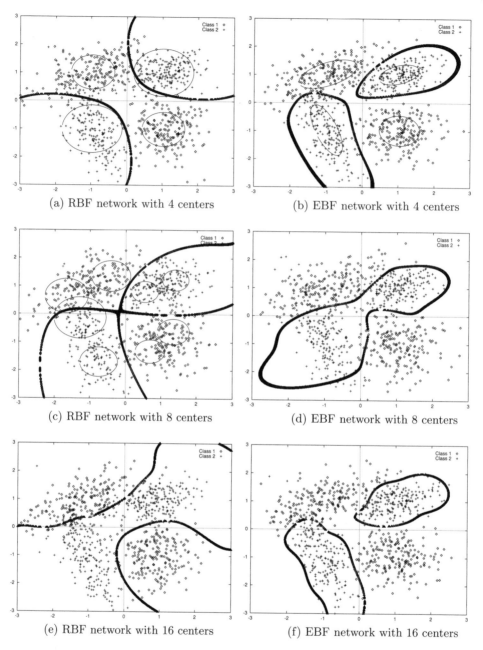

Figure 5.7. Decision boundaries (*thick curves*) and contour lines of basis functions (*thin circles or ellipses*) produced by RBF and EBF networks with various numbers of centers. EBF networks' function centers and covariance matrices were found by the EM algorithm. *Note*: Only 1,200 test samples were plotted.

5.5.2   Two-Stage Training for LBF Networks

The two-stage training approach just discussed can also be applied to train multi-layer LBF networks. Here a two-layer LBF network is used to illustrate the procedure. Recall from Section 5.4 that the outputs of a three-layer LBF network (i.e., $L = 2$) is

$$y_k(\mathbf{x}^{(n)}) = a_k(2) \qquad k = 1, \ldots, K \tag{5.5.11}$$

$$a_k(2) = f(u_k(2))$$

$$u_k(2) = \sum_{j=1}^{M} w_{kj}(2)a_j(1)$$

$$a_j(1) = f(u_j(1))$$

$$u_j(1) = \sum_{i=1}^{D} w_{ji}(1)x_i(0),$$

where $f(\cdot)$ is a nonlinear activation function.[2] In matrix form, Eq. 5.5.11 can be written as $\mathbf{Y} = \mathbf{XH}$, where $\mathbf{Y}$ is an $N \times K$ matrix, $\mathbf{X}$ is an $N \times D$ matrix, and $\mathbf{H}$ is an $D \times K$ matrix. The weight matrix $\mathbf{W}$ is the least-squares solution of the matrix equation

$$\mathbf{XH} = \mathbf{D}, \tag{5.5.12}$$

where $\mathbf{D}$ is an $N \times K$ target matrix containing the desired output vectors in its rows.

As $\mathbf{X}$ is not a square matrix, its left-inverse is

$$\mathbf{X}^{-L} = (\mathbf{X}^T\mathbf{X})^{-1}\mathbf{X}^T. \tag{5.5.13}$$

So ideally, this is

$$\mathbf{H} = \mathbf{X}^{-L}\mathbf{D}. \tag{5.5.14}$$

Assuming that the number of hidden units M is less than that of the input, then $\mathbf{H}$ can be approximated by a rank M-matrix factorization:

$$\mathbf{H} \approx \mathbf{W}(1)\mathbf{W}(2), \tag{5.5.15}$$

where $\mathbf{W}(1)$ and $\mathbf{W}(2)$ are, respectively, $D \times M$ and $M \times K$ weight matrices corresponding to Layer 1 and Layer 2.

Suppose that the singular value decomposition of $\mathbf{H} = \mathbf{X}^{-L}\mathbf{D}$ can be decomposed (approximately) into the product $\mathbf{X}^{-L}\mathbf{D} \approx \mathbf{U\Lambda V}^{\mathbf{T}}$, where $\mathbf{U}$ is a $D \times M$ column-orthogonal matrix, $\mathbf{\Lambda}$ is an $M \times M$ diagonal matrix containing the singular values, and $\mathbf{V}$ is an $M \times K$ orthogonal matrix. The weight matrix $\mathbf{W}(1)$ is given by $\mathbf{W}(1) = \mathbf{U}$, and the output of the hidden units will be $f(\mathbf{XW}(1))$, where $f(\cdot)$ stands for the nonlinear activation function. Having derived the initial values for

---

[2]For simplicity, the bias terms in Eq. 5.5.11 have been ignored.

the hidden units, the derivation for the upper-layer $\mathbf{W}(2)$ follows the same approach as that adopted in the RBF case. In other words, the kernel outputs are again computed to form the matrix equation (5.5.4). Then, the output weights are computed using SVD.

As a numerical example, let

$$
\mathbf{X} = \begin{bmatrix} 1 & 2 & 3 \\ 2 & 3 & 4 \\ 3 & 4 & 5 \\ -4 & -5 & 6 \end{bmatrix} \text{ and } \mathbf{D} = \begin{bmatrix} 7 & 8 \\ 9 & 10 \\ -11 & 12 \\ -13 & 14 \end{bmatrix}.
$$

Then,

$$
\mathbf{H} = \mathbf{X}^{-L}\mathbf{D} = \begin{bmatrix} -31.1111 & -1.6667 \\ 24.5556 & 1.3333 \\ -2.4444 & 2.3333 \end{bmatrix},
$$

which can be decomposed into

$$
\begin{aligned}
\mathbf{H} &= \mathbf{U}\mathbf{\Lambda}\mathbf{V}^T \\
&= \begin{bmatrix} -0.78 & -0.04 & -0.62 \\ 0.62 & 0.04 & -0.78 \\ -0.06 & 1.00 & 0.01 \end{bmatrix} \begin{bmatrix} 39.76 & 0 \\ 0 & 2.46 \\ 0 & 0 \end{bmatrix} \begin{bmatrix} 1.00 & 0.05 \\ -0.05 & 1.00 \end{bmatrix}.
\end{aligned}
$$

Hence, $\mathbf{W}(1)$ can be initialized to

$$
\mathbf{W}(1) = \begin{bmatrix} -0.78 & -0.04 & -0.62 \\ 0.62 & 0.04 & -0.78 \\ -0.06 & 1.00 & 0.01 \end{bmatrix}.
$$

5.6 Genetic Algorithm for Multi-Layer Networks

Another school of thought to train neural networks is to use an evolutionary search such as genetic algorithms [114, 243], evolutionary programming [100, 101], and evolution strategies [296, 325, 338]. This section presents a hybrid algorithm that combines the standard RBF learning and genetic algorithms to find the parameters of RBF networks.

Unlike a local search, an evolutionary search maintains a population of potential solutions rather than a single solution. Therefore, the risk of getting stuck in local optima is smaller. Each candidate solution represents one neural network in which the weights can be encoded as a string of binary [71, 381, 383] or floating point numbers [119, 240, 283, 334]. The performance of each solution is determined by the network error function, which is to be optimized by evolutionary search. Because gradient information is not required, evolutionary search is applicable to problems where gradient information is unavailable or to cases where the search surface contains many plateaus. However, the iterative process in an evolutionary search requires evaluation of a large number of candidate solutions; consequently, evolutionary search is usually slower than local search. The lack of fine-tuning operations in evolutionary search also limits the accuracy of the final solution. The focus here is on genetic algorithms (one of the evolutionary search approaches).

5.6.1   Basic Elements of GAs

Evolutionary search maintains a population of candidate solutions for a given problem. In genetic algorithm (GA) terminology, the candidate solutions are called *chromosomes*. New chromosomes are generated by a set of genetic operators. Each chromosome is associated with a fitness value that reflects the solution quality (performance). Based on the fitness values, the population of chromosomes is maintained by selection and reproduction processes.

As introduced by Holland [137], each chromosome is comprised of a number of genes, each of which can take a binary value of 0 or 1. A mapping function is therefore required to decode the binary strings into corresponding solutions. The search space of all binary strings is the genotype space, and the decoded solutions constitute the phenotype space. Depending on the characteristics of the problem to be solved, the mapping function can be complex and nonlinear. Consequently, a simple search surface in a solution space might be represented by a complex and rugged search surface in a genotype space, which could hinder the search process [243]. While binary representations are common in boolean optimization problems [72, 316], many practitioners [68, 243] prefer to use problem-oriented representations. In particular, it is common to represent a chromosome as a string of floating point numbers [90, 164]. With a problem-oriented representation, the correspondence between the genotype space and phenotype space is closer. This could reduce the complexity of the mapping function substantially.

5.6.2   Operation of GAs

As shown in Figure 5.8, the operation of the standard GA basically follows the structure of an evolutionary search. An initial population of size M is created by generating M chromosomes randomly. Based on the fitness values ($f(c_i)$, where $i = 1, 2, \ldots, M$), some chromosomes with better performance are selected to form an intermediate population. For example, in the proportional selection process proposed by Holland, the probability of selecting a chromosome c_k is determined by $f(c_k)/\sum_i f(c_i)$ (i.e., its fitness over the population's total fitness).

New chromosomes are generated from the selected chromosomes through genetic operations, where crossover and mutation operators are often used. In the standard GA proposed by Holland, each bit in a binary-encoded chromosome has a small probability of being flipped by mutation, and the crossover operator exchanges the binary substrings of two selected chromosomes to form new chromosomes. The M new chromosomes generated by the genetic operators constitute the new population for further processing. The iterative process is terminated when a satisfactory solution is found. It can also be terminated when the population has converged to a single solution. If the converged solution is an undesirable suboptimal solution,

```
procedure GA

begin
        Initialize a population of M chromosomes c_i, where i = 1, 2, ..., M.
        Evaluate the fitness of every chromosome in the population.
        repeat
            Select some chromosomes in the population to form an intermediate
                population by means of a selection and reproduction mechanism.
            Generate new chromosomes from the intermediate population by means
                of a set of genetic operators.
            Evaluate the fitness of the new chromosomes.
            Create a new population of chromosomes to replace the old ones.
        until termination condition reached
endproc GA
```

Figure 5.8. The procedure of the standard GA.

it is considered that premature convergence has occurred. Typically, premature convergence is a result of intense selection pressure (i.e., selection is heavily biased toward better-performing chromosomes) and insufficient exploration of search space [89, 204].

5.6.3 Applications of GAs to Evolve Neural Networks

Many attempts to apply GAs to construct and train neural networks have been made—see Schaffer et al. [335], Whitley [380], and Vonk et al. [362] for comprehensive reviews. Typically, GAs are used to determine the network weights. For example, Montana and Davis [254] have successfully applied GAs to train feedforward neural networks for sonar data classification. Promising results have also been reported in other studies (e.g., [19, 70, 381]).

GAs can also be applied to determine an appropriate network architecture. For instance, Miller et al. [246] represented each chromosome as a matrix that defines the network connection (i.e., each entry in the matrix determines the existence of a connection). The network connection can also be represented by using a more abstract encoding scheme in which variable-length chromosomes are used to specify the connections [127].

In addition, GAs can be used to find a suitable training mechanism and its associated parameters. To this end, Chalmers [48] encoded the dynamics of a training procedure in a chromosome. In the GANNET system developed by Robbins et al. [315], some of the genes in the chromosomes were used to specify the learning rate and the momentum of the BP algorithm.

Another application that is not common in the field of neural networks is to use

GAs for preprocessing and interpreting data. Typical examples include the selection of dominant features of input data [36, 49] and the selection of an appropriate training set [298]. Alternatively, GAs can be used to determine the input data patterns that produce a given network output [86].

Determination of RBF Centers by Genetic Algorithms

As mentioned in Section 5.4, the common approach to finding the parameters of RBF networks is to use unsupervised learning (e.g., the K-means algorithm and the K-nearest neighbor heuristic) to find the function centers and widths, and uses supervised learning (e.g., least-squares method) to determine output weights [255]. Although this approach is fast and a global minimum in output weight space is guaranteed to be found, the two-stage optimization process may lead to suboptimal solutions, especially when the K-means algorithm finds inappropriate function centers. An alternative approach is to replace the K-means algorithm with a genetic algorithm in an attempt to find the best locations for the function centers.

There are several approaches to embedding GA in the RBF training process. One approach uses space-filling curves to encode the function centers indirectly and evolve the parameters of the curves during the training process [378]. The success of this approach relies on the capability of the space-filling curves to capture the relevant properties of training data in the input space. Another approach evolves a population of genetic strings. Each of them encodes a single center and its associated width of an RBF network [379]. This method requires a shorter computation time as compared to the case where a single genetic string encodes the entire network. A more computation-intense, yet more powerful approach, is to evolve the centers, widths, and structure of the RBF networks simultaneously [26, 47]. It was found that this method is able to reduce the network complexity, which is particularly useful to the applications where runtime performance is an important issue.

This section describes a method that embeds a simple GA [114] into the RBF learning process. The result is referred to as the RBF-GA hybrid algorithm [224].[3] In the algorithm, each center of an RBF network is encoded as a binary string and the concatenation of the strings forms a chromosome. In each generation cycle, the simple GA is applied to find the center locations of the networks (chromosome), and the output weights are determined by a least-squares method. Therefore, the computation complexity of this approach falls between that of [378] and [26]. The proposed algorithm was applied to three artificial pattern classification problems to demonstrate that the best locations for the function centers may not necessarily be located inside the clusters of the training data.

Embedding GA in RBF Learning

To implement the RBF-GA hybrid algorithm, the following items should be derived:

  1. An appropriate genotype representation (encoding)

---

[3]Hereafter, the networks trained with this algorithm are referred to as RBF-GA networks.

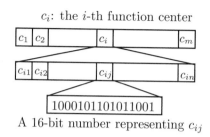

A 16-bit number representing c_{ij}

Figure 5.9. Structure of a chromosome

2. A fitness measure (performance evaluation)

3. A learning method for finding the output weights (RBF training)

4. A method for generating new genotypes from old ones (selection and reproduction)

Binary encoding is one of the common encoding methods for GAs because much of the existing GA theory is based on the assumption of fixed-length, fixed-order binary strings. This method can be classified into direct and indirect encoding [335]. For the former, little effort is required to decode the chromosome, and the transformation of genotypes into phenotypes is straightforward. The latter, on the other hand, requires more complicated decoding techniques.

Here, direct encoding is used to encode the function centers. Figure 5.9 shows the configuration of a binary string encoding m centers with each center being a point in an n-dimensional input space. Therefore, each chromosome is a concatenation of mn binary strings with each string being a 16-bit binary number. Each number is linearly mapped to a floating point number in $[U_{\min}, U_{\max}]$. In this work, $U_{\min}$ and $U_{\max}$ have been set to the minimum and maximum values of the training data in the input domain.

Fitness Measure

Training of neural networks usually involves the minimization of an objective function which, typically, is the mean-squared error (MSE) between the desired and actual outputs for the entire training set. However, GAs are usually applied to maximize a fitness function. Therefore, to apply GAs, a method is required to convert the objective function into a fitness function. A common approach is to use the transformation $f(X) = 1/(1 + \mathrm{MSE}(X))$, where X denotes the training set and f the fitness function. Moreover, it is common practice to use the linear scaling

$$g(X) = af(X) + b \tag{5.6.1}$$

as suggested by Goldberg [114] to obtain the scaled fitness $g(X)$. In this method, the values of a and b are chosen such that the scaled average fitness g_{av} is equal to the average fitness f_{av}, and the scaled maximum $g_{\max}$ is equal to a constant multiple of the average fitness (i.e., $g_{\max} = \gamma f_{\mathrm{av}}$). Therefore, after some mathematical manipulation, one has

$$a = \frac{f_{\mathrm{av}}(\gamma - 1)}{f_{\max} - f_{\mathrm{av}}} \quad \text{and} \quad b = f_{\mathrm{av}}(1 - a). \tag{5.6.2}$$

To ensure that $g(X)$ is always positive, check to see whether $g_{\min}(= a f_{\min} + b)$ is greater than 0. This leads to the following condition:

$$f_{\min} > \frac{\gamma f_{\mathrm{av}} - f_{\max}}{\gamma - 1}. \tag{5.6.3}$$

If Eq. 5.6.3 cannot be satisfied, set

$$a = \frac{f_{\mathrm{av}}}{f_{\mathrm{av}} - f_{\min}} \quad \text{and} \quad b = -\frac{f_{\mathrm{av}} f_{\min}}{f_{\mathrm{av}} - f_{\min}} \tag{5.6.4}$$

so that $g_{\min}$ will be equal to 0. The scaling ensures that the individuals with scaled average fitness g_{av} contribute one offspring on average during the selection process, while the best individual is expected to contribute a specified number (γ) of offspring. The advantage of this approach is that it can prevent individuals with small MSE (high fitness value) from dominating the population during the early stages of evolution. This can also improve selection resolution in the later stages of evolution when most individuals have high fitness values.

Selection and Reproduction

In this work, a simple genetic algorithm [114] has been used in the selection and reproduction processes. The operation of the simple GA is as follows. An initial population is created by randomizing the genetic strings. This is equivalent to randomizing the function centers of the corresponding RBF networks. In each generation, the algorithm creates a new population by selecting pairs of individuals from the current population and then allowing them to mate to produce new offspring. The new offspring are added to the new population. This process is continued until the size of the new population is the same as the current one. The evolution process is then repeated until a stopping criterion is met or the number of generations reaches a predefined maximum.

The roulette wheel selection method was used to pick an individual such that the fitter the individual, the higher the chance of selection for genetic operations. As a result, individual r has a probability p_r of being selected, where

$$p_r = \frac{g_r(X)}{\sum_k g_k(X)}. \tag{5.6.5}$$

The RBF-GA Hybrid Algorithm

The RBF-GA hybrid algorithm consists of ten steps that follow.

1. Specify the number of function centers m.

2. Set generation t to 0. Randomize an initial population $\mathcal{P}(t)$ of M (an even number) chromosomes $\{\mathcal{S}_i(t)\}_{i=1}^{M}$ with each chromosome being a $16mn$-bit binary string representing the m centers of an RBF network.

3. Determine the value of $U_{\max}$ and $U_{\min}$ based on the maximum and minimum values of the training data in the input domain.

4. Set $k = 1$. Decode the chromosomes to form M networks and compute the width of each center according to the K-nearest-neighbor heuristic with $K = 2$. Find the output weights of each network using singular value decomposition (SVD) [285]. Find the best individual and denote it as $\mathcal{S}^*(t)$.

5. Evaluate the scaled fitness value of each individual $\{\mathcal{S}_i(t)\}_{i=1}^{M}$ by using Eqs. 5.6.1 through 5.6.4. Compute their selection probabilities using Eq. 5.6.5.

6. Select two parents in $\mathcal{P}(t)$ according to their selection probabilities.

7. Apply crossover to the two selected parents with probability p_c to create two offspring. Next, apply mutation with probability p_m to every bit of the two offspring. Then add the resulting chromosomes to the new population $\mathcal{P}(t+1)$.

8. $k = k + 2$. If $k < M$, go to Step 6; otherwise, go to Step 9.

9. If all chromosomes in $\mathcal{P}(t+1)$ have a scaled fitness of less than that of $\mathcal{S}^*(t)$, replace the worst chromosome in $\mathcal{P}(t+1)$ by $\mathcal{S}^*(t)$.[4]

10. $t = t + 1$. If $t < t_{\max}$, go to Step 4; otherwise stop.

Experiments and Results

All experiments reported here were carried out on a SUN Sparc-10 workstation using a C++ genetic algorithm library (GAlib) [363]. The operation of the RBF-GA hybrid algorithm depends on the values of $t_{\max}$, p_c, p_m, M, and γ. In the experiments, $t_{\max} = 100$, $p_c = 0.6$, $p_m = 0.01$, $M = 30$, and $\gamma = 1.2$ have been set. These parameters typically interact with each other nonlinearly, and there are no conclusive results showing the best set of parameters. Therefore, the typical settings, which work well in other reports [249, 363], have been used.

Three problem sets, namely the XOR, overlapped Gaussian, and nonoverlapped Gaussian, have been investigated to evaluate the performance of the RBF-GA hybrid algorithm. For each problem, training of RBF networks using the K-means algorithm has also been performed so that comparisons can be made.

---

[4]This step implements elitism.

Table 5.3. RBF-GA hybrid and the standard RBF algorithms' performance when solving the XOR problem.

Algorithm	MSE	Training Time
RBF-GA	0.009	60 sec.
RBF	0.089	≈ 1 sec.

During the first experiments, a training set was created. The set consists of the input/output pairs $\{(x_1, x_2); y\}$ such that $y = r(x_1) \oplus r(x_2)$, where x_1 and x_2 belong to the set $\{0.0, 0.1, 0.2, 0.3, 0.7, 0.8, 0, 9, 1.0\}$, $y \in \{0.0, 1.0\}$, and $r(x)$ rounds the variable x to its nearest integer. The number of centers were set to four because there were four clusters in the input space. The effects of varying the number of centers were not investigated. Instead, the focus was on the differences in the performance of the RBF-GA hybrid and the standard RBF algorithms. Table 5.3 summarizes the MSE and training time (based on the CPU time of a SUN Sparc-10 workstation) of nine simulation runs of the two algorithms.

Table 5.3 shows that the MSE achieved by the RBF-GA hybrid algorithm is significantly smaller than that of the standard RBF learning algorithm. Figure 5.10 shows the locations of the function centers found by the RBF-GA hybrid algorithm and the K-means algorithm. One can deduce from this figure that the function centers found by the RBF-GA hybrid algorithm scatter around the input space, while most of the centers found by the K-means algorithm are located inside the input clusters. These results indicate that using the cluster centers as the function centers, as in standard RBF networks, may not necessarily give optimal performance.

Surprisingly, it was also noted that the networks trained by the RBF-GA hybrid algorithm provide high activation in the input space where no input data was present. This contrasts with the idea of using RBF networks with local perceptive fields in which networks are expected to produce low activation in regions without any training data. To further investigate this observation, two-dimensional Gaussian clusters were used in the experiments that followed.

In the second experiment, RBF networks were used to classify two overlapped Gaussian clusters into two classes. The mean vectors and covariance matrices of the two Gaussian clusters are shown in Table 5.4. The networks have two inputs, two function centers, and two outputs with each output corresponding to one class. The MSE, classification accuracy, and training time of the RBF-GA hybrid and the standard RBF algorithms are shown in Table 5.5.

Similar to the XOR problem, the MSE of the networks trained with the RBF-GA hybrid algorithm is lower than that of the standard RBF algorithm. This indicates that using the cluster centers found by the K-means algorithm can only provide a suboptimal solution. Figure 5.11 plots the decision boundaries, function centers, and the training data. A comparison between Figures 5.11(a) and 5.11(b) reveals

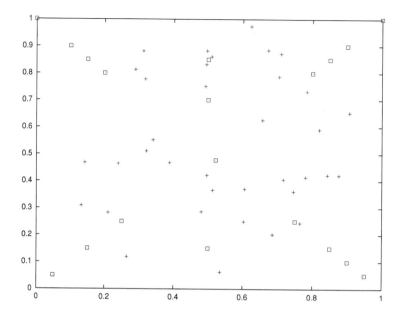

Figure 5.10. Function centers found by the RBF-GA hybrid algorithm (+) and the K-means algorithm ($\square$).

Table 5.4. Mean vectors and covariance matrices of the two overlapped Gaussian clusters (each with 1,000 data points).

Cluster 1 (Class 1)		Cluster 2 (Class 2)	
Mean vector	Cov. Matrix	Mean vector	Cov. matrix
$\begin{pmatrix} 2 \\ 2 \end{pmatrix}$	$\begin{pmatrix} 3 & 1 \\ 1 & 1 \end{pmatrix}$	$\begin{pmatrix} -1 \\ -1 \end{pmatrix}$	$\begin{pmatrix} 1 & 1 \\ 1 & 3 \end{pmatrix}$

that the decision boundary corresponding to the RBF-GA hybrid network is able to partially partition the two clusters—see the first quadrant of Figure 5.11(b)—resulting in a higher classification accuracy as shown in Table 5.5. This is because the GA is able to find appropriate locations for the function centers, which may not be located inside the input clusters. However, the function centers found by the K-means algorithm are always located inside the input clusters, as in Figure 5.11(a).

Although the RBF-GA hybrid network gives a lower MSE and higher classification accuracy on the training data, it requires longer computation time for the population to converge. In the experiment, 1,000 two-dimensional data points for

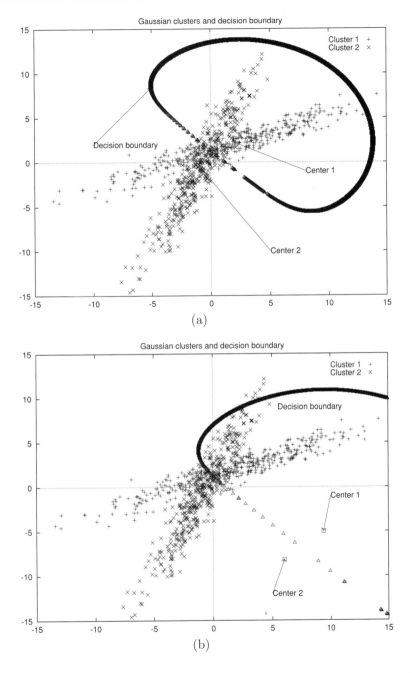

Figure 5.11. Decision boundaries and function centers found by (a) the standard RBF algorithm and (b) the RBF-GA hybrid algorithm. Note that in (b) only the centers and decision boundary of the best network were plotted.

Table 5.5. RBF-GA hybrid and standard RBF algorithms' performance in classifying the overlapped Gaussian cluster based on 10 simulation runs.

Algorithm	MSE	Classification Accuracy	Training Time
RBF-GA	0.157	66.5%	1814 sec.
RBF	0.223	64.8%	60 sec.

Table 5.6. Mean vectors and covariance matrices of the two nonoverlapped Gaussian clusters (each with 1,000 data points).

Cluster 1 (Class 1)		Cluster 2 (Class 2)	
Mean vector	Cov. matrix	Mean vector	Cov. matrix
$\begin{pmatrix} -2 \\ 2 \end{pmatrix}$	$\begin{pmatrix} 1 & 1 \\ 1 & 3 \end{pmatrix}$	$\begin{pmatrix} 2 \\ -2 \end{pmatrix}$	$\begin{pmatrix} 1 & 1 \\ 1 & 3 \end{pmatrix}$

each cluster were used. The standard RBF algorithm took about one minute to train a network while the RBF-GA hybrid algorithm required 30 minutes. This is because there are 30 chromosomes in the population and, in each generation, the SVD algorithm must be applied 30 times in order to find the output weights of all individuals.

In the third experiment, two nonoverlapped Gaussian clusters with mean vectors and covariance matrices, as shown in Table 5.6, were classified by RBF networks with two function centers.

The MSE, learning time, and classification accuracy are shown in Table 5.7, and Figure 5.12 shows the decision boundaries, function centers, and training data. Figure 5.12(b) demonstrates that the decision boundary formed by the RBF-GA hybrid network is able to separate the two clusters perfectly. On the other hand, the decision boundary created by the standard RBF network passes through both clusters, as shown in Figure 5.12(a). This leads to lower classification accuracy as shown in Table 5.7. The reason is that the RBF-GA hybrid algorithm considers all possible locations for the function centers in the input domain in order to minimize the MSE. Therefore, function centers outside the input clusters were found in this case, whereas function centers found by the K-means algorithm would always stay inside the Gaussian clusters.

In conclusion, this subsection has described a hybrid algorithm, which combines the standard RBF learning and genetic algorithms, to find the parameters of RBF networks. Unlike the K-means algorithm, the RBF-GA hybrid algorithm is able to find the function centers outside the input clusters. The results demonstrate

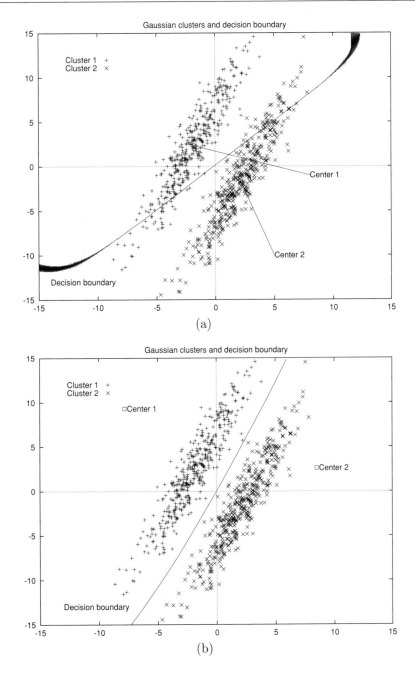

Figure 5.12. Decision boundaries and function centers found by (a) the standard RBF algorithm and (b) the RBF-GA hybrid algorithm. Note that in (b) only the centers and decision boundary of the best network were plotted.

Table 5.7. RBF-GA hybrid and standard RBF algorithms' performance in classifying the nonoverlapped Gaussian clusters (based on 10 simulation runs).

Algorithm	MSE	Classification Accuracy	Training Time
RBF-GA	0.021	100.0%	1884 sec.
RBF	0.075	89.5%	60 sec.

that, in some circumstances, the RBF networks with these center locations attain a higher recognition accuracy than that with centers located inside the input clusters.

Each generation of the RBF-GA hybrid requires the computation of the output weights of 30 networks and the evaluation of their fitness values. If the population takes many generations to evolve before an optimum is found, the hybrid algorithm will obviously require excessive training time. However, the main concern in practice is the runtime performance of the trained networks. Parallel implementation of GAs could be one possible solution to this problem.

In this work, only the locations of function centers are encoded in the genetic strings using direct binary encoding. The number of function centers (i.e., the network architecture) can also be evolved by GAs in future work. Moreover, local search methods, such as gradient descent, can be applied to fine-tune the center locations and the widths of basis functions during the course of Lamarckian evolution. This form of combining local search with GAs [185] could be applied to difficult problems where traditional RBF algorithms may fail to find a satisfactory solution.

5.7 Biometric Authentication Application Examples

Note that most biometric data can be adequately represented by a mixture of Gaussians, and also that RBF neural networks adopt basically the same type of Gaussian kernels. Therefore, RBF networks are more appealing for biometric applications than LBF networks. In fact, RBF and EBF networks have been used extensively in biometric applications. For instance, Brunelli and Poggio [38] used a special type of RBF network, called the "HyperBF" network, for face recognition and reported a 100% recognition rate on a 47-person database. In a speaker verification task, Mak and Kung [225] demonstrated that EBF networks perform substantially better than their RBF counterparts; and in a speaker identification task, Mak et al. [221, 222] found that RBF networks are superior to LBF networks in terms of both identification accuracy and training time.

In Er et al. [87], RBF neural networks are trained by a hybrid learning algorithm to drastically reduce the dimension of the search space in the gradient paradigm. On the widely used Olivetti Research Laboratory (ORL) face database, the proposed system reportedly achieved excellent performance in terms of both learning efficiency and classification error rates. In Zhang and Guo [401], a modular network structure based on an autoassociative RBF module is applied to face recognition applications based on both the XM2VTS and ORL face database. The scheme achieves highly accurate recognition on both databases. RBF neural networks have also been implemented on embedded systems to provide realtime processing speed for face tracking and identity verification [391].

In Cardinaux et al. [46], MLPs and GMMs were compared in the context of face verification. The comparison is carried out in terms of performance, effectiveness, and practicability. Apart from structural differences, the two approaches use different training criteria; the MLP approach uses a discriminative criterion, while the GMM approach uses a combination of maximum likelihood (ML) and maximum a posteriori (MAP) criteria. Experiments on the XM2VTS database have shown that for low-resolution faces the MLP approach has slightly lower error rates than the GMM approach; however, the GMM approach easily outperforms the MLP approach for high-resolution faces and is significantly more effective for imperfectly located faces. The experiments have also shown that computational requirements of the GMM approach can be significantly smaller than those of the MLP approach, with only a slight decrease in performance.

RBF networks have also been used for speaker verification. For example, in Jiang and Deng [165], a Bayesian approach is applied to speaker verification based on the Gaussian mixture model (GMM). In the Bayesian approach, the verification decision is made by comparing Bayes factors against a critical threshold. An efficient algorithm, using the Viterbi approximation, was proposed to calculate the Bayes's factors for the GMM. According to the reported simulation based on the NIST98 speaker verification evaluation data, the proposed Bayesian approach achieves moderate improvements over a well-trained baseline system using the conventional likelihood ratio test.

A close relative of the RBF clustering scheme is the family of vector quantization (VQ) clustering, since they are both based on the distance metric. In He et al. [129], an evaluation experiment was conducted to compare the codebooks trained by the Linde-Buzo-Grey (LBG), the learning vector quantization (LVQ), and the group vector quantization (GVQ) algorithms. (In GVQ training, speaker codebooks are trained and optimized for vector groups rather than for individual vectors.)

Many examples of applying GAs to face detection and recognition have been reported (e.g., references [15, 155, 156, 180, 275]). A face recognition method using support vector machines with the feature set extracted by genetic algorithms was proposed in Lee et al. [199]. By selecting the feature set with superior performance in face recognition, the use of unnecessary facial information can be avoided and memory requirements can be significantly reduced.

Problems

1. Consider the classical XOR problem defined in Eq. 5.3.4. Show that

   (a) the XOR data can be separated by the following elliptic decision boundary:
   $$6x_1^2 + 6x_2^2 + 8x_1x_2 - 10x_1 - 10x_2 + 3 = 0$$

   (b) they are separable by an even simpler hyperbolic decision boundary:
   $$4x_1x_2 - 2x_1 - 2x_2 + 1.5 = 0$$

   Verify your answers by using the Matlab function ezplot.m.

2. This problem again concerns the XOR problem:

   (a) Explain why multi-layer networks with nonlinear RBF kernels can be used for separating XOR data.

   (b) Repeat the same problem for multi-layer networks with nonlinear LBF kernels.

3. Compare multi-layer perceptrons and radial basis function networks in the context of

   (a) the number of possible hidden-layers

   (b) the representation of the hidden nodes' activation

   (c) the training time required

4. Given N input patterns, the output of a radial basis function network with D inputs, M hidden nodes, one bias term, and one output can be expressed as

   $$\mathbf{y} = \mathbf{\Phi}\mathbf{w},$$

   where $\mathbf{y} = [y_1, y_2, \ldots, y_N]^T$, $\mathbf{w} = [w_0, w_1, \ldots, w_M]^T$, and $\mathbf{\Phi}$ is an $(N \times M + 1)$ matrix with each row containing the hidden nodes' outputs. Assume that $N = 3$ and $M = 2$. Assume also that the desired output vector is

   $$\mathbf{d} = [1.0 \ \ 0.0 \ \ 1.0]^T \text{ and that } \mathbf{\Phi} = \begin{bmatrix} 1 & 0 & 0 \\ 1 & 1 & 0 \\ 1 & 1 & 1 \end{bmatrix}.$$

   Find the least-squares solution of $\mathbf{w}$ if the singular value decomposition of $\mathbf{\Phi}$ is given by $\mathbf{\Phi} = \mathbf{U}\mathbf{S}\mathbf{V}^T$, where

   $$\mathbf{U} = \begin{bmatrix} 0.33 & -0.74 & 0.59 \\ 0.59 & -0.34 & -0.74 \\ 0.74 & 0.59 & 0.33 \end{bmatrix}, \mathbf{V} = \begin{bmatrix} 0.74 & -0.59 & 0.33 \\ 0.59 & 0.33 & -0.74 \\ 0.33 & 0.74 & 0.59 \end{bmatrix},$$

   and $\mathbf{S} = \text{diag}\{2.25, 0.80, 0.56\}$. Hence, find the least-squares error

   $$E = \|\mathbf{\Phi}\mathbf{w} - \mathbf{d}\|^2.$$

5. When only a few training patterns are available, using SVD to determine the output weight matrix $\mathbf{W}$ of an RBF network is recommended (see Section 5.5). Explain why, if the amount of training data is large, it may not be necessary to use SVD. Instead, the *normal equation*

$$\mathbf{W} = (\mathbf{\Phi}^T \mathbf{\Phi})^{-1} \mathbf{\Phi}^T \mathbf{D}$$

of the linear least-squares problem

$$\min_{\mathbf{W}} \|\mathbf{\Phi}\mathbf{W} - \mathbf{D}\|^2$$

will be sufficient to find the output weights. In other words, show that the larger the training set, the more numerically stable the normal equation is. Equivalently, show that $\lambda_i^{(N+1)} \geq \lambda_i^{(N)}$, where $\lambda_i^{(N)}$ (resp. $\lambda_i^{(N+1)}$) is the i-th singular value for a data collection with N (resp. $N+1$) samples. *Hint*: If

$$\mathbf{C} = \left[\frac{\mathbf{A}}{\mathbf{B}} \right],$$

where all matrices are real matrices, show that

$$\mathbf{C}^T \mathbf{C} = \left[\mathbf{A}^T \ \mathbf{B}^T \right] \left[\frac{\mathbf{A}}{\mathbf{B}} \right] = \mathbf{A}^T \mathbf{A} + \mathbf{B}^T \mathbf{B}.$$

Since $\mathbf{B}^T \mathbf{B}$ is positive semidefinite, therefore, the eigenvalues of $\mathbf{C}^T \mathbf{C}$ are no smaller than that of $\mathbf{A}^T \mathbf{A}$ [116]. Equivalently, the singular values of $\mathbf{C}$ are no smaller than that of $\mathbf{A}$. (For a numerical example, see Problem 6.)

6. *Matlab Exercise.* It is well known that the SVD approach to solving least-squares problems requires a great deal more computation than using the normal equation. Therefore, SVD should only be used in a critical situation (i.e., if it can make a difference in numerical behavior). The following Matlab code shows that the more training patterns that are available, the more stable is a normal equation's numerical behavior. Therefore, there is less need for the computationally complex SVD algorithm. Verify this statement by using Matlab and varying the vectors $\mathbf{a}$, $\mathbf{b}$, $\mathbf{c}$, and $\mathbf{d}$. (See Figure 5.13.)

```
clear all; close all;
a=[5 6]'; b=[-3 4]'; c=[4 4]'; d=[-1 -4]';
A1 = [a*a']
svd(A1)
plot(svd(A1),'k');
axis([1 2 0 120]);
hold;
A2 = a*a' + b*b'
svd(A2)
```

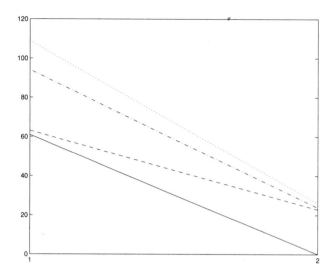

Figure 5.13. Results of the Matlab code shown in Problem 6. The figure shows the increases in the singular values when positive semidefinite matrices are added to a positive semidefinite matrix. The tick marks at the horizontal axis represent the two singular values (λ_1 and λ_2) and the vertical axis their numerical values. *Solid*: λ_i of $\mathbf{A}_1$; *dashed*: λ_i of $\mathbf{A}_1 + \mathbf{A}_2$; *solid dashed*: λ_i of $\mathbf{A}_1 + \mathbf{A}_2 + \mathbf{A}_3$; and *dotted*: λ_i of $\mathbf{A}_1 + \mathbf{A}_2 + \mathbf{A}_3 + \mathbf{A}_4$.

```
plot(svd(A2),'k--');
A3 = a*a' + b*b'+ c*c'
svd(A3)
plot(svd(A3),'k-.');
A4 = a*a' + b*b'+ c*c' + d*d'
svd(A4)
plot(svd(A4),'k:');
```

7. *Matlab Exercise.* The following Matlab code compares the numerical performance and stability of SVD and the normal equation for finding a least-squares
solution.

```
clear all; close all; echo off;
disp('Without dependent columns of A, press any key to start');
pause;
A=[1 2; 3 4; 5 6]; d=[-3 2 3]';
w=inv(A'*A)*A'*d;
y=A*w
E=(d-y)'*(d-y)
disp('Poor precision in w, press any key to start');pause;
```

```
w1 = w + [0.01 -0.01]'*w(1);
y1=A*w1
E1=(d-y)'*(d-y)
disp('With dependent columns of A, press any key to start');
pause;
A=[1 2; 2 4; 3.0000000001 6]; d=[-3 2 3]';
w=inv(A'*A)*A'*d;
y=A*w
E=(d-y)'*(d-y)
disp('Poor precision in w, press any key to start');pause;
w1 = w + [0.01 -0.01]'*w(1);
y1=A*w1
E=(d-y1)'*(d-y1)
disp('Solve the normal eq. by SVD, press any key to start');
pause;
[U,S,V]=svd(A);
invS = S;
for i=1:2,
  if(abs(S(i,i))< 0.0001)
    invS(i,i) = 0.0;
  else
    invS(i,i) = 1/S(i,i);
  end
end
w=V*invS'*U'*d;
y=A*w
E=(d-y)'*(d-y)
disp('Poor precision in w, press any key to start');pause;
w1 = w + [0.01 -0.01]'*w(1);
y1=A*w1
E=(d-y1)'*(d-y1)
disp('Ignore the second dimension of input data,...
      press any key to start');pause;
A=[1 ; 2 ; 3.0000000001]; d=[-3 2 3]';
w=inv(A'*A)*A'*d;
y=A*w
E=(d-y)'*(d-y)
```

(a) Explain why the error E is large when there are dependent columns in matrix $\mathbf{A}$ and the precision of $\mathbf{w}$ is poor.

(b) Discuss the geometrical interpretation of having a singular value of 0 in matrix $\mathbf{S}$. Hence, explain how the singular value decomposition approach helps avoid the numerical difficulty arising from the singularity of $\mathbf{A}^T\mathbf{A}$.

8. An elliptical basis function (EBF) network with M centers, K outputs, and no bias term has the form

$$y_k(\mathbf{x}) = \sum_{j=1}^{M} w_{kj}\phi_j(\mathbf{x}) \quad k = 1, \ldots, K,$$

where w_{kj} ($k = 1, \ldots, K$ and $j = 1, \ldots, M$) are the output weights and $\phi_j(\mathbf{x})$ is the j-th basis function output. An M-center GMM has the form

$$y(\mathbf{x}) = \sum_{j=1}^{M} \omega_j p_j(\mathbf{x}|\mu_j, \Sigma_j),$$

where $\Lambda = \{\omega_j, \mu_j, \Sigma_j\}_{j=1}^{M}$ are the GMM's parameters and $p_j(\cdot)$ is the j-th hidden node output. Compare the EBF network and the GMM in the context of their

   (a) hidden unit representation

   (b) output representation

   (c) training algorithm

9. A fully connected feed-forward multi-layer perceptron network has eight inputs, four outputs, and two hidden layers, one with six hidden nodes and the other with three nodes.

   (a) Draw a flow graph of this network.

   (b) For the retrieving phase, what is the computational complexity in terms of MACs (multiplier-accumulations)?

10. A radial basis function (RBF) network with 10 function centers is applied to approximate the function
$$f(x) = \frac{\sin(2\pi x)}{e^x}$$
shown in Figure 5.14. Suggest appropriate positions for the function centers and explain your suggestions. Verify your answer by using Matlab.

11. What are the advantage(s) and disadvantage(s) of using elliptical basis function networks instead of radial basis function networks for pattern classification?

12. The k-th output of an RBF network with K outputs has the form

$$y_k(\vec{x}) = \sum_{j=1}^{M} w_{kj}\phi_j(\vec{x}) + w_{k0} \qquad k = 1, \ldots, K.$$

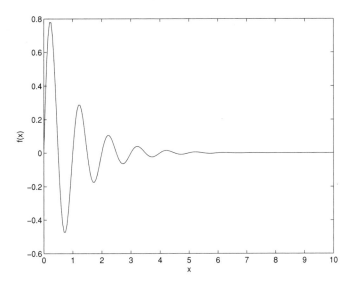

Figure 5.14. Function to be approximated by an RBF network.

For the case of Gaussian basis functions, there is either

$$\phi_j(\vec{x}) = \exp\left\{\frac{\|\vec{x} - \vec{\mu}_j\|^2}{2\sigma_j^2}\right\}$$

or

$$\phi_j(\vec{x}) = \exp\left\{-\frac{1}{2}(\vec{x} - \vec{\mu}_j)^T \Sigma_j^{-1}(\vec{x} - \vec{\mu}_j)\right\},$$

where $\vec{\mu}_j$ and Σ_j are the mean vectors and the covariance matrices, respectively, and σ_j is the function width associated with the j-th cluster.

(a) Give the conditions under which the preceding two equations are equivalent.

(b) If u_k and λ_k ($k = 1, \ldots, M$) are the orthonormal eigenvectors and eigenvalues of the matrix Σ_j^{-1}, respectively, that is,

$$\Sigma_j^{-1}\vec{u}_k = \lambda_k\vec{u}_k$$

show that

$$U^T\Sigma_j^{-1}U = D,$$

where

$$U = [\vec{u}_1\ \vec{u}_2 \ldots \vec{u}_M] \text{ and } D = \text{diag}\{\lambda_1, \lambda_2, \ldots, \lambda_M\}.$$

13. The outputs of an EBF network with I inputs, J hidden nodes, and K outputs are given by

$$y_k(\vec{x}) = \sum_{j=1}^{J} w_{kj}\phi_j(\vec{x}) + w_{k0} \qquad k = 1,\ldots,K,$$

where

$$\phi_j(\vec{x}) = \exp\left\{-\frac{1}{2}\sum_{i=1}^{I}\frac{(x_i - \mu_{ji})^2}{\sigma_{ji}^2}\right\}.$$

(a) Let the total squared error be

$$E = \frac{1}{2}\sum_{n=1}^{N}\sum_{k=1}^{K}\left[d_k(\vec{x}_n) - y_k(\vec{x}_n)\right]^2,$$

where $d_k(\vec{x}_n)$ is the desired output for the input vector $\vec{x}_n$. Derive expressions for

$$\frac{\partial E}{\partial w_{kj}} \quad k = 1,\ldots,K \text{ and } j = 1,\ldots,J$$

$$\frac{\partial E}{\partial \mu_{ji}}, \quad \frac{\partial E}{\partial \sigma_{ji}} \quad j = 1,\ldots,J, \text{ and } i = 1,\ldots,I$$

(b) Hence suggest a gradient-based learning rule for the EBF network.

(c) Suggest a method to ensure that the function widths σ_{ji} are always positive.

(d) Identify two disadvantages of the gradient-based learning rule derived in Problems 13b and 13c.

14. Assume that an RBF network with I inputs, J hidden nodes, and one output has a Gaussian-basis function of the form

$$z_j(\mathbf{x}) = \begin{cases} \exp\left\{-\frac{\|\mathbf{x}-\mathbf{c}_j\|^2}{2\sigma_j^2}\right\} & j = 1,\ldots,J \\ 1 & j = 0, \end{cases}$$

where $\mathbf{c}_j$ and σ_j are the j-th function center and function width, respectively.

(a) Express the output $y(\mathbf{x}_p)$ in terms of w_j, x_{pi}, c_{ji}, and σ_j, where $1 \leq i \leq I$, $0 \leq j \leq J$, $1 \leq p \leq N$, and N is the number of training pairs $\{\mathbf{x}_p, d(\mathbf{x}_p)\}$. Note that x_{pi} is the i-th input of the p-th pattern and c_{ji} is the position of the j-th center with respect to the i-th coordinate of the input domain.

(b) Show that the total squared error is

$$E = \frac{1}{2}(\mathbf{d} - Z\mathbf{w})^T(\mathbf{d} - Z\mathbf{w}),$$

where

$$Z = \begin{bmatrix} z_0(\mathbf{x}_1) & z_1(\mathbf{x}_1) & \cdots & z_J(\mathbf{x}_1) \\ z_0(\mathbf{x}_2) & z_1(\mathbf{x}_2) & \cdots & z_J(\mathbf{x}_2) \\ \vdots & \vdots & \ddots & \vdots \\ z_0(\mathbf{x}_N) & z_1(\mathbf{x}_N) & \cdots & z_J(\mathbf{x}_N) \end{bmatrix},$$

$$\mathbf{w} = \begin{bmatrix} w_0 \\ w_1 \\ \vdots \\ w_j \end{bmatrix}, \text{ and } \mathbf{d} = \begin{bmatrix} d(\mathbf{x}_1) \\ d(\mathbf{x}_2) \\ \vdots \\ d(\mathbf{x}_N) \end{bmatrix}.$$

Hence, show that

$$\mathbf{w} = (Z^T Z)^T Z^T \mathbf{d}.$$

(c) State the condition under which $(Z^T Z)^T Z^T \mathbf{d}$ is susceptible to numerical error.

15. An RBF network has n inputs, three hidden nodes, and one output. The output of the network is given by

$$y(\mathbf{x}) = w_1 \phi_1(\mathbf{x}) + w_2 \phi_2 \mathbf{x}() + w_3 \phi_3(\mathbf{x}).$$

where $\{\phi_i(\mathbf{x})\}_{i=1}^3$ are the activation of the hidden nodes and $\{w_i\}_{i=1}^3$ are the weights connecting the hidden nodes and the output node. Table 5.8 shows the relationship among $\phi_i(\mathbf{x}_p)$, $y(\mathbf{x}_p)$ and the desired output $d(\mathbf{x}_p)$ for input vectors $\{\mathbf{x}_p\}_{i=1}^3$.

Table 5.8.

Input Vector	$\phi_1(\mathbf{x})$	$\phi_2(\mathbf{x})$	$\phi_3(\mathbf{x})$	$d(\mathbf{x}_p)$	Actual Output
$\mathbf{x}_1$	0	0	0	1	$y(\mathbf{x}_1)$
$\mathbf{x}_2$	0	0	0	1	$y(\mathbf{x}_2)$
$\mathbf{x}_3$	1	1	1	0	$y(\mathbf{x}_3)$

(a) Determine the matrix A such that $A\mathbf{w} = \mathbf{y}$, where

$$\mathbf{w} = \begin{bmatrix} w_1 \\ w_2 \\ w_3 \end{bmatrix} \text{ and } \mathbf{y} = \begin{bmatrix} y(\mathbf{x}_1) \\ y(\mathbf{x}_2) \\ y(\mathbf{x}_3) \end{bmatrix}.$$

(b) Express the vector $\mathbf{w}^*$ that minimizes the error

$$E = \|A\mathbf{w} - \mathbf{d}\|^2$$

in terms of A, $\mathbf{w}$, and $\mathbf{d}$. Explain whether you can determine a numerical value of $\mathbf{w}^*$ by using your expression.

(c) If the singular decomposition of A is equal to USV^T, where

$$
U = \begin{bmatrix} 0 & 0 & -1 \\ 0 & 1 & 0 \\ 1 & 0 & 0 \end{bmatrix}, \quad S = \begin{bmatrix} 1.73 & 0 & 0 \\ 0 & 0 & 0 \\ 0 & 0 & 0 \end{bmatrix}, \quad \text{and}
$$

$$
V = \begin{bmatrix} 0.58 & -0.82 & 0 \\ 0.58 & 0.41 & -0.71 \\ 0.58 & 0.41 & 0.71 \end{bmatrix}
$$

suggest a least-squares solution for $\mathbf{w}^*$.

16. The probability density function of the k-th class is defined as $p(\mathbf{x}|C_k)$, where $\mathbf{x}$ denotes the vector in the feature space and C_k represents the k-th class.

    (a) Use Bayes's theorem to express the posterior probability $p(C_k|\mathbf{x})$ in terms of $p(\mathbf{x}|C_k)$ and $p(\mathbf{x})$, where $p(\mathbf{x}) = \sum_i p(\mathbf{x}|C_i)P(C_i)$ is the probability density function of $\mathbf{x}$.

    (b) Given that it is possible to use a common pool of M-basis functions $\{p(\mathbf{x}|j)\}_{j=1}^{M}$, labeled by an index j, to represent all of the class-conditional density; that is,

    $$
    p(\mathbf{x}|C_k) = \sum_{j=1}^{M} p(\mathbf{x}|j)P(j|C_k)
    $$

    show that $P(C_k|\mathbf{x}) = \sum_{j=1}^{M} P(C_k|j)P(j|\mathbf{x})$.

    (c) Hence, show that the basis function outputs and the output weights of an RBF network can be considered as posterior probabilities.

17. The posterior probability of class C_k in a K-class problem is given by

    $$
    P(C_k|\mathbf{x}) = \frac{p(\mathbf{x}|C_k)P(C_k)}{p(\mathbf{x})},
    $$

    where $p(\mathbf{x}|C_k) = \sum_{j=1}^{M} \pi_{kj} p(\mathbf{x}|\mu_{kj}, \Sigma_{kj}, C_k)$ is a GMM, $P(C_k)$ is the prior probability of class C_k, and $p(\mathbf{x})$ is the density function of $\mathbf{x}$.

    (a) Express $p(\mathbf{x})$ in terms of $p(\mathbf{x}|C_k)$ and $P(C_k)$.

    (b) Discuss the difference between $P(C_k)$ and π_{kj}.

    (c) If $P(C_k) \neq P(C_j) \ \forall k \neq j$, explain how you would use K GMMs to classify a set of unknown pattern $X = \{\mathbf{x}_1, \mathbf{x}_2, \ldots, \mathbf{x}_T\}$ into K classes.

18. The BP algorithm involves the computation of the gradient $\frac{\partial E}{\partial w_{ij}(l)}$, where E is the mean-squared error between the desired outputs and the actual outputs and $w_{ij}(l)$ is the weights at layer l.

(a) Based on the linear basis function in Eq. 5.2.1 and the BP algorithm (Eq. 5.4.3–Eq. 5.4.5), derive an algorithm for computing the error sensitivity of a multi-layer LBF network with respect to the network inputs x_i, $i = 1, \ldots, D$, where D is the input dimension.

(b) Repeat (a) for a multi-layer RBF network using Eq. 5.2.2 and the back-propagation rule for multi-layer RBF networks (Eq. 5.4.17–Eq. 5.4.19).

19. Derive the gradient of a Gaussian EBF in Eq. 5.4.10.

20. Assume that you are given a set of data with an XOR-type distribution: For one class there are two centroids located at (0,0) and (1,1), while for the other class there are two centroids located at (1,0) and (0,1). Each cluster can be statistically modeled by a uniform distribution within a disk of radius $r = 1$.

(a) Show that the two classes are not linearly separable.

(b) Derive an analytical solution (equations for the decision boundaries) for a multi-layer LBF network (no computer).

(c) Verify your solutions using Matlab and Netlab.

21. The implementation cost of multi-layer neural models depends very much on the computational complexity incurred in each layer of feed-forward processing. Compute the number of multiplications and additions involved when a 3×4 matrix is postmultiplied by a 4×12 matrix.

22. Is it true that the basis function of kernel-based networks must be in the form of a Gaussian function? Provide an explanation. *Note*: RBF networks are considered to be one type of kernel-based networks.

23. *Multiple Choice Questions:*

(a) The outputs of an RBF network are linear with respect to
     i. the inputs
     ii. the weights connecting the hidden nodes and the outputs
     iii. parameters of the hidden nodes
     iv. the outputs of the hidden nodes
     v. both (ii) and (iv)
     vi. none of the above

(b) The purpose of the function width σ in an RBF network is to
     i. improve the numerical stability of the K-means algorithm
     ii. allow the basis functions to cover a wider area of the input space so that better generalization can be achieved
     iii. ensure that the inference of the individual-basis functions will not overlap with each other

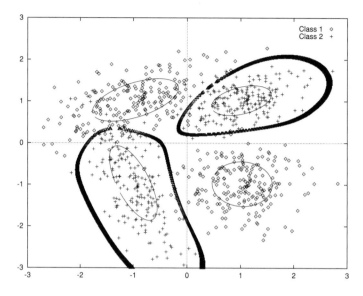

Figure 5.15. Decision boundaries (*thick curves*) and contours of basis function outputs (*thin ellipses*).

  iv. none of the above

(c) To train an RBF network, one should apply the K-means, least-squares, and K-nearest-neighbor algorithms in the following order:

  i. K-means, least squares, K-nearest neighbor

  ii. K-means, K-nearest-neighbor, least squares

  iii. Least squares, K-means, K-nearest neighbor

  iv. Least squares, K-means, and K-nearest-neighbor in any order

(d) To apply singular value decomposition to determine the output weights of an RBF network,

  i. decompose matrix A into $A = U\Lambda V^T$, where $A = \{a_{pi}\} = \{\Phi_i(\|\mathbf{x}_p - \mathbf{c}_i\|)\}$ contains the basis function outputs and $\Lambda = \text{diag}\{\lambda_1, \lambda_2, \ldots, \lambda_D\}$ contains the singular values

  ii. set $1/\lambda_j = 0$ if $\lambda_j \to 0$

  iii. set $\lambda_j = 0$ if $\lambda_j \to \infty$

  iv. both (i) and (ii)

  v. both (i) and (iii)

(e) Given the decision boundaries and the contours of basis function outputs of Figure 5.15, it is correct to say that

  i. The figure is produced by an RBF network with 2 centers.

    ii. The figure is produced by an RBF network with 4 centers.

    iii. The figure is produced by an EBF network with 4 centers.

    iv. The figure is produced by an EBF network with 2 centers.

(f) Based on the figure in the preceding question, it is correct to say that

    i. The covariance matrix corresponding to Class 1 (4th quadrant) is diagonal.

    ii. The covariance matrix corresponding to Class 2 (3rd quadrant) is not diagonal.

    iii. The covariance matrix corresponding to Class 1 (2nd quadrant) exhibits a larger variance along the horizontal axis.

    iv. All of the above.

(g) In the context of pattern classification, RBF outputs can be trained to give

    i. the posterior probabilities of class membership, $P(C_k|\mathbf{x})$

    ii. the likelihood of observing the input data given the class membership, $p(\mathbf{x}|C_k)$

    iii. the likelihood of observing the input data, $p(\mathbf{x})$

    iv. both (i) and (ii)

    v. (i), (ii), and (iii)

(h) Training of RBF networks is usually faster than training multi-layer perceptrons because

    i. Training in RBF networks is divided into two stages, where each stage focuses on a part of the optimization problem.

    ii. RBF networks have fewer parameters to be estimated.

    iii. The RBF networks' outputs are linear with respect to *all* of the networks' parameters.

    iv. None of the above.

(i) Training of RBF networks requires a large amount of memory because

    i. The K-means algorithm requires a great deal of memory.

    ii. The singular value decomposition requires to store an $N \times (M + 1)$ matrix in memory, where N is the number of patterns and M is the number of centers.

    iii. The size of RBF networks is usually considerably larger than that of MLPs.

    iv. None of the above.

Answers: a (v), b (ii), c (ii), d (iv), e (iii), f (iv), g (i), h (i), i (ii)

24. *Matlab Exercise.* Given two classes of patterns with distributions $p_1(x, y)$ and $p_2(x, y)$, where for p_1 the mean and variance are, respectively,

$$\mu_1 = [1.0, 1.0]^T \text{ and } \Sigma_1 = \begin{bmatrix} 2.0 & 0.5 \\ 0.5 & 1.0 \end{bmatrix}$$

and for p_2 the mean and variance are

$$\mu_1 = [-1.0, -1.0]^T \text{ and } \Sigma_1 = \begin{bmatrix} 1.0 & -1.0 \\ -1.0 & 2.0 \end{bmatrix},$$

write Matlab code to generate 1,000 patterns for each distribution. Create an RBF network and a GMM-based classifier to classify the data in these two classes. For each network type, find the classification accuracy and plot the decision boundary.

25. *Matlab Exercise.* Create 100 two-dimensional data patterns for two classes. Use an RBF BP Network to separate the two classes. Try to do clustering by using

(a) an RBF BP network with four hidden units

(b) an EBF BP network with four hidden units

Chapter 6

MODULAR AND HIERARCHICAL NETWORKS

6.1 Introduction

Multi-layer networks are perhaps the simplest and most popular supervised learning model and can be adopted for most biometric authentication applications. Structurally, a multi-layer network has full connectivity, as illustrated in Figure 6.1(a). More precisely, all hidden nodes of one lower layer are fully connected to all nodes in its immediate subsequent layer. In other words, the model adopts a flat network structure such that all synaptic weights of a layer are lumped together in one supernetwork. This type of network is also termed "all-class-one-network" (ACON) [186].

During the training phase, the fully connected network must be trained to simultaneously satisfy the specification given by all teachers. This results in two undesirable consequences. First, the number of hidden units required to meet the specification tends to become very large, especially for complex decision boundaries. Second, for every training pattern, all the synaptic weights must be trained. As a result, the training process is potentially influenced by conflicting signals from different teachers. Thus, the convergence rate of a fully connected multi-layer model degrades drastically with respect to network size, and the training time required may become excessively long.[1]

A more flexible and effective structure is a modular network built on many smaller modules. "An information processing network is said to be modular if the computation performed by the network can be decomposed into two or more modules (subsystems) that operate on distinct inputs without communicating with each other" [128]. The modular architecture follows the "divide-and-conquer" principle, solving a complex computation task by dividing it into simple subtasks and integrating the individual results in a postprocessing layer.

There are two prominent types of modular architecture: class- and expert-based.

---

[1]Note that this problem can be partially alleviated by restricting one group of neurons to be responsible for one class only, as exemplified by the RBF/EBF network shown in Figure 5.6.

1. *Class-Based Modular Networks.* The goal of a machine learning model is to determine the class to which an input pattern best belongs. Therefore, it is natural to use class-based modules as the basic partitioning units. In a class-based modular model, one subnet is designated to one pattern class, as illustrated in Figure 6.1(b). Each subnet specializes in distinguishing its own class from the others, so the number of hidden units is usually small. (This type of structure is also referred to as "one-class-one-network" (OCON) [186].) The design of class-based modular networks is briefly introduced in Section 6.2 and is covered in greater detail in Chapter 7.

2. *Expert-Based Modular Networks.* Another prominent type of modular structure is the expert-based modular model, shown in Figure 6.1(c), in which an individual module reflects one local expert's subjective viewpoint on how to distinguish different classes. In this type of structure, each subnet is a local classifier working only on local regions within the entire pattern space. The *gating network* is responsible for integrating the results from local classifiers and reaching a final decision. A typical example of such modular networks is the mixture-of-experts (MOE), illustrated in Section 6.3.

The two modular network types are similar in that both divide a complex classification problem into simpler subproblems and adaptively incorporate the results from individual modules into the final decision. The difference in the two structures is that the division used in the class-based structure is *pattern class oriented*, whereas the structure used in the expert-based networks is *pattern space oriented*.

Ultimately, the information made available by the functional modules in a modular network must be globally integrated to reach a final decision or recommendation. A hierarchical structure provides a systematic framework to integrate or fuse information from various local modules. Section 6.4 provides a comprehensive analysis of several important variants of hierarchical network structures. In particular, one prominent hierarchical model, the so-called hierarchical mixture-of-experts (HME), is explored in this chapter.

6.2 Class-Based Modular Networks

An important goal of pattern recognition is determining to which class an input pattern best belongs. Therefore, it is natural to consider class-level modules as the basic partitioning units, where each module specializes in distinguishing its own class from the others. Consequently, the number of hidden nodes designated to each class tends to be very small. The class-level modules are adopted by the OCON network. In contrast to expert-level partitioning, this OCON structure facilitates a global (or mutual) supervised training scheme. In global interclass supervised learning, any dispute over a pattern region by (two or more) competing classes can be effectively resolved by resorting to the teacher's guidance. Such a distributed processing structure is also convenient for network upgrading when there is a need to add or remove memberships. Finally, such a distributed structure is especially appealing to the design of the RBF networks.

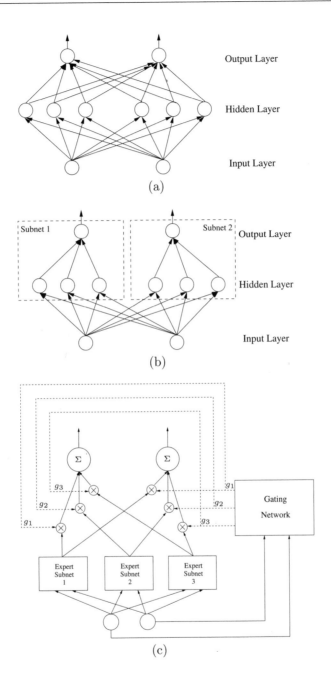

Figure 6.1. Different types of architectures for feed-forward neural networks: (a) ACON structure, (b) class-based, and (c) expert-based grouping structure.

6.2.1   Class-Based OCON Structure

The most popular class-based modular networks are generally based on the OCON-type structure. One prominent example of the class-based grouping structure is the Gaussian mixture model (GMM) classifier. Figure 6.2 depicts the architecture of a K-class classifier in which each class is represented by a GMM. Gaussian mixture models make use of semiparametric techniques for approximating probability density functions (pdf). The output of a GMM is the weighted sum of R component densities, as shown in Figure 6.2. Given a set of N independent and identically distributed patterns $X^{(i)} = \{\mathbf{x}_t; t = 1, 2, \ldots, N\}$ associated with class ω_i, the class likelihood function $p(\mathbf{x}_t|\omega_i)$ for class ω_i is assumed to be a mixture of Gaussian distributions, that is,

$$p(\mathbf{x}_t|\omega_i) = \sum_{r=1}^{R} P(\Theta_{r|i}|\omega_i)p(\mathbf{x}_t|\omega_i, \Theta_{r|i}), \qquad (6.2.1)$$

where $\Theta_{r|i}$ represents the parameters of the r-th mixture component, R is the total number of mixture components, $p(\mathbf{x}|\omega_i, \Theta_{r|i}) \equiv \mathcal{N}(\mathbf{x}; \mu_{r|i}, \Sigma_{r|i})$ is the probability density function of the r-th component, and $P(\Theta_{r|i}|\omega_i)$ is the prior probability (also called a mixture coefficient) of the r-th component. Typically, $\mathcal{N}(\mathbf{x}; \mu_{r|i}, \Sigma_{r|i})$ is a Gaussian distribution with mean $\mu_{r|i}$ and covariance $\Sigma_{r|i}$.

The training of GMMs can be formulated as a maximum-likelihood problem, where the mean vectors $\{\mu_{r|i}\}$, covariance matrices $\{\Sigma_{r|i}\}$ and mixture coefficients $\{P(\Theta_{r|i}|\omega_i)\}$ are typically estimated by the EM algorithm (see Chapter 3). More specifically, the parameters of a GMM are estimated iteratively by[2]

$$\mu_{r|i}^{(n+1)} = \frac{\sum_{t=1}^{N} P^{(n)}(\Theta_{r|i}|\mathbf{x}_t)\mathbf{x}_t}{\sum_{t=1}^{N} P^{(n)}(\Theta_{r|i}|\mathbf{x}_t)},$$

$$\Sigma_{r|i}^{(n+1)} = \frac{\sum_{t=1}^{N} P^{(n)}(\Theta_{r|i}|\mathbf{x}_t)[\mathbf{x}_t - \mu_{r|i}^{(n+1)}][\mathbf{x}_t - \mu_{r|i}^{(n+1)}]^T}{\sum_{t=1}^{N} P^{(n)}(\Theta_{r|i}|\mathbf{x}_t)}, \text{ and}$$

$$P^{(n+1)}(\Theta_{r|i}) = \frac{\sum_{t=1}^{N} P^{(n)}(\Theta_{r|i}|\mathbf{x}_t)}{N}, \qquad (6.2.2)$$

where n denotes the iteration index, $P^{(n)}(\Theta_{r|i}|\mathbf{x}_t)$ is the posterior probability of the r-th mixture ($r = 1, \ldots, R$), and T denotes matrix transpose. The posterior probability can be obtained by Bayes's theorem, yielding

$$P^{(n)}(\Theta_{r|i}|\mathbf{x}_t) = \frac{P^{(n)}(\Theta_{r|i})p^{(n)}(\mathbf{x}_t|\Theta_{r|i})}{\sum_{k=1}^{R} P^{(n)}(\Theta_{k|i})p^{(n)}(\mathbf{x}_t|\Theta_{k|i})} \qquad (6.2.3)$$

---

[2]To simplify the notation, ω_i in Eqs. 6.2.2 through 6.2.4 has been dropped.

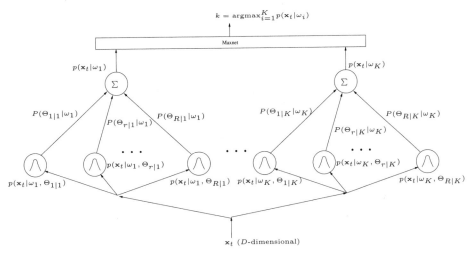

Figure 6.2. Architecture of a GMM-based classifier.

in which

$$p^{(n)}(\mathbf{x}_t|\Theta_{r|i}) = \frac{1}{(2\pi)^{\frac{D}{2}}|\Sigma_{r|i}^{(n)}|^{\frac{1}{2}}} \exp\left\{-\frac{1}{2}\left[\mathbf{x}_t - \mu_{r|i}^{(n)}\right]^T (\Sigma_{r|i}^{(n)})^{-1}\left[\mathbf{x}_t - \mu_{r|i}^{(n)}\right]\right\},$$

$$(6.2.4)$$

where D is the input dimension.

Unlike ACON, where an output node depends explicitly on all hidden nodes' outputs, the output of a GMM described before depends only on the hidden nodes of the corresponding class. This may restrict the classifier's ability to discriminate input patterns. One possible way of avoiding the complexity of ACON without sacrificing the modularity of the class-based grouping structure is to allow connections between the output nodes and all hidden nodes but to keep the hidden nodes class dependent. One typical example for this kind of structure is the multiclass RBF/EBF networks mentioned in Section 5.4. However, training time of this type of network is longer than that of the GMMs because connection weights from all hidden nodes must be trained for each output node.

6.2.2   ACON versus OCON Networks

Pandya and Macy [268] compared the performance of the ACON and OCON structures in handwritten character recognition. The authors observed that the OCON model achieves better training (99.5% vs 94%) and generalization (87% vs. 82%) accuracies, yet it requires only one-fourth of ACON's training time (cf. Table 6.1). Wang [369] uses an OCON model on computer-aided diagnosis for breast cancer and reports excellent experimental results.

One can argue that compared to the ACON structure, the OCON structure is slow in retrieving time when the number of classes is very large. This is not entirely

Table 6.1. The comparison of ACON and OCON structures in a handwritten character recognition problem. The training time is normalized so that the time for ACON structure is 1. (Adapted from Pandya and Macy [268].)

	Training Accuracy	Generalization Accuracy	Training Time
ACON	94.0%	82.0%	1.00
OCON	99.5%	87.0%	0.25

true. As mentioned earlier, when the number of classes is large; the number of hidden neurons in the ACON structure also tends to be very large; therefore, ACON is also slow. Since the computation time in either an OCON or an ACON structure increases as the number of classes grows, a linear increase of computation time (i.e., OCON) can be expected. In fact, for a neural network-based recognition system, the recognition time is usually just a small portion of the entire system time and it will not affect system performance too much, even if the class number grows very large. Take the 200-person PDBNN face recognition system (see Chapter 8) as an example. For a 320×240 grayscale image, the entire recognition process (including preprocessing, detection, localization, feature extraction, and recognition) takes about one second. In this one-second system time, recognition (the retrieving) only takes 100ms, which is only one-tenth of the system time. The retrieving time will become 50% of the system time when the number of people in the database exceeds 2,000. In this case, the system time becomes two seconds, which is still fast enough for most security applications.

6.3   Mixture-of-Experts Modular Networks

An expert-based modular network is built on smaller modules, each representing the behavior of a local or specially tailored pattern space. The most prominent expert-based modular network is the *mixture-of-experts* (MOE) [162]. The MOE exhibits an explicit relationship with statistical pattern classification methods. Given a pattern, each expert network estimates the pattern's conditional a posteriori probability on the (adaptively tuned or preassigned) feature space. Each local expert network performs multiway classification over K classes by using either K-independent binomial models, each being modeled after one and only one class, or one multinomial model for all classes. The corresponding output of the gating network represents the associated confidence of each expert. The final system output is the weighted sum of the estimated probabilities from all of the expert networks. Another example of the expert-based modular structure is the *committee machine* [262], which consists of a layer of elementary perceptrons (expert units) followed by a vote-taking perceptron (the gating unit) in the second layer.

6.3.1 Local Experts and Gating Network

With reference to Figure 6.3, the MOE comprises the following subsystems.

1. *Local Experts.* The design of modular neural networks hinges on the choice of local experts that serve the following functions:

 a. *Extract local features.* A *local expert* is adaptively trained to extract certain *local features* that are particularly relevant to its local decision. Sometimes, a local expert can be assigned a predetermined feature space.

 b. *Conduct local classification and make recommendations.* The local experts conduct pattern classification tasks only from the perspective of local regions of the pattern space. In short, based on local features, a local expert produces a *local recommendation* to the gating network.

 c. *Negotiate with the gating network.* For competitive learning models, the local expert must coordinate with the *gating network* to decide how much the local module should be adjusted in reaction to a particular training pattern.

The effectiveness of an expert-based modular network hinges on a proper designation of local experts. For example, one (LBF or RBF) hidden node may be devoted to extract a certain local feature of particular interest to an *expert*.

2. *Gating Network.* The gating network determines the rule that guides how to integrate recommendations collected from various local experts.[3] It estimates the proper weights to be used for the fusion of information, which should reflect the trustworthiness of a local expert's opinion on a given input pattern. A more trustworthy local expert will be assigned a larger (smaller) gating activation and in turn will be allocated a greater (lesser) influence in the final decision-making process. A gating network is often conveniently implemented as a softmax activation function [35, 236], and a probabilistic rule is often adopted to integrate recommendations from local experts. For competitive learning models, the gating network also has a major influence on how much individual modules should be trained in response to a new training pattern.

The expectation-maximization (EM) algorithm can be adopted to train the local experts and the confidence levels assigned by the gating network. The objective is to estimate the model parameters so as to attain the highest probability of the training set given the estimated parameters. For a given input $\mathbf{x}$, the posterior probability of generating class $\mathbf{y}$ given $\mathbf{x}$ using K experts can be computed as

$$P(\mathbf{y}|\mathbf{x}, \phi) = \sum_{j=1}^{K} P(\mathbf{y}|\mathbf{x}, \theta_j) a_j(\mathbf{x}), \qquad (6.3.1)$$

where $\mathbf{y}$ is a binary vector, $a_j(\mathbf{x})$ is the probability for weighting the expert outputs, and $P(\mathbf{y}|\mathbf{x}, \theta_j)$ is the output of the j-th expert network. For example, in a two-class

---

[3]In this sense, the *expert level* in neural networks is compatible with the *rule level* in fuzzy systems.

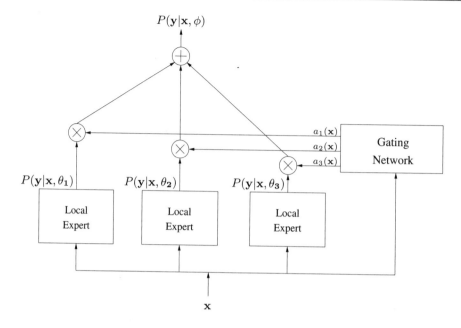

Figure 6.3. The baseline MOE architecture. An expert network estimates the pattern's conditional a posteriori probability. A baseline MOE comprises two subsystems: local experts and a gating network. The local experts are adaptively trained to extract certain local features, particularly relevant to their own local decisions, while the gating network computes the global weights to be applied to the local decisions. *Note*: For K-class classification problems, the output $P(\mathbf{y}|\mathbf{x}, \phi)$ is a K-dimensional vector with elements that sum to 1.

classification problem, $\mathbf{y}$ is either $[1\ 0]$ or $[0\ 1]$, ϕ is a parameter set comprising of $\{\mathbf{V}, \theta_j\}$, where $\mathbf{V} = \{\mathbf{V}_j, j = 1, \ldots, K\}$ is the parameter set for the gating network, and $\{\theta_j; j = 1, \ldots, K\}$ is the parameter set for the j-th expert network. The gating network of MOE can be a linear LBF network (say, a linear perceptron) or a nonlinear RBF network.

6.3.2   LBF MOE Networks

A simple example illustrates how an LBF MOE works. The output of a linear gating network [35, 236] can be expressed as:

$$a_j(\mathbf{x}) = \frac{\exp[b_j(\mathbf{x})]}{\sum_{k=1}^{K} \exp[b_k(\mathbf{x})]}, \tag{6.3.2}$$

where $b_j(\mathbf{x}) = \mathbf{v}_j^T\mathbf{x} + d_j$. Now, denote $\mathbf{V}_j = \{\mathbf{v}_j, d_j\}$ as the weight of the j-th neuron of the gating network. Suppose patterns from two classes occupy a two-dimensional space $\{\mathbf{x} = (x_1,\ x_2);\ 0 \le x_1 \le 1,\ 0 \le x_2 \le 1\}$. Class 1 patterns occupy the lower

quarter of the square in Figure 6.4, and Class 2 patterns take the remaining three
quarters of the space. A two-expert MOE classifier with the LBF local experts is
used:

$$P(\omega_1|\mathbf{x},\theta_1) = \frac{1}{1 + e^{\mathbf{w}_1^T\mathbf{x}+c_1}}$$

$$P(\omega_2|\mathbf{x},\theta_1) = 1 - P(\omega_1|\mathbf{x},\theta_1)$$

$$P(\omega_1|\mathbf{x},\theta_2) = \frac{1}{1 + e^{\mathbf{w}_2^T\mathbf{x}+c_2}}$$

$$P(\omega_2|\mathbf{x},\theta_2) = 1 - P(\omega_1|\mathbf{x},\theta_2)$$

$$a_1(\mathbf{x}) = \frac{e^{\mathbf{v}_1^T\mathbf{x}+d_1}}{\sum_{j=1}^2 e^{\mathbf{v}_j^T\mathbf{x}+d_j}}$$

$$a_2(\mathbf{x}) = \frac{e^{\mathbf{v}_2^T\mathbf{x}+d_2}}{\sum_{j=1}^2 e^{\mathbf{v}_j^T\mathbf{x}+d_j}},$$

where $P(\omega_1|\mathbf{x},\theta_j)$ is equal to $P(\mathbf{y} = [1\ \ 0]|\mathbf{x},\theta_j)$ in Eq. 6.3.1. Similarly, $P(\omega_2|\mathbf{x},\theta_j)$
is equal to $P(\mathbf{y} = [0\ \ 1]|\mathbf{x},\theta_j)$. One solution is

$$\mathbf{w}_1 = [-1\ \ 1]^T, c_1 = 0$$
$$\mathbf{w}_2 = [1\ \ 1]^T, c_2 = -1$$
$$\mathbf{v}_1 = [-200\ \ 0]^T, d_1 = 100$$
$$\mathbf{v}_2 = [0\ \ 0]^T, d_2 = 0.$$

For this solution, Expert 1 creates a decision boundary along the diagonal line
from (0,0) to (1,1), and Expert 2 creates a decision boundary along the antidiagonal
line from (1,0) to (0,1). The gating network gives Expert 1 higher confidence at
region $\{(x_1, x_2); 0 \le x_1 \le 0.5, 0 \le x_2 \le 1 \}$ and gives Expert 2 higher confidence
at the other half.

6.3.3   RBF MOE Networks

The MOE structure is natural for RBF networks. In an RBF network, each hidden
node represents a receptive field with the following normalized Gaussian activation
function:

$$a_j(\mathbf{x}) = \frac{\beta_j(\mathbf{x})}{\sum_{k=1}^K \beta_k(\mathbf{x})}, \tag{6.3.3}$$

where

$$\beta_k(\mathbf{x}) = \exp(-\|\mathbf{x} - \mu_k\|^2/2\sigma_k^2), \tag{6.3.4}$$

where $\mathbf{x}$ is the D-dimensional input vector and K is the number of hidden nodes.
The parameters μ_k and σ_k^2 denote the mean and the variance of the k-th Gaussian

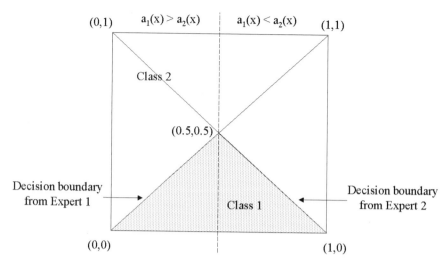

Figure 6.4. An example for MOE in which a local expert gets a higher weighting on its proclaimed feature space. In this example, the gating network assigns higher confidence to Expert 1 for region $\{(x_1, x_2); 0 \leq x_1 \leq 0.5, 0 \leq x_2 \leq 1\}$ and to Expert 2 for the remaining region.

function. The output $y(\cdot)$ can be computed as the weighted sum of the activation values

$$y(\mathbf{x}) = \sum_{j=1}^{K} w_j a_j(\mathbf{x}), \tag{6.3.5}$$

where w_j is the weight of the j-th Gaussian kernel.

The RBF network defined in Eq. 6.3.5 can be expressed in terms of the MOE structure. If the gating network output $a_j(\mathbf{x})$ in Eq. 6.3.1 is defined as the $a_j(\mathbf{x})$ in Eq. 6.3.3, and the $P(\mathbf{y}|\mathbf{x}, \theta_j)$ is defined to be a constant w_j, then the neural network becomes an MOE system with a radial-basis gating function and constant expert output.

The RBF MOE model has been applied to many applications. For example, in an application to sunspot time series, a Bayesian framework for inferring the parameters of an MOE model based on ensemble learning by variational free energy minimization proves to be very effective and performs significantly better than a single network [372]. MOE has also been shown to yield very good performance in automated cytology screening applications [158]. For broader application domains, MOE has also been extended to cope with a multiexpert-level tree structure, known as HME [167].

6.3.4   Comparison of MLP and MOE

The following is a comparison of MLPs and MOE in the context of their architecture, parameter specification, and learning algorithms.

Structural Similarity

As shown in Figures 6.1(a) and (c), MLPs and MOE have a similar structure in that they are both feed-forward networks. In the special case where the local experts of an MOE are single-layer perceptrons and the gating network's outputs are invariant with respect to the input vectors, the MOE is structurally equivalent to an MLP with output weights being equal to the gating network's outputs.

Difference in Parameter Specification

In terms of parameter specification, the output weights, hidden nodes' outputs, and network outputs in an MLP do not need to follow any probabilistic constraint. On the other hand, the training and operation of an MOE network require the gating network's outputs (which can be considered as the output weights of an MLP), experts' outputs, and the MOE's outputs to follow probabilistic constraints. More specifically, referring to Figure 6.3 the following conditions must be satisfied:

$$\sum_{j=1}^{M} a_j(\mathbf{x}) = 1, \quad \sum_{k=1}^{K} P(y_k|\mathbf{x}, \theta_j) = 1, \text{ and } \sum_{k=1}^{K} P(y_k|\mathbf{x}, \phi) = 1, \qquad (6.3.6)$$

where M denotes the number of experts, K denotes the number of classes, and y_k is the k-th component of the vector $\mathbf{y}$. Therefore, given the structural similarity between MOE and MLPs, one should consider the MOE a special type of MLP with some probabilistic constraints.

Algorithmic Comparison

The training algorithms of MLP and MOE have similarities, but they also have important differences. Their training algorithms are similar in that both of them iteratively update the network parameters in order to optimize an objective function. While the MLPs adopt an approximation-based learning scheme (back-propagation) to minimize the sum of squared errors between the actual outputs and desired outputs, the MOE adopts an optimization-based learning scheme (EM algorithm) to maximize the likelihood of model parameters given the training data. To a certain extent, the EM algorithm in MOE can be considered as a probabilistic back-propagation because the posterior probabilities in the E-step can be viewed as a set of optimized learning rates for the iterative M-step.

6.4   Hierarchical Machine Learning Models

The information made available by the functional modules in a modular network must be properly integrated to reach a final decision or recommendation. There are many possible techniques to integrate or fuse information from various local modules. The most popular approach uses some kind of hierarchical structure. For example, under a hierarchical model, a set of class-based modules can collectively form an expert subnet in expert-based type networks. Therefore, both the notions of expert modules and class modules play a vital role in the construction of such a hierarchical learning network.

According to the levels of hierarchy involved, hierarchical networks can be divided structurally into the following categories (Figure 6.5).

1. *One-level (i.e., structurally flat) networks.* This is exemplified by the traditional MLP that has a "single-network" structure shown in Figure 6.5(a).

2. *Two-level modular structures.* By adopting the divide-and-conquer principle, the task is first divided into modules, and then the individual results are integrated into a final and collective decision. Two typical modular networks are (1) mixture-of-experts (MOE), which uses the expert-level modules (cf. Figure 6.5(c)); and (2) the decision-based neural networks (DBNN) based on the class-level modules (cf. Figure 6.5(b)). See Section 6.3 and Chapter 7 for details.

3. *Three-level hierarchical structures.* To this end, the divide-and-conquer principle needs to be applied twice: one time on the expert-level and another on the class-level. Depending on the order used, this could result in two kinds of hierarchical networks: one with an experts-in-class construct and another a classes-in-expert construct, as depicted in Figures 6.5(d) and 6.5(e). These two hierarchical structures are further discussed in Section 6.4.2.

6.4.1 Hierarchical Mixture-of-Experts

A popular hierarchical learning model is the *hierarchical mixture-of-experts* (HME) network [167]. Many theoretical derivations of HME, which is basically a Bayesian belief network, have been proposed (e.g., [161, 167, 389, 390]). Given a pattern, each expert network estimates the pattern's conditional posterior probability on local areas of the input space, and the gating network computes the conditional probability, which represents the likelihood of the expert subnet producing the correct answer. The final output is the weighted sum of the estimated probabilities from all the expert networks.

Figure 6.6 illustrates the architecture of a K-output HME classifier. The experts and gating networks in the HME can take any form as long as the information about the probability-generating function for each expert is known. In most cases, the HME classifier adopts the generalized linear models (GLIM) [167, 373] for the expert and gating networks. The GLIM model is described next.

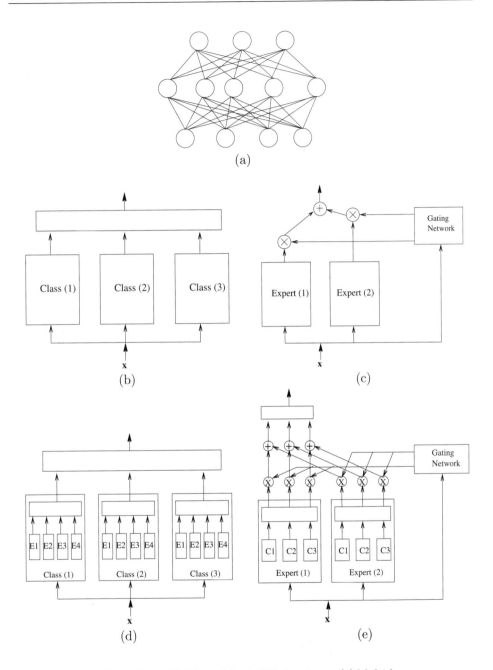

Figure 6.5. Several possible hierarchies in NN structures: (a) Multi-layer perceptrons, (b) decision-based neural network, (c) MOE network, (d) experts-in-class
network, and (e) classes-in-expert network.

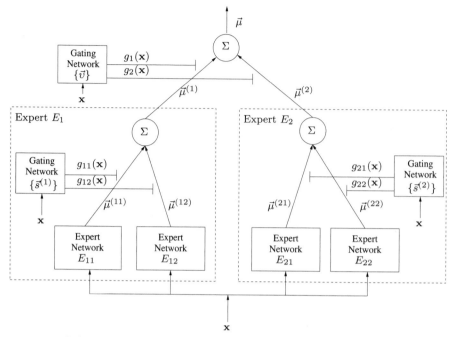

x : D-dimensional input vectors
$\vec{\mu}^{(ij)}$: K-dimensional output vectors from the lower-level expert networks
$\vec{\mu}^{(i)}$: K-dimensional output vectors from Expert E_i
$\vec{\mu}$: K-dimensional output vectors of the HME network

Figure 6.6. Architecture of an HME classifier.

To perform multiway classification over K classes, Jordan and Jacobs [167] and Waterhouse and Robinson [373] proposed that each expert network either adopts K-independent binomial models, each modeling only one class, or one multinomial model for all classes. For the former, the form of the binomial model in expert E_j is

$$\mu_k^{(j)} = \frac{1}{1 + \exp\left\{-(\vec{\eta}_k^{(j)})^T \mathbf{x}\right\}} \qquad k = 1, \ldots, K, \tag{6.4.1}$$

where $\mu_k^{(j)}$ is the k-th element of the output vector $\vec{\mu}^{(j)}$, and $\vec{\eta}_k^{(j)}$ is the parameter vector associated with output node k. This is clearly reminiscent of the single-layer perceptron, with logistic output functions.

As for the multinomial model, the output of the lower-level expert E_{ij} is

$$\mu_k^{(ij)}(\mathbf{x}) = \frac{\exp\left\{\left(\vec{\psi}_k^{(ij)}\right)^T \mathbf{x}\right\}}{\sum_r \exp\left\{\left(\vec{\psi}_k^{(ir)}\right)^T \mathbf{x}\right\}}, \tag{6.4.2}$$

where $\vec{\psi}_k^{(ij)}$ is the k-th vector of the weight matrix $\Psi^{(ij)}$ corresponding to Expert E_{ij} for class ω_i. This is equivalent to a single-layer perceptron with K outputs and a softmax output function. Richard and Lippmann [311] have shown that Eq. 6.4.2 can lead to the estimation of posterior probabilities. For simplicity's sake, the output of each expert network will henceforth be considered as the posterior probability estimated in a local area in the input space; that is, given an input vector $\mathbf{x}$, the k-th output in expert network E_{ij} is

$$\mu_k^{(ij)}(\mathbf{x}) = P(\omega_k|\mathbf{x}, E_{ij}). \tag{6.4.3}$$

The HME also uses an auxiliary network known as the *gating network*, which partitions the input space into regions corresponding to the various expert networks. The j-th output of g_{ij} of the lower-level gating network models the probability that Expert E_{ij} produces the correct answer, given the input vector $\mathbf{x}$—that is, $g_{ij}(\mathbf{x}) = P(E_{ij}|\mathbf{x})$. Jordan and Jacobs [167] implement the gating network by the softmax function:

$$P(E_{ij}|\mathbf{x}) = g_{ij}(\mathbf{x}) = \frac{\exp\left\{\left(\vec{s}_j^{(i)}\right)^T \mathbf{x}\right\}}{\sum_r \exp\left\{\left(\vec{s}_r^{(i)}\right)^T \mathbf{x}\right\}}, \tag{6.4.4}$$

where $\vec{s}_j^{(i)}$ is the j-th weight vector of the gating network corresponding to Expert E_i. By combining Eqs. 6.4.3 and 6.4.4, the k-th output of the Expert E_i in the HME network is the following:

$$\mu_k^{(i)} = P(\omega_k|\mathbf{x}, E_i) = \sum_j P(E_{ij}|\mathbf{x})P(\omega_k|\mathbf{x}, E_{ij}) = \sum_j g_{ij}(\mathbf{x})\mu_k^{(ij)}(\mathbf{x}). \tag{6.4.5}$$

Notice that both the expert outputs and network weights (gating outputs) are functions of input $\mathbf{x}$.

The higher-level gating network in the HME combines the outputs of the lower-level experts as follows:

$$\vec{\mu} = \sum_i g_i(\mathbf{x})\vec{\mu}^{(i)}, \tag{6.4.6}$$

where $\vec{\mu}$ is a K-dimensional output vector; and

$$g_i(\mathbf{x}) = P(E_i|\mathbf{x}) = \frac{e^{\vec{v}_i^T \mathbf{x}}}{\sum_r e^{\vec{v}_r^T \mathbf{x}}}, \tag{6.4.7}$$

where $\vec{v}_i$ is the i-th weight vector of the higher-level gating network. For K-class classification problem, $\mathbf{x}$ is assigned to the k^*-th class if

$$k^* = \arg\max_k \mu_k.$$

The training rules for the HME are based on the EM algorithm. The following details the procedures for estimating the parameters of the high-level expert. The extension to low-level experts is trivial. Suppose there are M experts and that each expert has K outputs. Define $\Upsilon^{(j)} = \{\vec{\eta}_i^{(j)}\}_{i=1}^K$ as the parameter set for the expert network E_j. Also define $V = \{\vec{v}_j\}_{j=1}^M$ as the parameter set for the gating network. The output vector of the whole HME network is defined as $\mathbf{y} = [y_1, y_2, \ldots, y_K]^T$. If the input pattern $\mathbf{x}$ belongs to class ω_i, then $y_i = 1$, $y_j = 0$, and $\forall j \neq i$. The likelihood function of the network output $\mathbf{y}$ is

$$p(\mathbf{y}|\mathbf{x}) = \sum_{j=1}^{M} g_j(\mathbf{x})p(\mathbf{y}|\mathbf{x}, E_j), \qquad (6.4.8)$$

where $p(\mathbf{y}|\mathbf{x}, E_j)$ is the output likelihood function conditional on the expert network E_j. The $p(\mathbf{y}|\mathbf{x}, E_j)$ is defined as follows:

$$p(\mathbf{y}|\mathbf{x}, E_j) = P(\omega_1|\mathbf{x}, E_j)^{y_1} P(\omega_2|\mathbf{x}, E_j)^{y_2} \cdots P(\omega_K|\mathbf{x}, E_j)^{y_K}. \qquad (6.4.9)$$

Since only one y_i is 1 (others are 0), the likelihood function of the *output* $\mathbf{y}$ is equal to the i-th posterior probability of the *input* $\mathbf{x}$, that is,

$$p(\mathbf{y}|\mathbf{x}, E_j) = P(\omega_i|\mathbf{x}, E_j) \quad \text{if } \mathbf{x} \in \omega_i. \qquad (6.4.10)$$

The EM algorithm is used to maximize the following likelihood function:

$$L = \sum_{t=1}^{N} \log p(\mathbf{y}(t)|\mathbf{x}(t)). \qquad (6.4.11)$$

Let $\mathbf{W}$ represents the set of parameters in the HME network ($\mathbf{W} = \{\{\Upsilon^{(j)}\}_{j=1}^M, V\}$). Given the current estimate $\bar{\mathbf{W}}$, the EM procedure for maximizing Eq. 6.4.11 consists of two steps.

1. *E-step*. First, for each pair $\{\mathbf{x}(t), \mathbf{y}(t)\}$, compute

$$h_j(t) = \frac{g_j(\mathbf{x}(t))p(\mathbf{y}(t)|\mathbf{x}(t), E_j)}{\sum_{r=1}^{M} g_r(\mathbf{x}(t))p(\mathbf{y}(t)|\mathbf{x}(t), E_r)} \qquad (6.4.12)$$

using the current estimates $\bar{\mathbf{W}} = \{\{\bar{\Upsilon}^{(j)}\}_{j=1}^M, \bar{V}\}$. A new set of objective functions is then formed:

$$\begin{aligned} Q_j^e(\Upsilon^{(j)}|\bar{\Upsilon}^{(j)}) &= \sum_{t=1}^N h_j(t) \log p(\mathbf{y}(t)|\mathbf{x}(t), E_j) \quad j = 1, \ldots, M \\ Q^g(V|\bar{V}) &= \sum_{t=1}^N \sum_{j=1}^M h_j(t) \log g_j(\mathbf{x}(t)), \end{aligned} \qquad (6.4.13)$$

where $\Upsilon^{(j)}$ and V are dependent variables.

2. *M-step.* Find a new estimate $\mathbf{W}' = \{\{\Upsilon^{(j)'}\}_{j=1}^M, V'\}$ with

$$
\begin{aligned}
\Upsilon^{(j)'} &= \arg\max_{\Upsilon^{(j)}} Q_j^e(\Upsilon^{(j)}|\bar{\Upsilon}^{(j)}), \quad j = 1, \ldots, M \\
V' &= \arg\max_V Q^g(V|\bar{V}).
\end{aligned}
\tag{6.4.14}
$$

Then, the old estimates $\{\{\bar{\Upsilon}^{(j)}\}_{j=1}^M, \bar{V}\}$ are replaced by the new ones $\{\{\Upsilon^{(j)'}\}_{j=1}^M, V'\}$ and the EM cycle is repeated.

Chapter 3 provides a detailed discussion on the EM algorithm.

Notice that due to the nonlinearity of the softmax functions, it is impossible to obtain an analytical solution of $\max_{\Upsilon^{(j)}} Q_j^e(\Upsilon^{(j)}|\bar{\Upsilon}^{(j)})$ and $\max_V Q^g(V|\bar{V})$. Therefore, Jordan and Jacobs [167] use a gradient ascent method called the *iteratively reweighted least-squares* (IRLS) algorithm in the M-step optimization.

Unlike the Jordan and Jacobs proposal, where the exact values of y_k are needed, this method requires only the class labels. Eq. 6.4.9 expresses the log-likelihood function conditional on the expert E_j as

$$
\log p(\mathbf{y}|\mathbf{x}, E_j) = \sum_{k=1}^K y_k \log P(\omega_k|\mathbf{x}, E_j).
\tag{6.4.15}
$$

By substituting Eq. 6.4.15 into Eq. 6.4.13, $Q_j^e(\Upsilon^{(j)}|\bar{\Upsilon}^{(j)})$ is rewritten as

$$
Q_j^e(\Upsilon^{(j)}|\bar{\Upsilon}^{(j)}) = \sum_{t=1}^N \sum_{k=1}^K h_j(t) y_k \log P(\omega_k|\mathbf{x}(t), E_j).
\tag{6.4.16}
$$

Since $y_i = 1$ if $\mathbf{x}(t) \in \omega_i$ and $y_i = 0$, if $\mathbf{x}(t) \notin \omega_i$, Eq. 6.4.16 becomes

$$
\begin{aligned}
Q_j^e(\Upsilon^{(j)}|\bar{\Upsilon}^{(j)}) = \quad &\sum_{\mathbf{x}(t)\in\omega_1} h_j(t) \log P(\omega_1|\mathbf{x}(t), E_j) \\
&+ \sum_{\mathbf{x}(t)\in\omega_2} h_j(t) \log P(\omega_2|\mathbf{x}(t), E_j) \\
&\vdots \\
&+ \sum_{\mathbf{x}(t)\in\omega_K} h_j(t) \log P(\omega_K|\mathbf{x}(t), E_j).
\end{aligned}
\tag{6.4.17}
$$

Then the derivatives of Q_j^e in Eq. 6.4.17 with respect to $\vec{\eta}_i^{(j)}$ are

$$
\begin{aligned}
\frac{\partial Q_j^e(\Upsilon^{(j)}|\bar{\Upsilon}^{(j)})}{\partial \eta_i^{(j)}} = \quad &\sum_{\mathbf{x}(t)\in\omega_i} h_j(t) \cdot \mathbf{x}(t) \cdot (1 - P(\omega_i|\mathbf{x}(t), E_j)) \\
&- \sum_{\mathbf{x}(t)\notin\omega_i} h_j(t) \cdot \mathbf{x}(t) \cdot P(\omega_i|\mathbf{x}(t), E_j),
\end{aligned}
\tag{6.4.18}
$$

where these relationships are used

$$
P(\omega_i|\mathbf{x}, E_j) = \mu_i^{(j)} = \frac{\exp\left\{\left(\vec{\eta}_i^{(j)}\right)^T \mathbf{x}\right\}}{\sum_k \exp\left\{(\vec{\eta}_k^{(j)})^T \mathbf{x}\right\}},
$$

$$\frac{\partial}{\partial \eta_i^{(j)}} \log P(\omega_i|\mathbf{x}, E_j) = (1 - P(\omega_i|\mathbf{x}, E_j))\mathbf{x},$$

and

$$\frac{\partial}{\partial \eta_i^{(j)}} \log P(\omega_k|\mathbf{x}, E_j) = -P(\omega_i|\mathbf{x}, E_j)\mathbf{x} \qquad \forall k \neq i.$$

The gradients for the gating network can be obtained in the same way. Eq. 6.4.18 shows that the gradient values are proportional to the estimation errors—$(1 - P(\omega_i|\mathbf{x}(t), E_j))$ in the first row or $-P(\omega_i|\mathbf{x}(t), E_j)$ in the second row—and that the ratio of the contribution of this expert to the final estimation results in $(h_j(t))$. Because the IRLS algorithm does not guarantee the updating process to converge to a global maximum, the learning algorithm used for the HME is actually not the EM, but the generalized EM (GEM) algorithm. Unlike the EM algorithm, in which convergence is guaranteed without setting learning parameters, some safeguard measures (e.g., appropriate choice of learning rate) are required for the IRLS algorithm to converge. Moreover, since the learning procedure for the HME has double training loops (at each M-step there are several iterations for IRLS), computation cost increases considerably.

6.4.2   Experts-in-Class and Classes-in-Expert Structures

An effective implementation of neural networks (NNs) hinges on (a combination of) *locally distributed* and *hierarchical* networks. Local and distributed processing is critical to fault tolerance and robustness in neural networks. A hierarchical design, on the other hand, often results in a more efficient network structure. Proper incorporation of *expert-level modules* and *class-level modules* into a hierarchical NN can improve computation and performance. A hierarchical NN comprises a variety of neural processing modules for which EM serves as a basic tool. In addition, the decision-based learning rule has proved to be effective in implementing a global (i.e., interclass) supervised training scheme. The structural designs and associated learning algorithms for decision-based neural networks are detailed in Chapter 7. There are many recent important applications involving fusion of information from completely different sources. The hierarchical NN structure can easily be extended to cope with multichannel information processing. Hierarchical neural networks with an embedded fusion agent offer an effective approach to channel fusion.

As mentioned in the previous sections, the MOE and HME adopt the *expert-level partitioning strategies*, whereas the DBNNs, GMMs, and EBFNs adopt the *class-level partitioning strategies*. A hierarchical NN configuration should allow the incorporation of both the expert-level and class-level modules. As discussed next, the selection of the *inner blocks* versus the *outer blocks* will lead to very distinctive hierarchical structures.

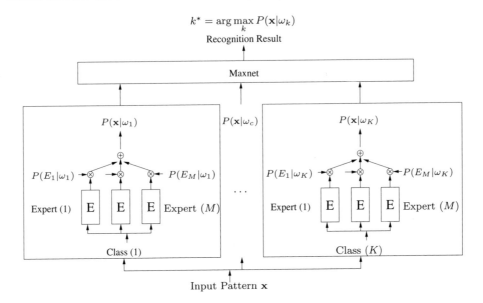

$$k^* = \arg\max_k P(\mathbf{x}|\omega_k)$$

Recognition Result

Figure 6.7. Experts-in-class hierarchy for K-class classification problems. When applications have several experts or information sources, an experts-in-class hierarchical scheme can be applied; gating networks are omitted in the drawing. Like the baseline DBNN model, the minimal updating principle can be adopted to train the parameters in the network. $P(E_j|\omega_c)$ serves as the confidence indicator for the j-th expert in class c. It is a trainable parameter, and its value is fixed during retrieval time.

Experts-in-Class Hierarchical Structures

Figure 6.7 depicts the experts-in-class structure. The inner blocks comprise expert-level modules, while the outer blocks focus on the class level. A typical example of this type of network is the hierarchical DBNN [211], which describes the class discriminant function as a mixture of multiple probabilistic distributions; that is, the discriminant function of the class ω_c in the hierarchical DBNN is a *class-conditional likelihood density* $p(\mathbf{x}|\omega_c)$:

$$p(\mathbf{x}|\omega_c) = \sum_{j=1}^{M} P(E_j|\omega_c)p(\mathbf{x}|\omega_c, E_j), \qquad (6.4.19)$$

where $p(\mathbf{x}|\omega_c, E_j)$ is the discriminant function of subnet c in expert j, and $p(\mathbf{x}|\omega_c)$ is the combined discriminant function for class ω_c. The expert confidence $P(E_j|\omega_c)$ can be learned from the following EM algorithm. Define $\alpha_j = P(E_j|\omega_c)$ and set the

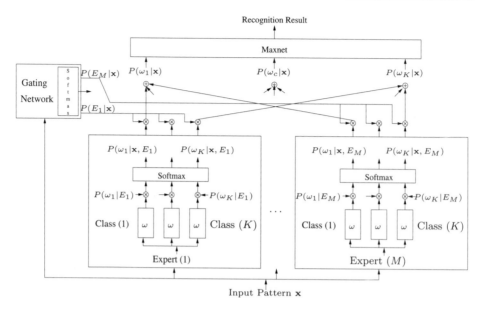

Figure 6.8. Classes-in-expert hierarchy for K-class classification problems. In this scheme, the expert weighting parameters ($P(E_j|\mathbf{x})$) are functions of the input pattern $\mathbf{x}$. The prior probability $P(\omega_i|E_j)$ could either be preassigned or estimated in a similar fashion as other prior probabilities such as $P(E_j|\omega_c)$ (cf. Eq. 6.4.20). This network can be viewed as a hybrid model of MOE and DBNN, where local expert networks serve as local classifiers. When applying such a hierarchical NN to channel fusion, each channel module can be regarded as a local expert with predetermined feature space.

initial value of $\alpha_j = 1/M$, $\forall j = 1, \ldots, M$. The result at the m-th iteration is

$$
h_j^{(m)}(\mathbf{x}_t) = \frac{\alpha_j^{(m)} p(\mathbf{x}_t|\omega_c, E_j)}{\sum_l \alpha_l^{(m)} p(\mathbf{x}_t|\omega_c, E_l)} \quad \text{and} \quad \alpha_j^{(m+1)} = \frac{1}{N} \sum_{t=1}^{N} h_j^{(m)}(\mathbf{x}_t). \quad (6.4.20)
$$

Each expert processes only the local features from its corresponding class. The outputs from different experts are linearly combined. The weighting parameters, $P(E_j|\omega_c)$, represent the confidence of expert E_j producing the correct answer for the object class ω_c. Once they are trained in the learning phase, their values remain *fixed* during the retrieving (identification) phase. By definition, $\sum_{j=1}^{M} P(E_j|\omega_c) = 1$, where K is the number of experts in the subnet ω_c, so it has the property of a probability function. In conjunction with the expert-level (or rule-level) hierarchy, each hidden node within one class subnet can be used to model a certain local expert (or rule) with a varying degree of confidence, which reflects its ability to interpret a given input vector. The locally unsupervised and globally supervised scheme described in Section 7.4.2 can be adopted to train the expert networks (cf. Figure 7.8).

Classes-in-Expert Hierarchical Structures

Figure 6.8 depicts the classes-in-expert structure. The inner blocks comprise class modules, and the outer blocks are the expert modules. Each expert has its own hierarchical DBNN classifier. The outputs of the hierarchical DBNNs are transformed to the posterior probabilities by softmax functions. In this scheme, the expert weighting $P(E_j|\mathbf{x})$ is a function of input pattern $\mathbf{x}$. Therefore, the importance of individual experts may vary with different input patterns observed.

This network adopts the posterior probabilities of electing a class given $\mathbf{x}$ (i.e., $P(\omega_c|\mathbf{x}, E_j)$)—instead of the likelihood of observing $\mathbf{x}$ given a class (i.e., $p(\mathbf{x}|\omega_c, E_j)$)—to model the discriminant function of each cluster. A new confidence $P(E_j|\mathbf{x})$ is also assigned, which stands for the confidence on Expert E_j when the input pattern is $\mathbf{x}$. Accordingly, the probability model is modified to become

$$P(\omega_c|\mathbf{x}) = \sum_{j=1}^{M} P(E_j|\mathbf{x})P(\omega_c|\mathbf{x}, E_j), \tag{6.4.21}$$

where

$$P(\omega_c|\mathbf{x}, E_j) = \frac{P(\omega_c|E_j)p(\mathbf{x}|\omega_c, E_j)}{p(\mathbf{x}|E_j)} \tag{6.4.22}$$

with

$$p(\mathbf{x}|E_j) = \sum_{i=1}^{K} P(\omega_i|E_j)p(\mathbf{x}|\omega_i, E_j). \tag{6.4.23}$$

The confidence $P(E_j|\mathbf{x})$ can be obtained by the following equation:

$$P(E_j|\mathbf{x}) = \frac{P(E_j)p(\mathbf{x}|E_j)}{\sum_{l=1}^{M} P(E_l)p(\mathbf{x}|E_l)}. \tag{6.4.24}$$

Eq. 6.4.22 is one type of the *softmax* function. Its weighting parameter $P(\omega_c|E_j)$, just like $P(E_j|\omega_c)$ in Eq. 6.4.19, is made to satisfy the property of a probability function (i.e., $\sum_c P(\omega_c|E_j) = 1$). Consequently, $P(\omega_c|E_j)$ can be recursively estimated by the EM iterations in Eq. 6.4.20. The term $P(E_j)$ can be interpreted as the *confidence-level* on Expert E_j, which can be learned by the EM iterations very similar to Eq. 6.4.20. The confidence level $P(E_j)$ provides the key factor affecting the fusion weight $P(E_j|\mathbf{x})$, which is the output of the softmax gating network. It can also be shown that $P(E_j)$ can be learned by Eq. 6.4.20 with slight modification. Unlike the experts-in-class approach, the fusion weights need to be computed for each testing pattern during the retrieval phase.

6.5 Biometric Authentication Application Examples

There are a number of examples in which a MOE has been successfully applied to face and speaker authentication systems. For example, in Gutta et al. [124], the mixture-of-experts is implemented based on the divide-and-conquer modularity principle, taking fully into account the granularity and locality of information. The proposed MOE consists of ensembles of local RBF experts, and a gating network is implemented by inductive decision trees and support vector machines (SVMs). The gating network is responsible for deciding which expert should be used to determine classification output. The learning model was successfully applied to pose classification and gender and ethnic classification of human faces.

In Zhang and Guo [401], a modular structure built on autoassociative RBF modules was applied to the XM2VTS and ORL face database and reportedly achieved satisfactory face recognition rates. Kuritaa and Takahashib [193] applied MOE to face recognition, in which viewpoint-dependent classifiers (experts) were independently trained based on facial images captured from different angles. During the recognition phase, classifier outputs from different viewpoints were combined by a gating network. Experiments were conducted on recognizing the face of 10 individuals, each taken from 25 different angles. Given a facial image taken at a particular angle, one of the 25 viewpoint classifiers could be properly selected. Nevertheless, a mixture of classifiers seemed to outperform a single classifier.

The HME model has been applied to system identification [167], speech recognition [373], motion estimation [375], face recognition, and speaker recognition. In the latter case, a modified hierarchical mixture-of-experts (HME) architecture was applied to text-dependent speaker identification [54]. The expectation-maximization (EM) algorithm (see Chapter 3) was adopted to train the HME model. A novel gating network was introduced to help incorporate instantaneous and transitional spectral information into text-dependent speaker identification. The authors further extended the modified HMEs to the fusion of diverse speech features—including different orders of LPCCs, delta LPCCs, MFCCs, and delta MFCCs (see Section 2.6.1)—in a speaker identification task [53]. The idea is based on the understanding that no unique speech feature is highly superior to others for all kinds of speech. Therefore, it is reasonable to expect that a probabilistic fusion of diverse features should achieve a better performance than a single feature. The fusion is carried out by dividing N expert networks into K groups for K-diverse features so that networks in the same group receive the same feature vector, while expert networks in different groups receive different feature vectors. Likewise, there are K gating networks, each of which receives a different set of feature vectors. The outputs of the gating networks are N-component weighting vectors used for weighing the contribution of the N expert networks. Simulation results demonstrate that using an HME to fuse different features achieves better performance than using an HME on a single feature or on a composite of features.

There are many other ways of generalizing the global hierarchical structure to integrate information from various local modules or from different sensors. Several successful face and speaker verification systems that apply the class-based OCON modular networks are proposed in Chapter 7. In Chapters 7 through 9, several enhanced hierarchical architectures are shown to be effective for numerous biometric applications—even for those applications with complex decision regions. More important, the hierarchical structure lends itself to a natural extension of fusion architectures amenable to multimodality and/or multisample sensor fusion. This is illustrated in Chapter 10.

Problems

1. The ACON and OCON differ significantly in the total number of synaptic weights. For convenience, all subnets of OCON are assumed to have a uniform size, say k. The number of hidden units in the ACON supernetwork is denoted as K. We denote the input dimension and output dimension as n and N respectively. Verify the following analyses.

   (a) When the number of outputs N is relatively small with respect to n, the ratio between the numbers of ACON weights and OCON weights is

   $$\frac{\text{number of ACON weights}}{\text{number of OCON weights}} = \frac{(N+n) \times K}{N \times n \times k} \approx \frac{l}{N},$$

   where

   $$l = \frac{K}{k} < N.$$

   Therefore, ACON should have fewer weights, and the two numbers of weights are compatible if $K \approx N \times k$.

   (b) When N is very large (compared with n), the ratio becomes

   $$\frac{\text{number of ACON weights}}{\text{number of OCON weights}} \approx \frac{l}{n}.$$

   When N is very large, it is plausible that

   $$n < l.$$

   Therefore, OCON should have an advantage in terms of the number of weights used.

2. Design an MOE model to classify the two distinctive regions shown in Figure 6.4. It is required that the model make proper use of two linear experts depicted in the figure.

3. A hierarchical neural network of the experts-in-class architecture has eight inputs, four outputs, and two experts.

   (a) Draw a flow graph of this network.

   (b) State the conditions under which the experts-in-class architecture of the network degenerates to the EBF-based classifier shown in Figure 5.6.

   (c) For the retrieval phase of the network in (b), determine the computational complexity in term of MACs (multiplier-accumulations).

(d) Repeat (a) and (b) if the network has a classes-in-expert architecture.

4. Construct a four-input and two-output hierarchical network with three levels

   (a) based on the experts-in-class architecture using two experts per class.

   (b) based on the classes-in-expert architecture, assuming three experts are used.

5. Compare the expert-based versus class-based modular networks in terms of overall computation costs.

   (a) Identify plausible application scenarios such that the expert-based modular networks hold a clear computational advantage over class-based models.

   (b) Conversely, describe possible scenarios under which the class-based models could become more advantageous than the expert-based models.

6. There are three learning paradigms for neural networks: (1) supervised learning, (2) unsupervised learning, and (3) combined supervised/unsupervised learning. Identify which learning paradigm would be most accurate in describing the following neural models:

   (a) EM

   (b) SVM

   (c) LBF multi-layer network

   (d) RBF multi-layer network

   (e) MOE

   (f) HME

   Give a brief justification for each case.

7. Design some new variants of hierarchical networks by incorporating as many as possible of the following design features:

   (a) Use of unsupervised learning paradigms such as principle components, K-means/EM clustering, and other competition-based learning rules.

   (b) Use of supervised learning paradigms such as reinforced/antireinforced learning and positive versus negative training strategies.

   (c) Soft versus hard memberships (i.e., soft vs. hard decision) in local levels.

   (d) Use of prior information.

Chapter 7

DECISION-BASED NEURAL NETWORKS

7.1 Introduction

In supervised training, training patterns are provided in input/teacher pairs, $[\mathcal{X}, \mathcal{T}] = \left\{ [\mathbf{x}^{(1)}, \mathbf{t}^{(1)}], [\mathbf{x}^{(2)}, \mathbf{t}^{(2)}], \ldots, [\mathbf{x}^{(N)}, \mathbf{t}^{(N)}] \right\}$. In a decision-based network, the teacher's values are exclusively the class membership or class symbols, as opposed to any quantitative or numeric values. Consequently, in training a decision-based neural network (DBNN), the teachers do not know (and need not know) the network's exact quantitative output values. More precisely, a decision-based network receives instruction from the teacher on the correct or preferred decision and performs the training process accordingly. This chapter introduces the (supervised) decision-based learning rule and shows how such a learning rule may be combined with the (unsupervised) EM clustering for training hierarchical networks. The results are several variants of the basic and probabilistic decision-based neural networks (PDBNNs) that show promise for biometric authentication.

7.2 Basic Decision-Based Neural Networks

In the binary classification problem, the pattern space is divided into two regions, and each class occupies its own region. In the clearly separable case, the two classes are separated by the *decision boundary*, defined as the hypersurface on which the two discriminant functions have equal scores. The objective of the learning phase is to determine the best discriminant functions, which in turn dictate decision boundaries.

A pioneering decision-based neural model is the *perceptron* originally proposed by Rosenblatt [322]. The (linear) perceptron was designed to separate two classes by a linear decision boundary (cf. Figure 7.1(a)). To deal with a more flexible decision boundary, the linear decision boundary is often replaced by more nonlinear boundaries. Figure 7.1(b) depicts nonlinear decision boundaries in the pattern space.

An example of a two-class decision-based network is illustrated in Figure 7.2(a),

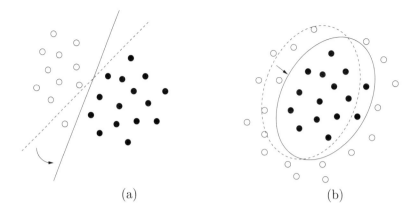

(a)   (b)

Figure 7.1. Adjustment of (a) linear and (b) nonlinear decision boundaries. The gradient-type adjustment of the weights cannot be explicitly reflected in the figure because it is not drawn on the **w**-plane.

where the input patterns are to be classified into two groups—*accept: (A)* or *reject: (R)*. The output of the first subnet $\mathbf{y}_A$ is a function of the input $\mathbf{x}$ and the weights $\mathbf{w}_A$:

$$\mathbf{y}_A = \phi_A(\mathbf{x}, \mathbf{w}_A).$$

This is the *discriminant function* of the subnet. Similarly, the second subnet has a discriminant function:

$$\mathbf{y}_R = \phi_R(\mathbf{x}, \mathbf{w}_R).$$

The classification is based on the values of the discriminant functions. More precisely, if

$$\phi_R(\mathbf{x}, \mathbf{w}_R) > \phi_A(\mathbf{x}, \mathbf{w}_A),$$

then the pattern is classified as "A"; otherwise, it is classified as "R." The teacher in Figure 7.2(a) points out the correct class for each training pattern—A or R. In the baseline DBNN, no training is needed if the decision is already correct. Only when the decision is incorrect do the weights ($\mathbf{w}_A$ and $\mathbf{w}_R$) require further updating. Once the network completes the learning phase, the network is ready for use in the *retrieving phase*, during which the network decides the class of unknown patterns based on the trained discriminant functions. As shown in Figure 7.2(b), the binary classification model can be reduced into one network with the discriminant function $\phi(\mathbf{x}, \mathbf{w}) = \phi_A(\mathbf{x}, \mathbf{w}_A) - \phi_R(\mathbf{x}, \mathbf{w}_R)$.

7.2.1 Decision-Based Learning Rule

For most pattern recognition applications, the ultimate goal is to correctly assign an input pattern to its proper class. Therefore, it is natural to make use of the class-

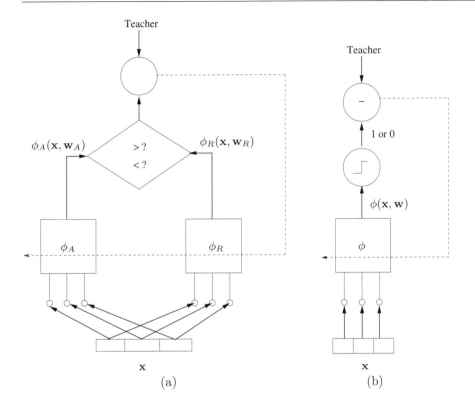

Figure 7.2. (a) Illustration of a two-class DBNN, which can be trained to determine whether to accept (A) or reject (R) an input. In decision-based training, the teacher indicates whether a correct classification has been made. The update strategy depends on the decision. (b) The binary classification model can be reduced into one network by defining a discriminant function of the form $\phi(\mathbf{x}, \mathbf{w}) = \phi_A(\mathbf{x}, \mathbf{w}_A) - \phi_R(\mathbf{x}, \mathbf{w}_R)$.

based modular structure. For multiclass classification, the design of a decision-based neural network follows two important guidelines:

- *OCON Network Structure.* For multiclass classification, K subnets are used for a K-class classification problem. Figure 7.3 shows a modular *one-class-one-network* (OCON) structure, where one class is assigned to each of the K subnets. The outputs of the subnets are the *discriminant functions.* The outputs will compete with each other, and the winner will claim the identity of the input pattern.

- *Decision-Based Learning Rule.* The teacher reveals only the correct or preferred classification for each training pattern. In decision-based learning, based on the teacher information, a class's discriminant function will either

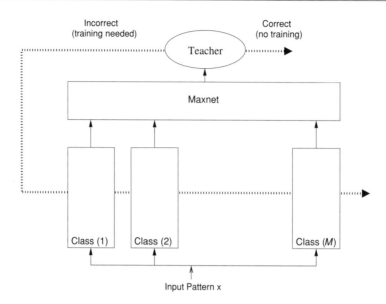

Figure 7.3. The baseline OCON structure and decision-based learning of a DBNN. In the DBNN, one subnet is designated to represent one class. The output scores of the subnets compete with each other, and the highest scoring subnet claims the identity of the input pattern. The decision-based learning rule is adopted for fine-tuning the decision boundaries between classes.

be reinforced or antireinforced upon the presentation of a new training pattern. The reinforced learning moves the network parameters, say $\mathbf{w}$, along a positive gradient direction, so the value of a discriminant function increases, enhancing the chance of the class's future selection when an input similar to the current training pattern is presented. In contrast, the antireinforced learning moves $\mathbf{w}$ along a negative gradient direction, thus lowering the value of discriminant function and thereby suppressing the chance of the class's selection when a similar pattern is presented in the future.

Decision-Based Credit-Assignment Principle

The decision-based learning rule is built on the following distributive credit-assignment principle:

- *When to update?* In the decision-based learning rule, weight updating is performed only when misclassification occurs.

- *Which subnets to update?* The learning rule is distributive and localized. It applies *reinforced learning* to the subnet corresponding to the correct class and *antireinforced learning* to the (unduly) winning subnet.

- *How to update?* Because the decision boundary depends on the discriminant function $\phi(\mathbf{x}, \mathbf{w})$, it is natural to adjust the boundary by adjusting the weight vector $\mathbf{w}$ either in the direction of the gradient of the discriminant function (i.e., reinforced learning) or opposite to that direction (i.e., antireinforced learning)

$$\Delta\mathbf{w} = \pm\eta\nabla\phi(\mathbf{x}, \mathbf{w}), \tag{7.2.1}$$

where η is a positive learning rate. The gradient vector of the function ϕ with respect to $\mathbf{w}$ is denoted as follows:

$$\nabla\phi(\mathbf{x}, \mathbf{w}) = \frac{\partial\phi(\mathbf{x}, \mathbf{w})}{\partial\mathbf{w}}.$$

The fundamental principle of decision-based neural networks has been studied previously by several researchers [105, 184, 262, 322]. The *decision-based learning rule* outlined below is based on a formulation proposed by Kung and Taur [191].

Decision-Based Training Algorithm

Suppose that $S = \{\mathbf{x}^{(1)}, \ldots, \mathbf{x}^{(N)}\}$ is a set of given training patterns, each corresponding to one of the L classes $\{\Omega_i, i = 1, \ldots, L\}$. Each class is modeled by a subnet with discriminant functions, say $\phi(\mathbf{x}, \mathbf{w}_i)$, $i = 1, \ldots, L$. Suppose that the n-th training pattern $\mathbf{x}^{(n)}$ is known to belong to class Ω_i and

$$\phi(\mathbf{x}^{(n)}, \mathbf{w}_j^{(n)}) > \phi(\mathbf{x}^{(n)}, \mathbf{w}_l^{(n)}) \quad \forall l \neq j. \tag{7.2.2}$$

That is, the winning class for the pattern is the j-th class (subnet).

1. When $j = i$, then the pattern $\mathbf{x}^{(n)}$ is already correctly classified and no update is needed.

2. When $j \neq i$, that is, $\mathbf{x}^{(n)}$ is misclassified, then the following update is performed:[1]

$$\begin{array}{ll} \text{Reinforced Learning:} & \mathbf{w}_i^{(n+1)} = \mathbf{w}_i^{(n)} + \eta\nabla\phi(\mathbf{x}, \mathbf{w}_i) \\ \text{Antireinforced Learning:} & \mathbf{w}_j^{(n+1)} = \mathbf{w}_j^{(n)} - \eta\nabla\phi(\mathbf{x}, \mathbf{w}_j) \end{array} \tag{7.2.3}$$

In the preceding decision-based learning rule, $\mathbf{w}_k^{(n+1)} = \mathbf{w}_k^{(n)}$ for all $k \neq i$ and $k \neq j$. In other words, only the weights of the rival subnet (which is "supposed to win") is adapted, and the weights of all other subnets remain unchanged. Just like the linear perceptron, the N training patterns can be repeatedly used for as many sweeps as required for convergence.

---

[1] The algorithm remains valid with the following modification: If there is more than one integer j', $j' \neq i$, such that $\phi(\mathbf{x}^{(n)}, \mathbf{w}_{j'}^{(n)}) > \phi(\mathbf{x}^{(n)}, \mathbf{w}_i^{(n)})$ then the antireinforced learning can be applied either to *all* such indices j' or to the highest winner j only.

Example 7.1: *RRF DBNN*

The following is an example of decision-based neural networks that uses radial basis function (RBF) as the discriminant function. An RBF discriminant function is a function of the distance between an input pattern $\mathbf{x}$ and a centroid $\mathbf{w}_l$. More specifically,

$$\phi(\mathbf{x}, \mathbf{w}_l) = -\frac{\| \mathbf{x} - \mathbf{w}_l \|^2}{2} \tag{7.2.4}$$

is used for the l-th subnet so that the centroid ($\mathbf{w}_l$) closest to the presented pattern is the winner. By applying the decision-based learning formula to Eq. 7.2.4 and noting that $\nabla \phi(\mathbf{x}, \mathbf{w}) = \mathbf{x} - \mathbf{w}$, the following learning rules can be derived:

$$\begin{aligned} \text{Reinforced Learning:} \quad & \mathbf{w}_i^{(n+1)} = \mathbf{w}_i^{(n)} + \eta(\mathbf{x} - \mathbf{w}_i^{(n)}) \\ \text{Antireinforced Learning:} \quad & \mathbf{w}_j^{(n+1)} = \mathbf{w}_j^{(n)} - \eta(\mathbf{x} - \mathbf{w}_j^{(n)}) \end{aligned} \tag{7.2.5}$$

The RBF decision-based learning is effective for many practical applications, especially for nearest-neighbor types of clustering.

Given a training pattern, there are two possible scenarios. One is that the pattern is already correctly classified by the network. In this case, there is no need to further update on the current network, and the learning process will proceed with the next training pattern. Another scenario is that the pattern is classified incorrectly to another winning class. In this case, two subnets must be updated. The score of the (incorrectly) winning class should be lowered by the antireinforced learning rule, while the score of the correct (but not winning) class should be enhanced by the reinforced learning rule. Since the training process is restricted to two subnets only, the learning rule is said to be based on a minimal updating principle, which is meant to avoid unnecessary overtraining and thus reduce unwanted side effects.

7.2.2 Comparison of MOE and DBNNs

The MOE and DBNNs share many similarities. For example, both modular structures are based on the divide-and-conquer modularity principle and both employ the EM algorithm in the training phase. Substantial differences also exist between them.

- *Network properties.* Each expert network in the MOE estimates the conditional posterior probabilities for all pattern classes. The output of a local expert is ready to make the classification based on its own local information and expert's perspective. This characteristic suggests that the *interclass* communication/decision exists in the local network level under the MOE framework. In contrast, each neuron in a DBNN estimates the *class*-conditional likelihood density. The interclass communication/decision does not occur until the final subnet output is formed. This makes possible the absence of interclass communication across the (class) modules of DBNNs. (In a sense, this achieves a truly distributive processing.)

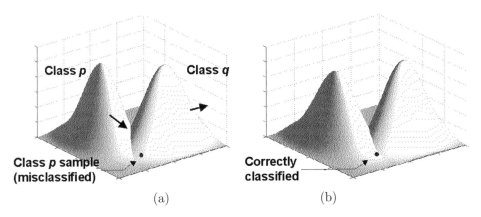

Figure 7.4. In DBNNs, the winning class is the one with the highest score function. The score functions of two competing classes (Class p and Class q) are depicted in this figure. In this example, the training pattern under consideration is marked as a dark dot. Class p is assumed to be the correct class for the dark dot.

- *Training strategies.* The training strategies of these two models are very different. During the MOE training, all training patterns have the power to update every expert. The influence from the training patterns on each expert is regulated by the gating network (which itself is under training) so that, as training proceeds, the training patterns will have more influence on the nearby experts and less influence on those far away. In contrast, unlike the MOE, the DBNNs make use of both unsupervised (EM-type) and supervised (decision-based) learning rules. The DBNNs use only the *misclassified* training patterns for its globally supervised learning. Moreover, unlike the MOE, which updates all experts, the DBNNs update only two subnets: the subnet preferred by the teacher and the incorrectly selected winning subnet.

Figure 7.4(a) shows a misclassification on the dot pattern because Class q shows a higher score than Class p. The DBNN learning rule adopts a minimal updating principle so that only the two classes immediately involved in the dispute should be updated. In Figure 7.4(a), the dot is incorrectly classified into Class q. The action required to correct such a mistake is twofold: (1) The reinforced learning rule is to be applied to Class p so as to enhance its score (e.g., by moving its centroid closer to the dark dot); (2) the antireinforced learning rule is to be applied to Class q, which results in a lower score (e.g., by moving its centroid further from the pattern). After multiple training sweeps, the boundary (respectively the pattern) should ultimately be placed on the correct side of pattern (respectively the boundary) as depicted in figure (b).

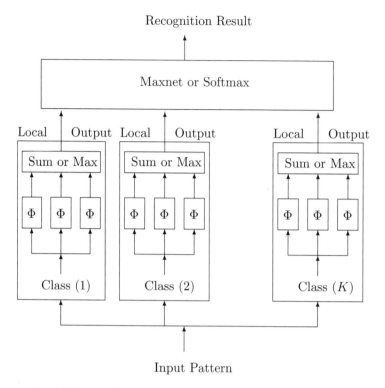

Figure 7.5. The hierarchical structure of decision-based neural networks. The basic OCON structure is adopted in the top level of the hierarchy. In the next sublevel, each subnet is further modeled by a GMM structure. The discriminant function (i.e., the output value) of a subnet is either the weighted sum or the maximum of the outputs of the subclusters within the subnet.

7.3   Hierarchical Design of Decision-Based Learning Models

The basic DBNN adopts a simple OCON structure, and K subnets are used for the classification of K categories. To facilitate a more flexible learning machine model, a two-level hierarchical network is constructed by embedding GMM into the OCON framework. More exactly, the top level still retains the basic OCON structure as shown in Figure 7.3, while in the second level each subnet is further modeled by a GMM structure, resulting in a more sophisticated hierarchical structure as shown in Figure 7.5.

7.3.1   Hybrid Supervised/Unsupervised Learning Scheme

To effectively train the hierarchical network, the learning rule should be expanded to cover two distinctive training stages. The first stage takes place in the *sublevel*, including (a) determination of the cluster structure for each class model and (b) estimation of data distributions for each induced cluster inside each class. The second stage then takes place in the *top level*, where the decision boundaries between the classes are fine-tuned by a reinforced or antireinforced supervised learning scheme.

In other words, the training of the hierarchical DBNN can be divided into two decoupled phases: *locally unsupervised* and *globally supervised* learning. During the locally unsupervised (LU) phase, each subnet is trained individually at the subnet level, and no mutual information is used across classes. Once the LU phase is completed, the training continues with the globally supervised (GS) phase, in which the teacher's preference is taken into account to further adjust the decision boundaries. The locally unsupervised globally supervised (LUGS) learning scheme can effectively combine the roles played by the supervised and unsupervised learning rules (and, in effect, complement one another) [192].

Locally Unsupervised Learning Phase

In the LU phase, each subnet is trained individually and involves the following two tasks.

1. *Determination of Subcluster Structure in Each Class.* An initial clustering can be found either by the K-means or EM methods. The K-means or VQ-type clustering algorithm offers a computationally simpler option to derive the centroids' initial locations. On the other hand, a maximum-likelihood estimation for each *class-conditional likelihood density* can be obtained via the computationally more demanding EM algorithm. Note that once the likelihood densities are obtained, the posterior probabilities can also be easily derived.

2. *Optimal Number of Subclusters for Each Class.* A critical task in the clustering phase is to determine an appropriate number of clusters for each class. If the cluster number is set too high, then noisy data or outlier patterns may excessively influence the cluster structure. If too few clusters are used, data distribution may be too coarsely or otherwise inadequately modeled. For the estimation of an optimal number of clusters, AIC and MDL represent two prominent criteria [5, 314].

Globally Supervised Learning Phase

After VQ or EM clustering, the decision-based learning rule can be applied to further fine-tune decision boundaries. Interclass mutual information (i.e., the globally supervised learning) is used to fine-tune decision boundaries. In this phase, DBNN applies a reinforced–antireinforced learning rule [192] or a discriminative learning rule [170] to adjust network parameters. In contrast to the criterion for EM clustering, which is based on the maximum-likelihood estimation, the objective of the DBNN learning aims to achieve a *minimum classification error*.

Discriminant Function. As shown in Figure 7.5, the local discriminant functions of the Gaussian subclusters are integrated to produce the output of a subnet. The integration units could be either (1) maxnet or (2) softmax. For the former, the local winner's discriminant function represents the subnet's output. For the latter, the subnet's output is expressed as a weighted sum of all of the local discriminant functions of the subclusters. More exactly,

$$\phi(\mathbf{x}, \Theta_l) = \sum_{s_l} P(s_l|l)\psi_l(\mathbf{x}, \mathbf{w}_{s_l}),$$

where the subcluster within the l-th subnet is indexed by s_l, and the parameter Θ_l is used to collectively represent all of the subcluster parameters $\{\mathbf{w}_{s_l}\}$ within the subnet. Typically, the weights must also meet the following probabilistic normalization constraint:

$$\sum_{s_l} P(s_l|l) = 1.$$

This is exemplified in the probabilistic decision-based neural networks (PDBNN), which is explored in full detail in Section 7.4.

Reinforced-Antireinforced Learning Rules. Suppose that the t-th training pattern $\mathbf{x}_t$ is known to belong to class Ω_i, and that the leading challenger is denoted as

$$j = \arg\max_{j \neq i} \phi(\mathbf{x}_t, \Theta_j),$$

where $\phi(\mathbf{x}, \Theta_j)$ is the likelihood function associated with class Ω_j. The decision-based learning rule is

$$
\begin{aligned}
\text{Reinforced Learning:} \quad & \Theta_i' = \Theta_i + \eta\nabla\phi(\mathbf{x}, \Theta_i) \\
\text{Antireinforced Learning:} \quad & \Theta_j' = \Theta_j - \eta\nabla\phi(\mathbf{x}, \Theta_j),
\end{aligned}
\tag{7.3.1}
$$

where η is a positive learning rate, and Θ_i' and Θ_j' are the updated parameters corresponding to classes i and j, respectively.

Example 7.2: *XOR Problem: A Numerical Example*

The RBF DBNN can be applied to find the decision boundary for an XOR-type problem, where four RBF clusters are distributed around the four original XOR centroids (with 30 data per cluster). Two subclusters are used to model each class. Ideally speaking, the "0-class" data should have two centroids located at (0, 0) and (1, 1), while the centroids for the "1-class" should be (1, 0) and (0, 1). In the simulation, the initial centroids are deliberately misplaced. Figure 7.6 shows how the DBNN learns to gradually adjust the centroids and decision boundaries as the adaptive iterations proceed.

7.3.2 Local Winners: Minimal Side-Effect Learning Rule

As previously mentioned, a maxnet could be adopted to select a local winner from the local discriminant functions to represent the output of the subnet. For notational convenience, here is an introduction to the notion of local winner and global

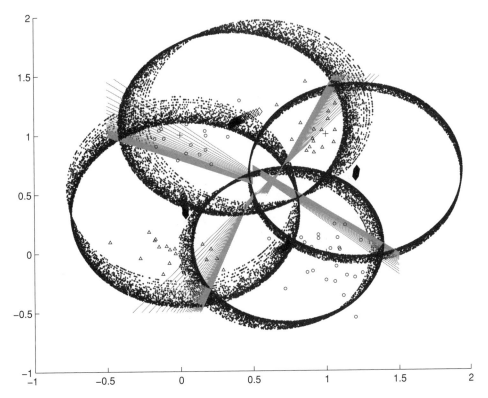

Figure 7.6. Iteration results based on decision-based learning algorithms for the XOR-type problem. Three different types of objects in the figure are all moving: (1) The trajectories of the four centroids (marked in ◊) move slowly toward the ideal centroids; (2) the territories (*big circles*) associated with of each of the four centroids are migrating in counterclockwise fashion; and (3) the linear decision boundaries (*blade-shaped*, resembling an electric fan) are also moving counterclockwise. As a result, the classification errors gradually decrease as iterations continue. For example, there are approximately 10 △-shaped patterns inside the lower-left blade. This implies that the 10 patterns were originally misclassified and that their classification is being corrected during decision-based learning iterations.

winner. The local winner is the winner among the subclusters within, say, the i-th subnet, the index of which is denoted as $\hat{s}_i$:

$$\hat{s}_i = \arg\max_{s_i} \psi_i(\mathbf{x}, \mathbf{w}_{s_i}).$$

The output value of a subnet is the maximum among the local subcluster outputs. In other words, the discriminant function of the subnet is defined to be the discriminant function of its local winner:

$$\phi(\mathbf{x}, \Theta_i) \equiv \psi_i(\mathbf{x}, \mathbf{w}_{\hat{s}_i}).$$

As depicted in Figure 7.5, the local winners continue the competition and the final winner is selected by the maxnet in the upper level. The subnet with the highest output wins the final classification. For example, the j-th subnet is declared the *global winner* if

$$\psi_j(\mathbf{x}, \mathbf{w}_{\hat{s}_j}) > \psi_k(\mathbf{x}, \mathbf{w}_{\hat{s}_k}) \quad \forall \ k \neq j.$$

In this case, the input pattern is classified into the j-th class.

Decision-Based Learning Algorithm with Local Winners

The learning rule largely follows the basic decision-based learning rule in Eq. 7.2.3. Suppose there are multiple subclusters in each class and the index of the local winners is labeled as, say $\hat{s}_i$. Suppose that a pattern $\mathbf{x}$ should belong to class Ω_i, but unfortunately the j-th subnet ($j \neq i$) is incorrectly selected as the global winner instead of the i-th subnet. Since misclassification occurs, the responsible local winners must be adjusted by the following updating algorithm:

$$\begin{aligned} \text{Reinforced Learning:} \quad & \mathbf{w}'_{\hat{s}_i} = \mathbf{w}_{\hat{s}_i} + \eta \nabla \psi(\mathbf{x}, \mathbf{w}_{\hat{s}_i}) \\ \text{Antireinforced Learning:} \quad & \mathbf{w}'_{\hat{s}_j} = \mathbf{w}_{\hat{s}_j} - \eta \nabla \psi(\mathbf{x}, \mathbf{w}_{\hat{s}_j}) \end{aligned} \qquad (7.3.2)$$

In other words, the algorithm involves local winners only. The reinforced learning is applied to the local winner of the subnet preferred by the teacher, and antireinforced learning is applied to the local winner of the ineligible winning subnet.

Comparison of LVQ and RBF DBNN

The RBF DBNNs have a strong resemblance to the well-known Learning VQ (LVQ) algorithm, proposed by Kohonen [184]. The RBF DBNN (with local winner) is especially close to the LVQ2 learning scheme. These two schemes share two attributes because they both (1) adopt a reinforced/antireinforced learning rule and (2) adopt a radial basis subcluster scheme. Therefore, both learning algorithms are based on Eq. 7.2.5 (i.e., the updating amount is proportional to the distance between the pattern and the target centroid).

The difference between DBNN and LVQ2 is in their credit assignment principles. The DBNN only learns when there is a misclassification. Moreover, DBNN and LVQ2 target different updating subclusters. The LVQ2 updates the local winner of the globally winning class and the local winner of the runner-up class. The DBNN, on the other hand, updates the local winner of the globally winning class and the local winner of the correct (or teacher-preferred) class. There are two scenarios when the preferred class is not selected as the winner. First, if the correct class, although not selected as the winner, is selected as runner-up, then DBNN and LVQ2 coincide in terms of which subnets need further updating. However, if the correct class is not even selected as runner-up, then DBNN and LVQ2 choose to train different subnets.

7.4   Two-Class Probabilistic DBNNs

A probabilistic decision-based neural network (PDBNN) (e.g., described in [208, 210–212]) is a probabilistic variant of its predecessor—the decision-based neural network (DBNN) [192]. This section describes the network structure and learning rules of the PDBNN for binary classification problems. The extension to multiclass pattern classification is presented in Section 7.5.

Consider an the example of binary classification (hypothesis testing) problems based on an independent and identical distributed (i.i.d.) observation sequence:

- Hypothesis Ω_0: $\mathbf{x}_n$ is defined by the probability distribution $p(\mathbf{x}|\Omega_0)$ for $n = 1, 2, \ldots, N$.

- Hypothesis Ω_1: $\mathbf{x}_n$ is defined by the probability distribution $p(\mathbf{x}|\Omega_1)$.

A conventional approach to binary classification can be divided into three main steps that: (1) estimate the likelihood densities $p(\mathbf{x}|\Omega_0)$ and $p(\mathbf{x}|\Omega_1)$ by statistical inference methods (parametric or nonparametric [85]); (2) compute the posterior probabilities $P(\Omega_0|\mathbf{x})$ and $P(\Omega_1|\mathbf{x})$ by the Bayes rule,

$$P(\Omega_i|\mathbf{x}) = \frac{p(\mathbf{x}|\Omega_i)P(\Omega_i)}{\sum_{j=0}^{1} p(\mathbf{x}|\Omega_j)P(\Omega_j)};$$

and (3) form a decision rule based on these posterior probabilities:

$$\begin{aligned} \mathbf{x} &\in \Omega_1 \quad \text{if } P(\Omega_1|\mathbf{x}) \geq P(\Omega_0|\mathbf{x}) \\ \mathbf{x} &\in \Omega_0 \quad \text{if } P(\Omega_1|\mathbf{x}) < P(\Omega_0|\mathbf{x}). \end{aligned} \qquad (7.4.1)$$

An equivalent way of forming a decision rule is based on the *likelihood ratio*:

$$\begin{aligned} \mathbf{x} &\in \Omega_1 \quad \text{if } \frac{p(\mathbf{x}|\Omega_1)}{P(\mathbf{x}|\Omega_0)} \geq T \\ \mathbf{x} &\in \Omega_0 \quad \text{if } \frac{p(\mathbf{x}|\Omega_1)}{p(\mathbf{x}|\Omega_0)} < T, \end{aligned} \qquad (7.4.2)$$

where the threshold T is equal to $P(\Omega_0)/P(\Omega_1)$. It can be further modified if certain risk factors are taken into consideration. Analogous to Eqs. 7.4.1 and 7.4.2, there are two types of neural network structures for binary classification problems. The first one has two output neurons, with each output designated to approximate the probability distribution of one hypothesis. The second type of structure has only one output neuron, in which the discriminant function computes the likelihood ratio. In general, PDBNNs adopt the first type of network structure; that is, one of the two subnets in the PDBNN is used for estimating the data distribution of one of the two hypotheses.

Two types of binary classification problems exist: balanced and unbalanced.

1. In a *balanced* binary classification problem, the distributions of two classes have similar complexities. One example of the balanced type is gender

classification based on facial features. For this type of two-class classification problems, it is sensible to consider using *two* subnets in the classifier, one for each class.

2. In a *unbalanced* binary classification problem, the distribution of one class is much more complex than that of the other class. One example of an unbalanced classification problem is face detection. For such kinds of unbalanced two-class problems, it is often more advisable to adopt only *one* subnet in the PDBNN classifier.

The probability distribution of all nonface patterns in the world is much more complex than the distribution of human face patterns. Estimating the distribution of nonface patterns would exhaust many resources (neurons) but still achieve less than satisfactory results. For this kind of unbalanced problem, PDBNNs emphasize obtaining a good estimate of the likelihood density of the face class only and treating the other hypothesis as the "complement" of the first hypothesis. The decision rule is as follows:

$$\mathbf{x} \in \Omega_1 \quad \text{if } p(\mathbf{x}|\Omega_1) \geq T$$
$$\mathbf{x} \in \Omega_0 \quad \text{if } p(\mathbf{x}|\Omega_1) < T, \qquad (7.4.3)$$

where the hypothesis Ω_1 represents the face class and Ω_0 represents the nonface class. The threshold T is learned by discriminative learning rules: reinforced/antireinforced rules. If a face pattern falls out of the decision region made by the threshold, the threshold value will be decreased so that the face region can include the pattern. In contrast, if a nonface pattern is found in the face region, the threshold value is increased so that the pattern may be rejected. In order to reduce the effect of outlier patterns, a window function can be applied so that only the neighboring patterns are involved in the threshold adjustment. It has been proven that the discriminant learning rules can achieve a minimum classification error rate [170]. When the window function is present, not all of the nonface patterns are used for training. Only the patterns that "look like face patterns" are included in the training set, thus yielding a more efficient yet accurate learning scheme. One simple example illustrating the power of this one-subnet PDBNN is depicted in Figure 7.7.

Similar to DBNNs, PDBNNs use the decision-based learning scheme (see Section 7.3). In the LU phase, PDBNNs use the positive training patterns to adjust the subnet parameters by an unsupervised learning algorithm, and in the GS phase they use only the misclassified patterns for reinforced/antireinforced learning. The negative patterns are used only for the antireinforced training of the subnet. The decision boundaries are determined by a threshold, which can also be trained by reinforced/antireinforced learning. The detailed description of a PDBNN detector (a two-class classifier) follows.

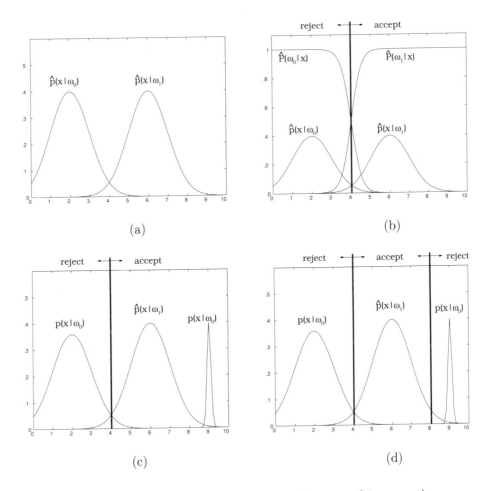

(a)   (b)

(c)   (d)

Figure 7.7. An example demonstrating the advantage of incorporating an adaptive decision-threshold in a two-class problem. (a) The estimated probability distributions of hypothesis ω_0 ($\hat{p}(\mathbf{x}|\omega_0)$) and hypothesis ω_1 ($\hat{p}(\mathbf{x}|\omega_1)$); here $\hat{p}(\mathbf{x}|\omega_0) = \mathcal{N}(2;1)$ and $\hat{p}(\mathbf{x}|\omega_1) = \mathcal{N}(6;1)$. (b) The decision boundary (*thick straight line* at $\mathbf{x} = 4$) is generated by estimated posterior probabilities $\hat{P}(\omega_0|\mathbf{x})$ and $\hat{P}(\omega_1|\mathbf{x})$. Prior probabilities are assumed to be equal—that is, $P(\omega_0) = P(\omega_1)$; the input pattern is accepted if $\hat{P}(\omega_1|\mathbf{x}) \geq \hat{P}(\omega_0|\mathbf{x})$, otherwise it is rejected. (c) Decision making based on posterior probabilities may incur an additional misclassification rate in the test phase due to poor density estimation; here the true distribution of hypothesis ω_0 is $p(\mathbf{x}|\omega_0) = \frac{9}{10}\mathcal{N}(2;1) + \frac{1}{10}\mathcal{N}(9;0.1)$. (d) The decision boundary is created by $\hat{p}(\mathbf{x}|\omega_1)$ only; here the input pattern is accepted if $\hat{p}(\mathbf{x}|\omega_1) \geq 0.054$, otherwise it is rejected. Note that the misclassification rate is reduced.

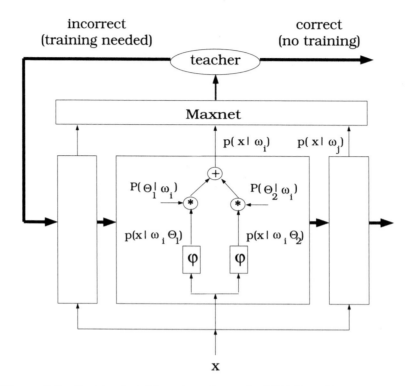

Figure 7.8. Decision-based learning scheme for PDBNNs. Unlike the HME, each subnet in the PDBNN estimates the class-conditional likelihood density function. The neurons estimate the likelihood densities in some local areas, and the subnet output is the weighted sum of the neuron outputs. PDBNN has a two-phase learning scheme. In the first phase, network parameters are trained by the unsupervised learning algorithm (e.g., EM). In the second phase, the obtained decision boundaries are further adjusted by supervised learning rules.

7.4.1　Discriminant Functions of PDBNNs

One major factor differentiating PDBNNs from DBNNs (or other RBF networks) is that PDBNNs must follow probabilistic constraints. That is, the subnet discriminant functions of PDBNNs are designed to model log-likelihood functions. The reinforced and antireinforced learning is applied to *all* the clusters of the global winner and the correct winner, with a weighting distribution proportional to the degree of possible responsibility (measured by the likelihood) associated with each cluster.

Given a set of identical, independent distributed (i.i.d.) patterns

$$X^+ = \{ \mathbf{x}_t \mid t = 1, 2, \ldots, N\},$$

the class-likelihood function $p(\mathbf{x}_t|\Omega)$ for class Ω (e.g., face class) is assumed to be a mixture of Gaussian distributions. Define

$$p(\mathbf{x}_t | \Omega, \mathbf{w}_r)$$

as one of the Gaussian distributions that comprise $p(\mathbf{x}_t | \Omega)$; that is,

$$p(\mathbf{x}_t | \Omega) = \sum_{r=1}^{R} P(\mathbf{w}_r | \Omega) p(\mathbf{x}_t | \Omega, \mathbf{w}_r),$$

where

$$p(\mathbf{x}_t | \Omega, \mathbf{w}_r) = \mathcal{N}(\mu_r, \Sigma_r),$$

$P(\mathbf{w}_r | \Omega)$ is the prior probability of cluster r, and $\mathbf{w}_r$ denotes the parameters of the r-th cluster in the subnet. By definition

$$\sum_{r=1}^{R} P(\mathbf{w}_r | \Omega) = 1.$$

The discriminant function of a subnet can be expressed in a form of log-likelihood functions:

$$\phi(\mathbf{x}_t, \Theta) = \log p(\mathbf{x}_t | \Omega) = \log \left[\sum_r P(\mathbf{w}_r | \Omega) p(\mathbf{x}_t | \Omega, \mathbf{w}_r) \right], \qquad (7.4.4)$$

where the classifier is parameterized by

$$\Theta \equiv \{ T, \mu_r, \Sigma_r, P(\mathbf{w}_r | \Omega); \ \forall \ r \}. \qquad (7.4.5)$$

In Eq. 7.4.5, T is the threshold of the subnet and is trained in the GS learning phase. The overall system diagram leading to such a form of discriminant function is illustrated in Figure 7.8. Just like the original DBNN, only the teacher's class preference is needed. In other words, no quantitative value is explicitly required from the teacher.

Elliptic Basis Function

In most general formulation, the basis function of a cluster should be able to approximate the Gaussian distribution with a full-rank covariance matrix. However, for those applications that deal with high-dimensional data but provide only a smaller number of training patterns, such a general covariance formulation is often discouraged because it involves training of a large number of covariance parameters using a relatively small database. A natural compromise is to assume that all of the features are uncorrelated but not all features are of equal importance. That is, suppose that $p(\mathbf{x} | \Omega, \mathbf{w}_r)$ is a D-dimensional Gaussian distribution with uncorrelated features

$$p(\mathbf{x} | \Omega, \mathbf{w}_r) = \frac{1}{(2\pi)^{D/2} \prod_{d=1}^{D} \sigma_{rd}} \exp \left\{ -\frac{1}{2} \sum_{d=1}^{D} \frac{(x_d - \mu_{rd})^2}{\sigma_{rd}^2} \right\} \sim \mathcal{N}(\mu_r, \Sigma_r), \quad (7.4.6)$$

where $\mathbf{x} = [x_1, x_2, \ldots, x_D]^T$ is the input pattern, $\mu_r = [\mu_{r1}, \mu_{r2}, \ldots, \mu_{rD}]^T$ is the mean vector, and the diagonal matrix $\Sigma_r = \text{diag}\{\sigma_{r1}^2, \sigma_{r2}^2, \ldots, \sigma_{rD}^2\}$ is the covariance matrix.

To approximate the density function in Eq. 3.2.16, the elliptic basis function (EBF) serves as the basis function for each cluster:

$$\psi(\mathbf{x}, \Omega, \mathbf{w}_r) = -\frac{1}{2}\sum_{d=1}^{D} \beta_{rd}(x_d - \mu_{rd})^2 + \theta_r, \qquad (7.4.7)$$

where $\theta_r = -\frac{D}{2}\ln 2\pi - \sum_{d=1}^{D}\ln\sigma_{rd}$. After passing an exponential activation function, $\exp\{\psi(\mathbf{x}, \Omega, \mathbf{w}_r)\}$ can be viewed as the same Gaussian distribution described in Eq. 3.2.16, except for a minor notational change: $\frac{1}{\beta_{rd}} = \sigma_{rd}^2$.

7.4.2   Learning Rules for PDBNNs

Recall that the training scheme for PDBNNs follows the LUGS principle. The locally unsupervised phase for the PDBNNs can adopt one of the unsupervised learning schemes (e.g., VQ, K-means, EM, etc.). As for the globally supervised learning, the decision-based learning rule is adopted.

Unsupervised Training for LU Learning

The network parameter values are initialized in the LU learning phase. The following are two unsupervised clustering algorithms, either one of which can be applied to the LU learning.

- *K-means and vector quantization.* The K-means method adjusts the centroid of a cluster based on the distance $\|\mathbf{x} - \mu_r\|^2$ of its neighboring patterns $\mathbf{x}$. Basically, the K-means algorithm assumes the covariance matrix of cluster i is $\sigma_i^2 I$. An iteration of the K-means algorithm in the LU phase contains two steps: (1) All input patterns are examined to determine their closest cluster centers, and (2) each cluster centroid is moved to the mean of its neighboring patterns.

- *Expectation-Maximization (EM).* In the LU learning phase of PDBNNs, K-means or VQ are used to determine initial positions of the cluster centroids and EM is used to learn the cluster parameters of each class. Chapter 3 explores the EM algorithm in detail.

One limitation of the K-means algorithm is that the number of clusters needs to be decided before training. An alternative is to use the vector quantization (VQ) algorithm. VQ can adaptively create a new neuron for an incoming input pattern if it is determined (by a vigilance test) to be sufficiently different from the existing clusters. Therefore, the number of clusters may vary from class to class. Notice that one must carefully select the threshold value for the vigilance test so that the clusters are neither too large to incur high reproduction error nor too small to lose generalization accuracy. (Like K-means, the VQ algorithm assumes the covariance matrix of cluster i is $\sigma_i^2 I$.)

Supervised Training for GS Learning

There is one major difference between PDBNNs and traditional statistical approaches: after maximum-likelihood estimation, the PDBNN has one more learning phase—the GS learning, which minimizes the classification error. The GS learning algorithm is used after the PDBNN finishes the LU training. In the globally supervised training phase, teacher information is used to fine-tune decision boundaries. When a training pattern is misclassified, the LVQ-type reinforced or antireinforced learning technique is applied [184]. The GS training is detailed next.

Assume a set of training patterns $X=\{\mathbf{x}_t; t = 1, 2, \ldots, M\}$. Now further divide the data set X into (1) the "positive training set" $X^+ = \{\mathbf{x}_t; \mathbf{x}_t \in \Omega, t = 1, 2, \ldots, N\}$ and (2) the "negative training set" $X^- = \{\mathbf{x}_t; \mathbf{x}_t \notin \Omega, t = N + 1, N + 2, \ldots, M\}$. Define an energy function

$$E = \sum_{t=1}^{M} l(d(t)), \tag{7.4.8}$$

where

$$d(t) = \begin{cases} T - \phi(\mathbf{x}_t, \Theta) & \text{if } \mathbf{x}_t \in X^+ \\ \phi(\mathbf{x}_t, \Theta) - T & \text{if } \mathbf{x}_t \in X^-. \end{cases} \tag{7.4.9}$$

The discriminant function $\phi(\mathbf{x}_t, \Theta)$ is defined in Eq. 7.4.4. T is the threshold value. The *penalty function* l can be either a piecewise linear function

$$l(d) = \begin{cases} \zeta d & \text{if } d \geq 0 \\ 0 & \text{if } d < 0, \end{cases} \tag{7.4.10}$$

where ζ is a positive constant, or a sigmoidal function

$$l(d) = \frac{1}{1 + \exp^{-d/\xi}}. \tag{7.4.11}$$

Figure 7.9 depicts these two possible penalty functions. The reinforced and antireinforced learning rules for the network are:

$$\begin{aligned} \text{Reinforced Learning:} \quad & \Theta^{(j+1)} = \Theta^{(j)} + \eta l'(d(t))\nabla\phi(\mathbf{x}_t, \Theta^{(j)}) \\ \text{Antireinforced Learning:} \quad & \Theta^{(j+1)} = \Theta^{(j)} - \eta l'(d(t))\nabla\phi(\mathbf{x}_t, \Theta^{(j)}) \end{aligned} \tag{7.4.12}$$

If the misclassified training patterns are from the positive training set (e.g., face or eye), reinforced learning will be applied. If the training patterns belong to the negative training (i.e., nonface or noneye) set, then only the antireinforced learning rule will be executed, since there is no "correct" class to be reinforced.

The gradient vectors in Eq. 7.4.12 are computed:

$$\frac{\partial \phi(\mathbf{x}_t, \Theta)}{\partial \mu_{rd}}\bigg|_{\Theta=\Theta^{(j)}} = h_r^{(j)}(t) \cdot \beta_{rd}^{(j)}(x_d(t) - \mu_{rd}^{(j)})$$

$$\frac{\partial \phi(\mathbf{x}_t, \Theta)}{\partial \beta_{rd}}\bigg|_{\Theta=\Theta^{(j)}} = h_r^{(j)}(t) \cdot \frac{1}{2}\left(\frac{1}{\beta_{rd}^{(j)}} - (x_d(t) - \mu_{rd}^{(j)})^2\right), \qquad (7.4.13)$$

where $h_r^{(j)}(t)$ is the conditional posterior probability

$$h_r^{(j)}(t) = \frac{P^{(j)}(\mathbf{w}_r|\Omega)p^{(j)}(\mathbf{x}_t|\Omega, \mathbf{w}_r)}{\sum_k P^{(j)}(\mathbf{w}_k|\Omega)p^{(j)}(\mathbf{x}_t|\Omega, \mathbf{w}_k)}, \qquad (7.4.14)$$

and $\mu_{rd}^{(j)}$ and $\beta_{rd}^{(j)}$ are defined in Eqs. 3.2.16 and 7.4.7, respectively. As to the conditional prior probability $P(\mathbf{w}_r|\Omega)$, since the EM algorithm can automatically satisfy the probabilistic constraints $\sum_r P(\mathbf{w}_r|\Omega) = 1$ and $P(\mathbf{w}_r|\Omega) \geq 0$, it is applied to update $P(\mathbf{w}_r|\Omega)$ in the GS phase so that the influences of different clusters are regulated. Specifically, at the end of the epoch j, one computes

$$P^{(j+1)}(\mathbf{w}_r|\Omega) = (1/N)\sum_{t=1}^{N} h_r^{(j)}(t). \qquad (7.4.15)$$

7.4.3 Threshold Updating

The threshold value of PDBNNs can also be learned by reinforced/antireinforced learning rules. Since the increment of the discriminant function $\phi(\mathbf{x}_t, \Theta)$ and the decrement of the threshold T have the same effect on the decision-making process, the direction of the reinforced and antireinforced learning for the threshold is the opposite of the one for the discriminant function. For example, given an input $\mathbf{x}_t$, if $\mathbf{x}_t \in \Omega$ but $\phi(\mathbf{x}_t, \Theta) < T$, then T should reduce its value. On the other hand, if $\mathbf{x}_t \notin \Omega$ but $\phi(\mathbf{x}_t, \Theta) > T$, then T should increase:

$$\begin{array}{lll} \text{Reinforced Learning:} & T^{(j+1)} = T^{(j)} - \eta l'(d(t)) \text{ if } \mathbf{x}_t \in \Omega & \\ \text{Antireinforced Learning:} & T^{(j+1)} = T^{(j)} + \eta l'(d(t)) \text{ if } \mathbf{x}_t \notin \Omega & (7.4.16) \end{array}$$

7.5 Multiclass Probabilistic DBNNs

Similar to the previous two-class (one-subnet) probabilistic DBNNs, the PDBNNs adopt a modular structure for a multiclass classification problem: one subnet is designated to one object class. The output of a subnet represents the class's likelihood function. The outputs of the subnets compete with each other, and the winner can claim the identity of the input pattern. For the multiclass probabilistic DBNNs, a mixture of Gaussians is often adopted as the class-likelihood function $p(\mathbf{x}_t|\Omega_i)$.

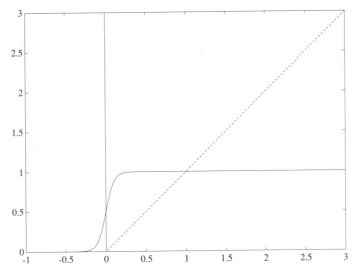

Figure 7.9. Linear penalty function (*solid line*) and sigmoidal penalty function (*dashed line*). From J. S. Taur, "Structural Study on Neural Networks for Pattern Recognition," Ph.D. Thesis, Princeton University, 1993; used with permission.

Thus, the discriminant function and the network parameters are expressed according to a probabilistic format. For example, the discriminant function is expressed as:

$$\phi(\mathbf{x}_t, \Theta_i) = \log p(\mathbf{x}_t|\Omega_i) = \log\left[\sum_r P(\mathbf{w}_{r|i}|\Omega_i)p(\mathbf{x}_t|\mathbf{w}_{r|i}, \Omega_i)\right], \qquad (7.5.1)$$

where the i-th subnet is parameterized by

$$\Theta_i \equiv \{T_i, \mu_{r|i}, \Sigma_{r|i}, P(\mathbf{w}_{r|i}|\Omega_i); \ \forall \ r\}.$$

The overall system diagram is depicted in Figure 7.10. Because the output of PDBNNs is of probability form, the discriminant function can also be adopted as a convenient indicator of the confidence level of the recognition result. This is of vital importance for many applications such as sensor fusion.

There are two important characteristics in the multiple subnet PDBNNs: (1) They adopt the OCON structure as discussed in Section 6.1 and (2) they produce FARs and FRRs. These characteristics are further explored next.

7.5.1 Structure of Multiclass PDBNNs

The PDBNNs adopt the one-class-one-network (OCON) structure, where one subnet is designated to one class only (see structure of a PDBNN recognizer in Figure 7.10). Each subnet specializes in distinguishing its own class from the rest of the

$$k = \begin{cases} \mathrm{argmax}_j\{\phi(x(t), \mathbf{w}_j)\} & \text{if } \phi(x(t), \mathbf{w}_j) - T_j > 0 \quad \forall j = 1, \dots, K \\ \text{unknown class} & \text{if } \phi(x(t), \mathbf{w}_j) - T_j \leq 0 \quad \forall j = 1, \dots, K \end{cases}$$

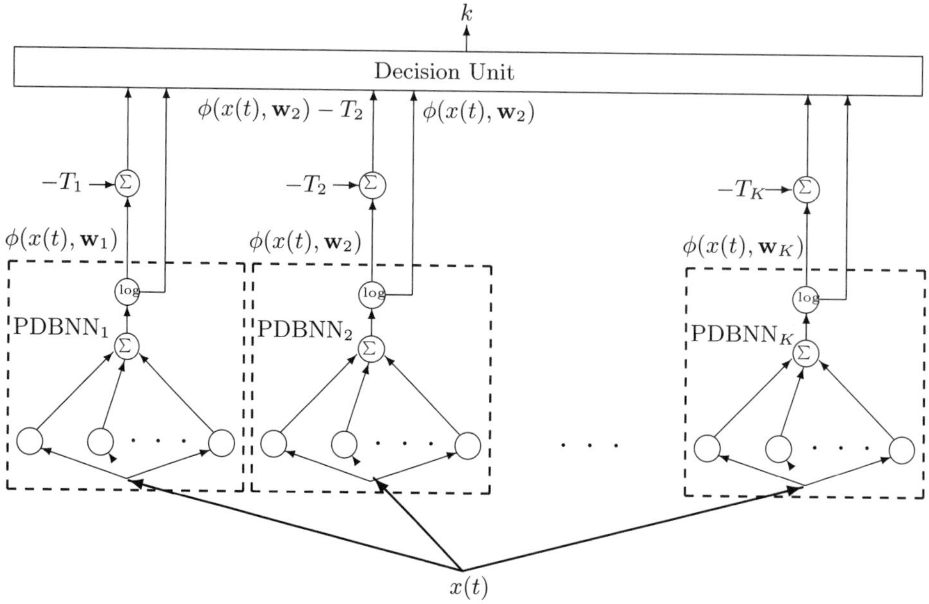

Figure 7.10. Structure of a multiclass PDBNN. Each class is modeled by a subnet. The subnet discriminant functions are designed to model the log-likelihood functions given by Eq. 7.5.1. All network weights are in probabilistic format.

classes. The number of hidden units in each subnet can be kept reasonably small. As mentioned in Section 6.1, the OCON structure is superior to the ACON structure in many aspects; and the OCON structure is particularly advantageous for face recognition applications. For example, the OCON structure makes the network amenable to *incremental training* by permitting a simple network upgrade on any new addition (or withdrawal) of class memberships. Moreover, thanks to the distributive property inherent in the OCON structure, the PDBNN (face or speaker) recognizer can be easily adapted to (face or speaker) verification systems. Individual information and features can be stored, either in centralized computer systems or in users' magnet cards, and this information can be individually retrieved for verifying users' identities.

7.5.2   False Acceptance and False Rejection

False rejection and false acceptance rates are important not only for detection problems but also for several recognition problems. For example, it is more dangerous for a face recognition security system to falsely give an entry permit to an intruder than to mistakenly recognize one personnel member for another member. As illustrated in the following paragraphs, the OCON structure and probabilistic discriminant functions enable PDBNNs to produce low false rejection rates (FRR) and false acceptance rates (FAR).

Statistical theory of pattern classification shows that the decision boundaries generated by Bayes's posterior probabilities produce minimum classification error. For K-class classification problems, the Bayes decision boundaries divide the feature space into K different regions. One simple example is shown in Figure 7.11(a), which depicts two Gaussian distributions with decision boundaries depicted by dotted lines. In many classification applications (e.g., OCR), the classifiers are designed, either implicitly or explicitly, to approach the Bayes decision boundaries so that the minimum classification error can be reached [311, 324]. However, since the Bayes decision theory is based on the assumption that all data classes are "known," the Bayes decision rule may not be suitable for those object recognition problems that deal with "unknown" data classes as well. For this type of recognition problem, false rejection and false acceptance rates need to be taken care of; Figure 7.11(b) provides an example. The asteroid points represent the unknown data points derived from an unknown class. The job of a classifier is now not only to decide which data class this input pattern belongs to but also to trigger an alarm if this input pattern is not from either one of the data classes. Obviously, the Bayes classifier shown in the figure is not capable of triggering an alarm caused by unknown data.

There are two approaches for tackling this recognition problem: thresholding and approximation of data distribution. In the thresholding approach, the input pattern is considered as class i not only when the Bayes conditional posterior of class i for the pattern is the highest among all of the classes but also when the value of the posterior is higher than a certain threshold (see Figure 7.10). With this thresholding mechanism, a certain amount of outlier data can be rejected. However, as Figure 7.11(c) shows, because all conditional posterior probabilities sum to one, some data points, though far away from data clusters, still possess high posterior values. As a result, thresholding cannot reject these data.

In the second approach, the distribution of unknown data is approximated by a probability density function. That is, all unknown data is categorized into a large object class called an *unknown class*; its distribution is estimated just as in normal object classes. This approach is feasible, but the distribution of the unknown data is usually much more complicated than that of normal class data. One example is face recognition for security systems. The objective of such systems is to reject all people who are not in the database. Therefore, the size of the unknown class, or intruder class, is close to the population of the entire world. Intuition suggests that the distribution of the unknown face class is very complicated.

The PDBNN is suitable for tackling the false rejection and false acceptance problems mentioned before because it focuses only on the distributions of individual classes rather than partitioning classes globally. As a result, no additional intruder class subnet is needed. Also, since density value usually drops as the distance to class centroids increases, the decision regions tend to be more locally conserved, thus

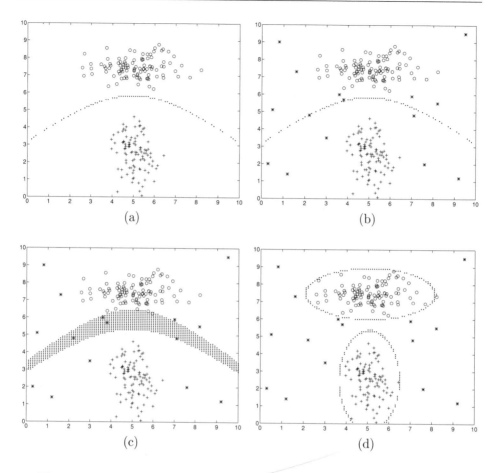

Figure 7.11. An example of the false acceptance and false rejection issues in pattern recognition problems. Three types of patterns appear in this figure: ○ and + represent patterns from two different classes, and * represents intruder patterns; the *dotted lines* represent the decision boundaries. (a) A Bayes decision boundary between two classes. (b) Same as (a) but with intruder patterns that are classified to either class ○ or class +. (c) Rejection region is formed by increasing the decision threshold; 5 intruder patterns are rejected by using the threshold but 10 are still misclassified. (d) Decision boundaries generated by a PDBNN; the decision regions are locally preserved, and most intruder patterns are now rejected. (Refer to other parts of this section for a detailed description.)

a lower FAR is to be expected. Numerous experiments indicate this characteristic. Figure 7.11(d) shows the decision boundaries made by PDBNNs.

7.5.3   Learning Rules for Multiple Subnet PDBNNs

The multiple subnet PDBNNs follow the same learning rules as the one-subnet PDBNNs in the LU learning phase. Each subnet performs K-means, VQ, or EM to adjust its parameters. In the GS learning phase, only the misclassified patterns are used. The GS learning rules for multiclass PDBNNs are a straightforward extension of the one-class version mentioned in Section 7.4.2. The following are two types of misclassified patterns.

1. The first type is from the positive dataset. A pattern of this type is used for (1) reinforced learning of the threshold and discriminant function of the class that it actually belongs to, and (2) antireinforced learning of the thresholds and discriminant functions of the classes that have higher discriminant function values than the true class has.

2. The second type of misclassified patterns is from the negative data set. Examples of negative training patterns are intruder face patterns and nonface patterns. Since there is no subnet representing the negative data, a pattern of this type is used only for antireinforced learning of the threshold and discriminant function of the class to which it is misclassified.

GS Learning for Subnet Parameters

The data adaptive scheme used by the GS learning of the multiclass PDBNNs is an extension of the learning scheme shown in Section 7.4.2. This section shows the block adaptive gradient version of the GS learning. At the beginning of each GS iteration, the still-in-training PDBNN classifies all input patterns $\mathbf{x}_t$'s. $\mathbf{x}_t$ is classified to class Ω_j if $\phi(\mathbf{x}_t, \Theta_j) > \phi(\mathbf{x}_t, \Theta_k), \forall k \neq j$ and $\phi(\mathbf{x}_t, \Theta_j) \geq T_i$, where T_i is the threshold for subnet i. Three data sets are created according to the classification results:

- $D_1^i = \{\mathbf{x}_t; \mathbf{x}_t \in \Omega_i, \mathbf{x}_t \text{ is classified to } \Omega_i\}$ (correctly classified set)

- $D_2^i = \{\mathbf{x}_t; \mathbf{x}_t \in \Omega_i, \mathbf{x}_t \text{ is misclassified to other class } \Omega_j\}$ (false rejection set)

- $D_3^i = \{\mathbf{x}_t; \mathbf{x}_t \notin \Omega_i, \mathbf{x}_t \text{ is misclassified to } \Omega_i\}$ (false acceptance set)

Only patterns in the false rejection set (D_2^i) and the false acceptance set (D_3^i) need to be used to train the subnet Ω_i. In the beginning of the GS learning, the intermediate parameter $h_{r|i}^{(j)}(t)$ is computed as in Eq. 7.4.14. Also, the cluster prior probabilities $P(\mathbf{w}_{r|i}|\Omega_i)$ are updated by the same rule as in Eq. 7.4.15. To update the mean $\mu_{r|i}$ and the diagonal covariance matrix $\Sigma_{r|i}$, a gradient ascent approach is applied. At GS iteration j, $\mu_{r|i}$ and $\Sigma_{r|i}$ are updated by the following:

$$\mu_{r|i}^{(j+1)} = \mu_{r|i}^{(j)} + \eta_\mu \sum_{t, \mathbf{x}_t \in D_2^i} h_{r|i}^{(j)}(t) \Sigma_{r|i}^{-1(j)} \left[\mathbf{x}_t - \mu_{r|i}^{(j)} \right]$$

$$-\eta_\mu \sum_{t,\mathbf{x}_t \in D_3^i} h_{r|i}^{(j)}(t) \Sigma_{r|i}^{-1(j)} \left[\mathbf{x}_t - \mu_{r|i}^{(j)}\right]$$

$$\Sigma_{r|i}^{(j+1)} = \Sigma_{r|i}^{(j)} + \frac{1}{2}\eta_\sigma \sum_{t,\mathbf{x}_t \in D_2^i} h_{r|i}^{(j)}(t) \left[H_{r|i}^{(j)}(t) - \Sigma_{r|i}^{-1(j)}\right]$$

$$-\frac{1}{2}\eta_\sigma \sum_{t,\mathbf{x}_t \in D_3^i} h_{r|i}^{(j)}(t) \left[H_{r|i}^{(j)}(t) - \Sigma_{r|i}^{-1(j)}\right], \tag{7.5.2}$$

where $H_{r|i}^{(j)}(t) = \Sigma_{r|i}^{-1(j)}[\mathbf{x}_t - \mu_{r|i}^{(j)}][\mathbf{x}_t - \mu_{r|i}^{(j)}]^T \Sigma_{r|i}^{-1(j)}$, and η_μ, η_σ are user-defined positive learning rates.

GS Learning for Thresholds

The thresholds T_i in the multiple subnet PDBNN are also trained by the reinforced and antireinforced rules, just like the one-subnet case. Recall that the energy function to be minimized is

$$E = \sum_{t=1}^{M} l(d_i(t)) \qquad i = 1, \ldots, K, \tag{7.5.3}$$

where

$$d_i(t) = \begin{cases} T_i - \phi(\mathbf{x}_t, \Theta_i) & \text{if } \mathbf{x}_t \in X^+ \\ \phi(\mathbf{x}_t, \Theta_i) - T_i & \text{if } \mathbf{x}_t \in X^- \end{cases} \tag{7.5.4}$$

and the penalty function l is depicted in Figure 7.9 (*dotted line*). The sets X^+ and X^- are the positive training set and the negative training set, respectively. Therefore, an updating rule similar to Eq. 7.4.16 can be easily derived.

Convergence Properties of GS Learning

The reinforced and antireinforced updating rules defined in Eqs. 7.4.13 and 7.4.16 are just the negative derivative of E with respect to the cluster means, cluster variances, and the threshold. The EM algorithm used for updating the conditional prior probabilities has proven to have a negative projection on the gradient [389]. If the patterns are separable and the learning rate is properly selected such that no new misclassified patterns will be incurred, the energy function decreases iteration by iteration since these parameters are moved along the negative gradient direction (or at least a negative projection). Also, since the energy function defined in Eq. 7.5.3 is always positive, the algorithm stops after finite steps. In most cases, PDBNNs can demonstrate fast convergence and high performance in many practical applications. However, in some nonseparable cases, PDBNNs can get trapped between local minima. One example follows.

The nonseparable case shown in Figure 7.12(a) has one cluster in the PDBNN. From Eqs. 7.4.12 and 7.4.16, the reinforced and antireinforced learning rules for one-subnet PDBNN are:

$$\text{Reinforced Learning:}\quad \Theta^{(j+1)} = \Theta^{(j)} + \eta l'(d(t))\nabla\phi(\mathbf{x}_t, \Theta^{(j)})$$
$$\text{Antireinforced Learning:}\quad \Theta^{(j+1)} = \Theta^{(j)} - \eta l'(d(t))\nabla\phi(\mathbf{x}_t, \Theta^{(j)})$$

The gradient vectors for the block adaptive updating scheme of the one-cluster, one-subnet PDBNN are as follows:

$$\left.\frac{\partial E}{\partial \mu_d}\right|_{\Theta=\Theta^{(j)}} = \beta_d^{(j)}\left[\sum_{\mathbf{x}_t \in D_2} l'(d(t))(x_d(t) - \mu_d^{(j)}) - \sum_{\mathbf{x}_t \in D_3} l'(d(t))(x_d(t) - \mu_d^{(j)})\right]$$

$$\left.\frac{\partial E}{\partial \beta_d}\right|_{\Theta=\Theta^{(j)}} = \frac{1}{2}\left[\sum_{\mathbf{x}_t \in D_2} l'(d(t))(\frac{1}{\beta_d^{(j)}} - (x_d(t) - \mu_d^{(j)})^2)\right.$$

$$\left. - \sum_{\mathbf{x}_t \in D_3} l'(d(t))(\frac{1}{\beta_d^{(j)}} - (x_d(t) - \mu_d^{(j)})^2)\right]$$

$$\left.\frac{\partial E}{\partial T}\right|_{\Theta=\Theta^{(j)}} = \sum_{\mathbf{x}_t \in D_3} l'(d(t)) - \sum_{\mathbf{x}_t \in D_2} l'(d(t)), \qquad (7.5.5)$$

where the energy function E is defined in Eq. 7.5.3, and D_2, and D_3 are the falsely rejected data set and the falsely accepted data set. Because there is only one cluster, the subscript r is removed and the conditional prior $P(\mathbf{w}|\Omega)$ and the conditional posterior $h^{(j)}(t)$ are always equal to 1.

Suppose the number of patterns in D_2 and D_3 are p and q, respectively, and suppose the penalty function described in Eq. 7.4.10 is used. Eq. 7.5.5 can be rewritten to a more useful form:

$$\left.\frac{\partial E}{\partial \mu_d}\right|_{\Theta=\Theta^{(j)}} = \zeta\beta_d^{(j)}\left[\sum_{t=1}^{p} x_d^+(t) - \sum_{t=1}^{q} x_d^-(t) + (q-p)\mu_d^{(j)}\right]$$

$$\left.\frac{\partial E}{\partial \beta_d}\right|_{\Theta=\Theta^{(j)}} = \frac{1}{2}\zeta\left[\sum_{t=1}^{q}(x_d^-(t) - \mu_d^{(j)})^2 - \sum_{t=1}^{p}(x_d^+(t) - \mu_d^{(j)})^2 + (p-q)\frac{1}{\beta_d^{(j)}}\right]$$

$$\left.\frac{\partial E}{\partial T}\right|_{\Theta=\Theta^{(j)}} = \zeta(q-p), \qquad (7.5.6)$$

where $x_d^+(t)$ is the d-th component of the t-th pattern in D_2, $x_d^-(t)$ is the d-th component of the t-th pattern in D_3, and ζ is a positive constant. The learning algorithm will stop if the gradient vectors are 0. Eq. 7.5.6 shows that the gradient vectors are 0 if and only if $p = q$, because of $\frac{\partial E}{\partial T}$. Furthermore, when $p = q$, the vector $\frac{\partial E}{\partial \beta_d}$ can be written as

$$\left.\frac{\partial E}{\partial \beta_d}\right|_{\Theta=\Theta^{(j)}} = \frac{1}{2}\zeta\left\{\sum_{t=1}^{q}\left[\left(x_d^-(t) - \mu_d^{(j)}\right)^2 - \left(x_d^+(t) - \mu_d^{(j)}\right)^2\right]\right\}. \qquad (7.5.7)$$

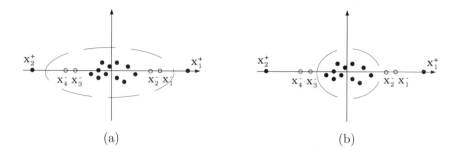

Figure 7.12. (a) A local minimum for the reinforced and antireinforced learning rule; (b) another local minimum. The energy might be lower than (a).

Notice that the square distance from the cluster center to a misclassified negative pattern is smaller than the square distance to any misclassified positive pattern (otherwise it would be correctly rejected). Therefore, if $p = q$, $\frac{\partial E}{\partial \beta_d}$ is zero only when $p = q = 0$. The preceding discussion demonstrates that the gradient vectors for GS learning are zero when and only when there are no misclassified patterns.

Consider again the example in Figure 7.12(a). The black circles represent the positive training patterns, and the white circles are the negative training patterns. The dotted line represents the decision boundary determined by threshold T and discriminant function $\phi(\mathbf{x}_t, \Theta)$. There are two falsely rejected patterns ($\mathbf{x}_1^+$ and $\mathbf{x}_2^+$) and four false accepted patterns ($\mathbf{x}_1^-$, $\mathbf{x}_2^-$, $\mathbf{x}_3^-$, and $\mathbf{x}_4^-$). In this example it is impossible for a one-cluster PDBNN to correctly separate all the patterns. Because both the mean of the misclassified positive patterns and that of the misclassified negative patterns are equal to the cluster center (which is 0), the gradient vector for cluster center $\frac{\partial E}{\partial \mu_d}$ is 0. Also, the locations of the misclassified patterns are carefully designed such that $\frac{\partial E}{\partial \beta_d}$ is 0. The only non-zero gradient vector is $\frac{\partial E}{\partial T}$. Because the number of misclassified negative patterns is larger than that of misclassified positive patterns, the threshold value increases and the decision boundary is therefore pushed inward, as shown in Figure 7.12(b). Now, because there are still misclassified patterns ($\mathbf{x}_1^+$ and $\mathbf{x}_2^+$) and the remaining misclassified patterns are in the positive set, the decision boundary is again pushed outward, going back to a situation similar to Figure 7.12(a). Therefore, in this example, the PDBNN oscillates between the two local minima.

Several remedies can solve the oscillation problem. First, the sigmoidal penalty function (cf. Eq. 7.4.11) could replace the linear penalty function. The misclassified patterns falling out of the window will be considered outliers. Also, the learning rate could be changed from a constant (η) to a function dependent on the number of patterns that have been presented (i.e., $\eta(t)$). As time goes by, the learning rate decreases. Therefore, the learning eventually converges. Finally, some kind of "terminating function" could be designed by monitoring either the error count or the energy value. If oscillation happens, the learning process terminates.

7.6　Biometric Authentication Application Examples

In Lin et al. [211], the DBNN was applied to all three modules of face recognition systems: (1) face detector, (2) eye localizer, and (3) face recognizer. The face detector finds the location of a human face in an image, then the eye localizer determines the positions of both eyes in order to generate meaningful feature vectors. The facial region proposed contains eyebrows, eyes (eyeglasses allowed), and nose. The face recognition system was successfully applied to two public databases (FERET and ORL) and one in-house (SCR) database. Chapter 8 further describes how to apply the probabilistic DBNN to various face detection and recognition applications. In addition to automated face recognition, the probabilistic DBNN has also been demonstrated to be promising for other applications including intelligent multimedia processing, medical image quantification, and bioinformation data mining (e.g., see [190, 192, 211, 346, 347, 366]).

Problems

1. A pioneering decision-based neural model is the perceptron originally proposed by Rosenblatt [322], which has linear discriminant functions:

$$y = \phi(\mathbf{x}, \Theta) = \sum_{j=1}^{P} w_j x_j + \theta.$$

   For convenience, consider the threshold value θ as an additional weight parameter. Denote $\theta = w_{P+1}$, then

$$\Theta = [w_1 \ w_2 \ldots w_P \ \theta]^T$$

   and

$$\mathbf{z} = [x_1 \ x_2 \ldots x_P \ 1]^T.$$

   That is, $\mathbf{z}$ is the augmented pattern $\mathbf{x}$. Now the linear discriminant function can be rewritten as

$$y = \phi(\mathbf{x}, \Theta) = \Theta^T \mathbf{z}$$

   Show that, according to Eq. 7.2.3, the linear perceptron learning rule is

$$\Theta^{(m+1)} = \Theta^{(m)} \pm \eta \mathbf{z}^{(m)}$$

   where η is a positive learning rate.

2. Show that if the training patterns of two classes are linearly separable, then the linear perceptron learning algorithm

$$\mathbf{w}^{(n+1)} = \begin{cases} \mathbf{w}^{(n)} & \text{if } \mathbf{w}^{(n)} \cdot \mathbf{x} + w_0 \geq 0 \text{ and } \mathbf{x} \in \text{Class 1} \\ \mathbf{w}^{(n)} - \eta \mathbf{x} & \text{if } \mathbf{w}^{(n)} \cdot \mathbf{x} + w_0 \geq 0 \text{ and } \mathbf{x} \in \text{Class 2} \\ \mathbf{w}^{(n)} + \eta \mathbf{x} & \text{if } \mathbf{w}^{(n)} \cdot \mathbf{x} + w_0 < 0 \text{ and } \mathbf{x} \in \text{Class 1} \end{cases}$$

   converges to a correct solution in a finite number of iterations.

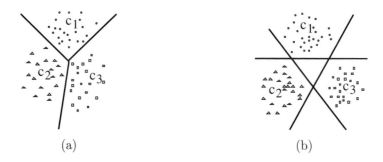

Figure 7.13. The clusters shown are (a) linearly separable, and (b) strongly linearly separable.

3. *Multiclass Linear Separability.* A practical "linear separability" definition is described as follows: Suppose that $S = \{\mathbf{z}^{(1)}, \ldots, \mathbf{z}^{(M)}\}$ is a set of given training patterns, with each element $\mathbf{z}^{(i)} \in R^N$ belonging to one of the L classes $\{\Omega_i, i = 1, \ldots, L\}$. The set is said to be *linearly separable* when (and only when) there exists weight vectors Θ_i, $i = 1, \ldots, L$, such that for any pattern $\mathbf{z} \in \Omega_i$, $\Theta_i^T \mathbf{z} \geq \Theta_j^T \mathbf{z}$ for all $j \neq i$.

   (a) Show that when there are three linear separable classes in a two-dimensional vector space, there always exists a decision boundary with appearance similar to the one shown in Figure 7.13(a).

   (b) Show that three linear separable classes may not have a decision boundary like Figure 7.13(b).

   (c) Suppose that there are four linear separable classes in a two-dimensional vector space. Sketch a plausible decision boundary.

   (d) Suppose that there are three linear separable classes in a three-dimensional vector space. Sketch a plausible decision boundary (surface).

4. *Multiclass Linear Perceptron.* The basic perceptron can be extended to the problem of classifying multiple (e.g., L) classes. For this purpose, the following important features are incorporated into the general DBNN:

   - One subnet is designated for one class.

   - The linear discriminant function of the subnets are denoted by $\phi(\mathbf{x}, \Theta_i)$, for $i = 1, \ldots, L$.

   - A maxnet is used to select the winning class.

   Following the two-class perceptron learning rule, design a corresponding learning rule for multiple-class perceptrons. *Hint:* Suppose that $S = \{\mathbf{z}^{(1)}, \ldots, \mathbf{z}^{(M)}\}$

is a set of given training patterns, with each element $\mathbf{z}^{(m)} \in R^N$ belonging to one of the L classes $\{\Omega_i, i = 1, \ldots, L\}$, and that the discriminant functions are $\phi(\mathbf{z}, \Theta_i) = \Theta_i^T \mathbf{z}$ for $i = 1, \ldots, L$. Suppose that the m-th training pattern $\mathbf{x}^{(m)}$ presented is known to belong to class Ω_i and that the winning class for the pattern is denoted by an integer j; that is, for all $l \neq j$,

$$\Theta_j^T \mathbf{z} > \Theta_l^T \mathbf{z}.$$

If $j \neq i$, this indicates that $\mathbf{z}^{(m)}$ is misclassified, and the following update is executed:

- Reinforced Learning: $\Theta_i^{(m+1)} = \Theta_i^{(m)} + \eta \mathbf{z}^{(m)}$
- Antireinforced Learning: $\Theta_j^{(m+1)} = \Theta_j^{(m)} - \eta \mathbf{z}^{(m)}$

The other weights remain unchanged: $\Theta_l^{(m+1)} = \Theta_l^{(m)}$ for all $l \neq i$ and $l \neq j$.

5. *Multiclass Linear Perceptron Convergence Theorem.* Show that if a given multiclass training set is linearly separable, then the linear perceptron learning algorithm (in the previous problem) converges to a correct solution after a finite number of iterations.

6. Suppose that a DBNN of the class-in-expert architecture has eight inputs, four outputs, and three experts per class.

   (a) Draw a flow graph of this network.

   (b) Show that in the special case where each expert in the network is a Gaussian density function, the class-in-expert architecture degenerates to the GMM-based classifier shown in Figure 6.2.

   (c) For the retrieving phase of the network in (b), calculate the computational complexity in term of MACs (multiplier-accumulations).

7. As depicted in Figure 7.9, different degrees of error are associated with each decision (e.g., marginally erroneous, erroneous, and extremely erroneous). The technique imposes a proper penalty function on all the "bad" decisions as well as the "marginally correct" ones. This consideration leads to the notion of a *fuzzy decision neural network* (FDNN) [176, 345].

Following the notations used in DBNNs, suppose that $S = \{\mathbf{x}^{(1)}, \ldots, \mathbf{x}^{(M)}\}$ is a set of given training patterns and the discriminant function for the class Ω_i is denoted $\phi(\mathbf{x}, \Theta_i)$, for $i = 1, \ldots, L$. For DBNNs, the *winner* class is denoted as Ω_j, where $j = \arg\max_j \phi_j(\mathbf{x}^{(m)}, \Theta_j)$. In contrast, for FDNNs, an alternative notation is adopted. Instead, Ω_j now denotes the leading challenger among all the classes excluding the correct class Ω_i; that is,

$$j = \arg\max_{j \neq i} \phi_j(\mathbf{x}^{(m)}, \Theta_j). \tag{7.6.1}$$

For a training pattern, a measure of misclassification can be introduced:

$$d = d^{(m)}(\mathbf{x}^{(m)}, \Theta) = -\phi_i(\mathbf{x}^{(m)}, \Theta_i) + \phi_j(\mathbf{x}^{(m)}, \Theta_j), \qquad (7.6.2)$$

where Θ denotes all the involved weight vectors. In fact, a more general measure was proposed in Katagiri et al. [176] as

$$d = -\phi_i(\mathbf{x}^{(m)}, \Theta_i) + \left\{ \frac{1}{M-1} \sum_{j, j \neq i} \phi_j(\mathbf{x}^{(m)}, \Theta_j)^{\gamma} \right\}^{1/\gamma}, \qquad \gamma > 0, \quad (7.6.3)$$

where ϕ is assumed to be positive.

It is plausible to modify Eq. 7.6.3 into

$$d_k(\mathbf{x}, \mathbf{w}_k) = -\phi_k(\mathbf{x}, \mathbf{w}_k) + \left\{ \frac{1}{M} \sum_j \phi_j(\mathbf{x}, \mathbf{w}_j)^{\gamma} \right\}^{1/\gamma}, \qquad \gamma > 0 \quad (7.6.4)$$

(a) Show that when $\gamma \to \infty$, Eq. 7.6.4 asymptotically approximates Eq. 7.6.2.

(b) Show that when Eq. 7.6.3 or Eq. 7.6.4 is adopted as the penalty function, it incurs an extra computational cost so that all subnets (instead of the key winner and loser only) must be updated on presentation of each training pattern.

Chapter 8

BIOMETRIC AUTHENTICATION BY FACE RECOGNITION

8.1 Introduction

Machine recognition of human faces from still or video images has attracted a great deal of attention in the psychology, image processing, pattern recognition, neural science, computer security, and computer vision communities. Its popularity results from its wide application ranging from static matching of controlled format photographs, such as passports, credit cards, driving licenses, and police photos, to real-time matching of video images for surveillance, access control, and security of public areas such as airports. Face recognition is probably one of the most *nonintrusive* and *user-friendly* biometric authentication methods currently available; a screensaver equipped with face recognition technology can automatically unlock the screen whenever the authorized user approaches the computer.

After many years of research and development, face recognition technology has reached a point where large-scale deployment is possible. For example, face recognition tools were used to scan the faces of people in crowds at the January 2001 Super Bowl in Tampa, Florida. Since the September 11 terrorist attacks, face recognition has attracted increasing attention from the media and law enforcement agencies. The nonintrusiveness of this seemingly mature technology has led many government and security agencies to consider it a viable means of enhancing proactive measures in public areas. For instance, in July 2002, the Virginia Beach Police Department linked Identix Inc.'s FaceIt recognition system to its beachfront cameras [153]. In addition, customs authorities at Sydney's airport adopted Cognitec's FaceVACS system to automatically process face-to-passport checks of Quantas Air crew members.

While such large-scale trials demonstrate the promise of face recognition technology, several reports question its reliability. For example, a study by the American Civil Liberties Union (ACLU) on an Identix face recognition system at the

Palm Beach, Florida, airport [139][1] shows that the system fails to detect suspected criminals (airport employees posing as criminals) 53% of the time—a disappointing performance that is just slightly better than a coin toss. A similar test at Boston's Logan Airport in 2002 showed that both the Identix and Viisage systems failed to identify positive matches 38% of the time.

The results of these two airport trials clearly indicate the weakness of current face recognition systems. Some of the major obstacles include system vulnerability to pose variation (e.g., rotation, translation, scaling, occlusion), facial expression, facial hair, makeup, background complexity, lighting conditions, and parameter variation of acquisition devices (e.g., noise level, distortion, f number). In most cases, variations between images of the same faces due to the influence of the preceding sources are much larger than image variations due to changes in facial identity. This makes face recognition a challenging problem. This chapter describes several image preprocessing procedures commonly used by face recognition systems in an attempt to reduce the influence of change in lighting conditions and head poses. The results show that facial images can be detected only under limited conditions even with such preprocessing procedures. In fact, achieving true illumination and pose invariants for three-dimensional deformable object recognition is still an unsolved problem for computer vision research. Until face recognition technology overcomes this obstacle, it cannot reliably perform automatic surveillance and security control without human assistance.

Most face recognition algorithms follow the standard recognition system structure described in Chapter 2. The first stage of the algorithm is the *feature extractor*, which transforms the input image into features that are the most representative of the face and the most invariant due to environmental changes. Section 8.2 describes several popular facial feature extraction techniques. Once the features are extracted, the *pattern classifier* uses them to determine the identity of the input patterns. Facial pattern classifiers are covered in Section 8.3.

The second half of the chapter emphasizes applications for surveillance and access control. In this type of application, one or more cameras are installed to constantly feed video streams into a realtime face recognition system. Illumination may change drastically at different times of day, but a lamp can be used to mitigate the effects. Image background can be stationary most of the time, but the setting is usually complicated. If necessary, at the access point the camera can take multiple captures of frontal views of a person's face; sometimes partial cooperation from the tester is expected. Before the *face recognizer* begins to identify the testee, the real-time face recognition system requires the *face detector* to identify the presence and the locations of human faces among the complicated image scene. After a face is detected, the *eye localizer* is then applied to pinpoint the locations of two eyes, which are used to align and normalize the facial image. Section 8.4 discusses face detection and eye localization techniques, and Section 8.5 provides a system implementation example. This system applies probabilistic decision-based neural

---

[1]Quote from the press release.

networks (PDBNNs) to all three major modules—face recognizer, face detector, and eye localizer.

8.2   Facial Feature Extraction Techniques

Ideal facial feature extraction techniques are able to (1) eliminate the influence of all environmental, non-identity-related variations and (2) maximize the difference between faces that belong to distinctive people. Numerous methods have been proposed that would attempt to extract such features to facilitate the recognition process. These methods can be classified into two categories: *feature-invariant approaches* and *template-based approaches* [38].

8.2.1   Feature-Invariant Approaches

This type of algorithm looks for structural features that exist even when the pose, viewpoint, or lighting conditions vary. Some prior knowledge may be used to develop rules to capture the relationship between facial features. Kelly [178] pioneered the work by using various features including width of the head; distance between the eyes, top of the head to eyes, eyes to the nose; and distance from eyes to the mouth. One of the recent image-invariant approaches is the local ordinal structure of brightness distribution between different parts of a human face [340]. The scheme checks the invariants for positive occurrences at all image locations. Yow and Cipolla [396] used a multistage feature extraction method for face detection. The first stage applies a second derivative Gaussian filter to the raw image. The filter kernel is of elliptical shape with a three-to-one aspect ratio designed for responding to many oval-shaped facial features, such as eyes and mouth. The second stage examines the edges around the interesting points that the Gaussian filter detects and groups those edges into regions. Measurements of a region's characteristics, such as edge length, edge strength, and intensity variance, are computed and stored as a "feature vector" for that facial feature.

One advantage of feature-invariant approaches is that they tend to have better tolerance toward variations of pose and facial expression. Because individual facial features are treated and examined locally, this category of feature extractor often accompanies a wire-frame type of pattern classifier such as active contour or dynamic link graph algorithms. For example, Manjunath et al. [230] used elastic graphs to link local facial features extracted by a Gabor filter set. Because wire-frame recognizers are suitable for incorporating the three-dimensional structural information of human heads, they have a better chance of recognizing faces using a nonfrontal view.

On the other hand, the major drawback of feature-based algorithms is their expensive computational requirements. For instance, Wiskott and Von der Malsburg's dynamic link system [384] takes 10 to 15 minutes of a SPARC 10's CPU time to recognize one face from a gallery of 111 models. In addition, the focus of feature-invariant approaches on facial details often requires higher resolution images than template-based approaches.

8.2.2 Template-Based Approaches

This type of algorithm designs one or several standard face "templates" (usually frontal) either manually or by learning from examples in the image database. The face templates are used to scan through the incoming test images. If the template matcher returns a high correlation value when it scans a certain region in the test image, the algorithm determines that this region contains a human face. Compared to feature-invariant methods, template-matching approaches have the advantage of being simple to implement. However, studies have shown that template matching is inadequate when dealing with variation. Supplementary techniques, such as multiresolution, multiscale subtemplates, and deformable templates, have subsequently been proposed to achieve scale, pose, and shape invariance. To obtain other invariance (e.g., lighting conditions, facial expression), template matching requires either more sophisticated statistical assistance or a powerful pattern classifier.

One important issue for statistical template matching is the *curse of dimensionality* [85]. As the dimension of the extracted feature vector increases, the number of training samples required for correct classification grows exponentially. A straightforward facial image input feature of 120×120 pixels results in a feature dimension of 14,000. An efficient preprocessing mechanism is therefore necessary to select a set of limited yet salient features. With the help of efficient feature extraction, the curse of dimensionality will be alleviated, and the accompanying classifier will run faster and consume less memory.

Principal Component Analysis

The most practical, systematic method of discovering low-dimensional representations of signals by relying on their statistical regularities is called *principal component analysis* (PCA). PCA assumes that the probability density of the input ensemble in the space of feature vectors is significantly non-zero only in a low-dimension linear subspace, which is parameterized with a linear expansion in the eigenvectors of the correlation matrix of the ensemble. PCA's power stems from its ease of computability and general applicability; so far it has been used in many real-world problems, including face recognition. Pentland et al. [272] used PCA to produce a representation of two-dimensional faces called *eigenfaces*. Let a face image $I(\mathbf{x})$ be a two-dimensional $m \times m$ array of intensity values. $I(\mathbf{x})$ can also be viewed as a vector of dimension m^2 and $\mathbf{x}$ is the two-dimensional grid coordinate. Denote a training set (*ensemble*) of N face images by $\{I^i(\mathbf{x}), i = 1, 2, \ldots, N\}$. PCA extracts a hierarchical orthonormal basis of the linear subspace that the input ensemble spans. This is done by diagonalizing the correlation matrix of the ensemble

$$R(\mathbf{x}, \mathbf{y}) \equiv \frac{1}{N} \sum_{i=1}^{N} I^i(\mathbf{x}) \left(I^i(\mathbf{y})\right)^T = \sum_{r=1}^{N} \Psi_r(\mathbf{x}) \lambda_r \left(\Psi_r(\mathbf{y})\right)^T \qquad (8.2.1)$$

to produce the orthonormal set of eigenfaces $\Psi_r(\mathbf{x}), r = 1, 2, \ldots, N$ and their respective eigenvalues λ_r, ordered in the natural hierarchy of decreasing magnitude.

The PCA representation

$$I(\mathbf{x}) = \sum_{r=1}^{N} A_r \Psi_r(\mathbf{x}) \tag{8.2.2}$$

with $A_r = \int \Psi_r(\mathbf{x}) \, (I(\mathbf{x}))^T \, \mathrm{d}\mathbf{x}$ is decorrelated in the sense that

$$\langle A_r, A_q \rangle = \lambda_r \delta_{rq}. \tag{8.2.3}$$

It has the property of best reconstruction—truncation in the expansion with $K < N$ has a minimum mean-squared error where

$$I^{rec}(\mathbf{x}) \equiv \sum_{r=1}^{K} A_r \Psi_r(\mathbf{x}). \tag{8.2.4}$$

With proper alignment of all of the facial images in the ensemble, most of the energy is compacted in the first K eigenfactors, where K is much smaller than $m \times m$. For example, Pentland [272] compressed the original facial image with $128 \times 128 = 16,384$ pixels onto the subspace spanned by the first 40 significant eigenfaces and reached a 95% recognition rate among the 200 faces chosen from a group of 7,562 facial images (3,000 individuals). This 16,384-to-40 feature dimension reduction greatly alleviated the burden of its corresponding pattern classifier.

Fisher's Linear Discriminant

The error that PCA minimizes during dimension reduction is the "reconstruction" error of the whole ensemble, not the recognition error between the object classes within the ensemble. Fisher's linear discriminant (FLD) [20], on the other hand, tries to address the classification issue. With C different classes in an ensemble, FLD finds the optimal $C - 1$ dimensional subspace for classification. The objective of FLD is to maximize the ratio of the "between-class" scatter matrix and the "within-class" scatter matrix [85]. Let the between-class scatter matrix be defined as

$$S_B = \sum_{j=1}^{C} n^j (\bar{I}^j - \bar{I})(\bar{I}^j - \bar{I})^T \tag{8.2.5}$$

and the within-class scatter matrix be defined as

$$S_W = \sum_{j=1}^{C} \sum_{k=1}^{n^j} (I^k - \bar{I}^j)(I^k - \bar{I}^j)^T, \tag{8.2.6}$$

where $\bar{I} = (1/N) \sum_{k=1}^{N} I^k$ is the mean image of the ensemble, $\bar{I}^j = (1/n^j) \sum_{k=1}^{n^j} I^{k,j}$ is the mean image of the j-th class, n^j is the number of samples in the j-th class, $N = \sum_{j=1}^{C} n_j$ is the number of images in the ensemble, and C is the number of classes in the ensemble. The optimal subspace, $E_{optimal}$, is determined as follows:

$$E_{optimal} = \arg\max_E \frac{|E^T S_B E|}{|E^T S_W E|} = [e_1, e_2, \ldots, e_{C-1}], \tag{8.2.7}$$

where $[e_1, e_2, \ldots, e_{C-1}]$ is the set of generalized eigenvectors of S_B and S_W corresponding to the $C - 1$ largest generalized eigenvalues λ_i, $i = 1, 2, \ldots, C - 1$; that is,

$$S_B E_i = \lambda_i S_W E_i. \tag{8.2.8}$$

Thus, the feature vectors P for any query face images I in the most discriminant sense can be calculated as:

$$P = E_{optimal}^T \cdot I. \tag{8.2.9}$$

PCA and FLD can be combined in cascade. PCA could be used to reduce the feature dimension from m^2 to K and then FLD could be applied on the K-dimensional PCA subspace to further reduce the dimension to $C - 1$. The following two conditions need to be satisfied when combining PCA and FLD:

- From Eq. 8.2.6, the rank of S_W is smaller than or equal to the minimum of K and $C(n^j - 1)$. To prevent S_W from becoming singular, the value of K should be no more than $N - C$.

- From Eq. 8.2.5, the rank of S_B is smaller than or equal to the minimum of K and $C - 1$. Accordingly, there are at most $C - 1$ non-zero generalized eigenvectors. In other words, the FLD transforms the K-dimensional space into $(C - 1)$-dimensional space to classify C classes of faces. If K is smaller than $C - 1$, FLD does not reduce feature dimension.

Note that FLD is a linear transformation maximizing the ratio of the determinant of the projected samples' between-class scatter matrix to the determinant of the projected samples' within-class scatter matrix. The results are globally optimal only for linearly separable data. The linear subspace assumption is violated for face data with a great deal of overlap. Moreover, the separability criterion is not directly related to classification accuracy in the output space. Several researchers have indicated that the FLD method achieved the best performance on the training data but generalized poorly to new individuals [83].

Local Feature Analysis

Local feature analysis (LFA) [271] is the core technique of FaceIt,[2] one of the most popular commercial face recognition systems. Unlike PCA, which simply expands the two-dimensional image into a one-dimensional vector, LFA tries to take advantage of the facial image's "topographical" property. LFA's first step is to design a preprocessing filter to enhance any local facial features that may contribute the greatest variations among different facial images. The idea of LFA is to use the values of several "significant" locations of the filtered image to constitute a much lower-dimension ($V << m^2$) feature vector for classification. LFA can be viewed as a *feature-based template matching* approach; it uses a template, on which several focus points represent the features.

---

[2] *http://www.identix.com/newsroom/whatisfaceit.html.*

Define the output image after the LFA filtering to be

$$O(\mathbf{x}) \equiv \int K(\mathbf{x}, \mathbf{y}) I(\mathbf{x}) \mathrm{d}\mathbf{y}. \tag{8.2.10}$$

LFA defines the filter kernel $K(\mathbf{x}, \mathbf{y})$ to be a topographical one that projects input signals to the subspace spanned by the eigenvectors

$$K(\mathbf{x}, \mathbf{y}) = \sum_{r,s=1}^{K} \Psi_r(\mathbf{x}) Q_{rs} (\Psi_s(\mathbf{y}))^T, \tag{8.2.11}$$

where $\Psi_r(\mathbf{x})$ is the eigenface in PCA (Eq. 8.2.1), and Q_{rs} is a priori an arbitrary matrix. Minimizing the correlation of the output image by minimizing

$$E = \int |\langle O(\mathbf{x}), O(\mathbf{y}) \rangle - \delta(\mathbf{x}, \mathbf{y})|^2 \mathrm{d}\mathbf{x}\mathrm{d}\mathbf{y} \tag{8.2.12}$$

with respect to the matrix Q_{rs} results in Q_{rs} being given by

$$Q_{rs} = \frac{1}{\sqrt{\lambda_r}} U_{rs}, \tag{8.2.13}$$

where U_{rs} is any orthogonal matrix satisfying $U^T U = I$ and λ_r is the eigenvalue corresponding to $\Psi_r(\mathbf{x})$. The simplest choice for U is $U_{rs} = \delta_{rs}$. With this choice, the LFA outputs become

$$O(\mathbf{x}) = \int \sum_{r=1}^{K} (\Psi_r(\mathbf{x}))^T \frac{1}{\sqrt{\lambda_r}} \Psi_r(\mathbf{y}) I(\mathbf{y}) \mathrm{d}\mathbf{y} = \sum_{r=1}^{K} \frac{A_r}{\sqrt{\lambda_r}} \Psi_r(\mathbf{x}), \tag{8.2.14}$$

where A_r is defined in Eqs. 8.2.2 and 8.2.3, and their correlation can be easily computed using the orthonormality of the vectors $\Psi_r(\mathbf{x})$:

$$\langle O(\mathbf{x}), O(\mathbf{y}) \rangle = \sum_{r=1}^{K} (\Psi_r(\mathbf{x}))^T \Psi_r(\mathbf{y}) \equiv P(\mathbf{x}, \mathbf{y}). \tag{8.2.15}$$

The function $P(\mathbf{x}, \mathbf{y})$ is an interesting object; it can easily be recognized as the projection operator onto the subspace spanned by the PCA eigenvectors.

To summarize, given the eigenvectors $\Psi_r(\mathbf{x})$ with eigenvalues λ_r, the following two functions can be constructed:

$$K(\mathbf{x}, \mathbf{y}) = \sum_{r=1}^{K} \Psi_r(\mathbf{x}) \frac{1}{\sqrt{\lambda_r}} (\Psi_r(\mathbf{y}))^T \tag{8.2.16}$$

$$P(\mathbf{x}, \mathbf{y}) = \sum_{r=1}^{K} (\Psi_r(\mathbf{x}))^T \Psi_r(\mathbf{y}) \tag{8.2.17}$$

$K(\mathbf{x}, \mathbf{y})$ is the LFA filter kernel, and $P(\mathbf{x}, \mathbf{y})$ turns out to be the residual correlation of the outputs. To reconstruct the input image directly from the output $\{O(\mathbf{x})\}$, Eq. 8.2.14 can be rearranged to obtain A_r and substituted into the reconstruction formula (Eq. 8.2.4)

$$I^{rec}(\mathbf{x}) = \sum_{r=1}^{K} \int \sqrt{\lambda_r} \Psi_r(\mathbf{y}) O(\mathbf{y}) (\Psi_r(\mathbf{x}))^T \mathrm{d}\mathbf{y} = \int K^{(-1)}(\mathbf{x}, \mathbf{y}) O(\mathbf{y}) \mathrm{d}\mathbf{y}, \quad (8.2.18)$$

where the "inverse" kernel is

$$K^{(-1)}(\mathbf{x}, \mathbf{y}) = \sum_{r=1}^{K} \Psi_r(\mathbf{x}) \sqrt{\lambda_r} (\Psi_r(\mathbf{y}))^T. \quad (8.2.19)$$

It can be shown that the reconstruction error $||I - I^{rec}||^2$ for the LFA representation is exactly equal to that for the PCA representation [271].

Now that the filter and its inverse are designed, the next LFA step is to use only a few significant points on the output image to represent the whole output image, then use this representation to reconstruct the input. Penev and Atick [271] proposed a deterministic algorithm to incrementally produce (in no more than K steps) a "significant point set" M. The algorithm starts with an empty set $M^{(0)}$ and at each step adds a point to M, chosen according to the following criterion. Two things are required at the m-th step. First, given the current set $M^{(m)}$, an attempt is made to reconstruct $O(\mathbf{x})$ by

$$O^{rec}(\mathbf{x}) = \sum_{m=1}^{|M|} a_m(\mathbf{x}) O(\mathbf{x}_m). \quad (8.2.20)$$

If reconstructing $O(\mathbf{x})$ is successful, then by Eq. 8.2.18 the sample image $I(\mathbf{x})$ can be reconstructed. The optimal linear prediction coefficients $a_m(\mathbf{x})$, chosen to minimize the average reconstruction mean-squared error on $O(\mathbf{x})$

$$E = \langle ||O^{err}(\mathbf{x})||^2 \rangle \equiv \langle ||O(\mathbf{x}) - O^{rec}(\mathbf{x})||^2 \rangle, \quad (8.2.21)$$

are

$$a_m(\mathbf{x}) = \sum_{l=1}^{|M|} P(\mathbf{x}, \mathbf{x}_l) P^{-1}(\mathbf{x}_l, \mathbf{x}_m). \quad (8.2.22)$$

Then, at the $(m+1)$-st step the point that has the maximum reconstruction error $O^{err}(\mathbf{x}_{m+1})$ is chosen and M is added to it. Points are added to M until the reconstruction error goes below some acceptable level or until K of them have been chosen. In Penev and Atick [271], satisfactory results can be reached at $|M| = 64$ when the total number of facial images in the ensemble is 1,039, the number of pixels per image is 3,840, and the number of PCA components is 400.

Three statistical techniques for template matching have been discussed: (1) PCA uses eigenvalue decomposition or singular value decomposition to find a significant low-dimension linear subspace, (2) FLD reduces the dimension by using the information of class discrimination, and (3) LFA takes advantage of the image's localized similarity. One crucial criterion to ensure the effectiveness of all of these methods is that all facial images, either in training sets or under testing, need to be properly aligned to a predefined "grid." A slight translational or rotational displacement can easily spread the energy to higher dimensions in the feature space and render PCA useless. The most commonly used facial features for image alignment are the eyes. The position of the eyes makes images invariant to translation and rotation. The nose and mouth can also be used for scale invariance. Section 8.4 discusses automatic algorithms to detect human faces and facial features from still images or video streams.

8.3 Facial Pattern Classification Techniques

As discussed in Section 2.2.6, the pattern classifier and feature extractor are often a complementary pair in the biometric identification paradigm. A powerful classifier needs to be adopted to separate the features from a simple extractor. Neural network techniques are widely applied and suitable for face recognition algorithms. Instead of recognizing a face by following a set of human-designed rules, neural networks learn the underlying rules from a given collection of representative examples. This ability to automatically learn from examples makes neural network approaches attractive and exciting. Moreover, it is well known that neural networks are very robust and adaptive. Therefore, for applications with many variation factors, neural networks seem to be a good remedy.

Constructing a neural model is very much dependent on the intended application and is crucial for successful recognition. For example, the model for gender classification [115] is different than that used for facial expression classification [293]. For face detection, multi-layer perceptrons [343] and convolutional neural networks [323] have been applied. For face verification, the dynamic link architecture [194] is applied, using Gabor wavelets as features. Another example is the Cresceptron [376], which is a multiresolution pyramid structure similar to Fukushima's Neocognitron. Support vector machines [132] have been applied to recognize faces that are rotated as much as 40 degrees.

This chapter focuses on probabilistic decision-based neural network (PDBNN) [209, 210] and its applications to face recognition. PDBNN enjoys the merits of both neural network and statistical approaches and it inherits the modular structure from its predecessor, the decision-based neural network (DBNN) [192]. For each person to be recognized, a PDBNN devotes one of its subnets to the representation of that particular person. Such a structure makes system maintenance easy. The updating of a PDBNN-based security system for any personnel change in an organization is relatively straightforward—just a simple addition or deletion of the subnets (using a localized training process). A centralized system, in contrast, would require global updating. PDBNN's modular structure also facilitates portable ID devices; it is simple to implement a PDBNN-based biometric ID on a wallet-size smart card system. The smart card records only the parameters of the subnet in the PDBNN that represent the cardholder. In terms of statistical merits, PDBNN adopts the form of probability density as its discriminant function, which yields far fewer false acceptance cases even if there are not many "negative examples" in the training phase (see Section 7.5.2). This characteristic is highly desirable in systems that are frequently under intruder attack. The remaining sections of this chapter will demonstrate PDBNN's ability to handle face detection, eye localization, and face recognition all at once.

8.4   Face Detection and Eye Localization

Face (and eye) detection can be viewed as a two-category classification problem. Given an image patch, a face detector decides whether the patch contains a human face (class ω_1) (class ω_0). Misclassification happens when the face detector misses a facial image patch (false rejection) or mistakenly raises a flag on a normal, nonfacial image (false acceptance). According to Bayes's decision theory, the decision rule for the two-category classification problem can be designed by the likelihood ratio (see Eq. 7.4.2):

$$\mathbf{x} \in \omega_1 \quad \text{if } \frac{p(\mathbf{x}|\omega_1)}{p(\mathbf{x}|\omega_0)} \geq T$$
$$\mathbf{x} \in \omega_0 \quad \text{if } \frac{p(\mathbf{x}|\omega_1)}{p(\mathbf{x}|\omega_0)} < T, \tag{8.4.1}$$

where T is the threshold and $p(\mathbf{x}|\omega_i)$ is the likelihood density of class ω_i. For typical binary classification problems, the PDBNN allocates two subnets to estimate the likelihood densities of both categories. For detection problems, however, the PDBNN adopts a simpler structure:

$$\mathbf{x} \in \omega_1 \quad \text{if } p(\mathbf{x}|\omega_1) \geq T$$
$$\mathbf{x} \in \omega_0 \quad \text{if } p(\mathbf{x}|\omega_1) < T. \tag{8.4.2}$$

Figure 8.1 shows this type of PDBNN. The simplification comes from the assumption that the likelihood density of the "nonface" class is uniform across the whole sample space. The single-subnet PDBNN estimates only the distribution of the face class samples. The decision threshold T controls the operating point on the detector's ROC (or DET) curve; a larger T value reduces the false acceptance rate (FAR) but may increase the false rejection rate (FRR). For applications such as automatic security systems, the decision threshold is usually set in favor of a lower FAR because the face detection function (and, similarly, the eye localization function) plays a "gatekeeper" role for processes that follow). This allows the face recognition engine to remain dormant until a "very possible" face pattern is detected from the continuous video stream.

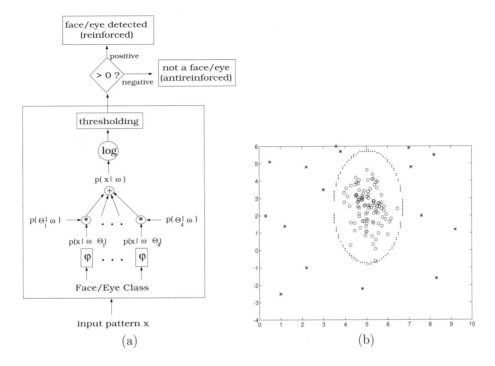

Figure 8.1. PDBNN-based face (or eye) detector. (a) Schematic diagram; (b) an example of pattern detection by PDBNN. ○ = patterns belonging to the target object; * = "nonobject" patterns.

8.4.1 Face Detection

Face detection is a difficult computer vision problem. Chapter 2 discussed many variation factors affecting detection accuracy. To simplify the task, many research groups focus only on developing detection algorithms for frontally viewed faces. For applications such as ATM access control, one can assume that rightful users will show their faces to the camera in an upright and mostly unoccluded fashion. Another useful application for frontal face detection is screening of police department photos of suspected criminals.

Although facial orientation may be constrained, frontal face detection remains a problem. Consider the picture on the left of Figure 8.2. It may look like a low-resolution image of a smiling face, but in fact it is just an enlarged image segment from the stadium crowd scene on the right. The face detector needs to carefully comb through arbitrary scenes to screen out false facial patterns like this one while also remaining responsive to the "true" face patches.

Sung and Poggio [343] developed a distribution-based system for face detection. Their system consisted of two components: distribution-based models for face and nonface patterns and a multi-layer perceptron classifier. In the training phase, each

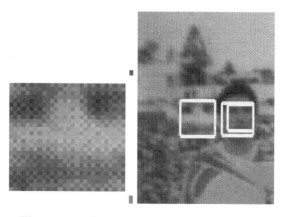

Figure 8.2. A pattern that looks like a face.

face and nonface example was first normalized and processed to a 19×19 pixel image and treated as a 361-dimensional feature vector. Next, the patterns were grouped into six face and six nonface clusters using the K-means algorithm. During the testing phase, two distance measures were applied. The first was the normalized Mahalanobis distance between the test image pattern and the centers of the proto-type clusters, measured within a low-dimension subspace spanned by the cluster's 75 largest eigenvectors. The second distance component was the Euclidean distance between the test pattern and its projection onto the 75-dimensional subspace. The last step was to use a multi-layer perceptron (MLP) network to classify face pat-terns from nonface patterns using the 12 pairs of distances to each face and nonface cluster. A database of 47,316 windowed patterns, 4,150 face patterns, and the rest nonface was used to train the classifier using the standard back-propagation algorithm.

Rowley et al. [323] also used a neural network model for face detection. A 20×20 window was used to scan the entire image. Each windowed pattern was preprocessed (normalization, histogram equalization) then fed into a three-layer convolutional neural network. Nearly 1,050 face samples of various sizes, orientations, positions, and intensities were used to train the network. The locations of the eyes, tip of the nose, and center of the mouth were labeled manually to normalize the faces. Multiple neural networks were trained, each focused on different sets of training images. In the testing phase, the decisions from these networks were merged and arbitrated by a simple scheme such as voting or logic operators (AND/OR). Rowley et al. reported that such an arbitration scheme was less computationally expensive than Sung and Poggio's system [343] and enjoyed higher detection rates on a test set of 144 faces in 24 images.

Lin et al. [211] proposed a one-class PDBNN model for face detection. There were four clusters (neurons) in the network; the number of clusters was determined based on empirical results. The input dimension of the network was $12 \times 12 \times 2 =$

Figure 8.3. Some faces detected by the PDBNN detector.

Table 8.1. Performance of the PDBNN face detector. *Note*: A database of 473 frontal facial images and most faces were detected within 5-pixel displacement.

Displacement from True Face Location	0 to 5 pixels	5 to 10 pixels	> 10 pixels
Percentage of test patterns	98.5%	1.5%	0 %

288. (12×12 image intensity plus x- and y-directional gradient feature vectors— compare to Figures 8.9(c) and 8.9(d)). The network weighting parameters and thresholds were trained by the procedures described in Chapter 7.

Table 8.2. Comparing PDBNN-based face detector with other face detection algorithms

System	Missed Faces	False Detection
Ideal	0 of 155	0 in 2,709,734
Rowley et al. [323]	34	3
Sung and Poggio [343]	36	5
PDBNN	43	6

The training database contained 92 annotated images (each image generated approximately 25 *virtual training patterns*). The images were taken in normal indoor lighting conditions with a cluttered background (refer to Figure 8.8). The image size was 320×240 pixels, and the face size was approximately 140×100 pixels. The variation of head orientation was about 15 degrees toward the four directions (up, down, right, left). The testing database contained 473 images taken under similar conditions. The testing performance was measured by the error (in terms of pixels) between detected face location and true location. To make it size invariant, errors were normalized with the assumption that the distance between both eyes is 40 pixels, which is the average distance in the annotated images. Among all testing images, 98.5% of the errors were within 5 pixels and 100% were within 10 pixels in the original high-resolution image (which is less than a 1-pixel error in the low-resolution images).

Table 8.1 lists the detection accuracy of the PDBNN face detector. For most face recognition applications, a 10-pixel error is acceptable since the main purpose of face detection is to restrict the searching area for eye localization. As for processing speed, the PDBNN face detector down-scaled the images (320×240 pixels) by a factor of 7. Working with low-resolution images (search range approximately 46×35 pixels, search step 1 pixel, and block size 12×12 pixels), the PDBNN pattern detector was reported to detect a face within 200ms on the SPARC II workstation.

The trained PDBNN face detector was applied to other image databases. Figure 8.3 shows its detection result on some sample images. The algorithm performed fairly well on images with different size faces (from the anchorwoman to the pedestrians) and in different lighting conditions.

Sung and Poggio's database was also used for comparison. The pictures in this database, which contained 155 faces, were from a wide variety of preexisting sources. Three face detection algorithms (PDBNN, Rowley et al., and Sung and Poggio) were applied in this comparison. Under the similar false acceptance performance, three nonface patterns were falsely detected as faces by Rowley et al., five by Sung and Poggio, and six by PDBNN. The false acceptance rates of all three were below 10^{-5}—Rowley et al. missed 34 faces, Sung and Poggio missed 36 faces, and the PDBNN face detector missed 43 faces (see Table 8.2).

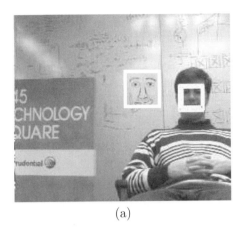
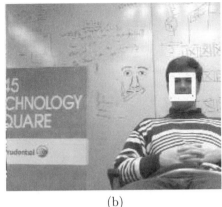

(a)   (b)

Figure 8.4. (a) Faces detected by Sung and Poggio's algorithm. (b) Face detected by PDBNN. Note that the artificial drawing (on the board) was marked as a face by (a) but not by (b).

Two reasons explain why PDBNN had an inferior performance. First, compared to the huge number of training samples used by both groups (4,000 in Sung and Poggio [343] and 16,000 in Rowley et al. [323]), PDBNN's training set only consisted of 92 images. A more interesting comparison would be made if a comparable-sized training database for PDBNN were available. Second, PDBNN did not detect the "artificial faces" (e.g., faces on poker cards, handdrawn face; cf. Figure 8.4). Since the PDBNN face detector was mainly used in surveillance and security applications, this "discrimination" may actually be beneficial.

8.4.2 Eye Localization

The eye localization module is activated whenever the face detector discovers facial patterns from the incoming image. Since the purpose of eye localization is to normalize facial patterns into a format the recognizer can accept, eye locations need to be pinpointed with much higher precision than face location.

The eye localizer must overcome several challenges. First of all, eye shape changes whenever people blink. Also, eyes are often occluded by eyeglass glare. Wu et al. [386] proposed a statistical inference method to "remove" eyeglasses by artificially synthesizing the face image without them. Although this method is not intended for face recognition applications, it would be interesting to determine whether the synthesized images could improve recognition accuracy.

Another challenge for the eye localizer is that the eye region usually occupies only a small portion of the captured image. In other words, the eye localizer deals with a smaller amount of image information but needs to generate more precise detection results. In Lin et al. [211], the PDBNN eye localizer used more examples

Table 8.3. Performance of eye localization. *Note*: Most eyes were detected with error of less than 3 pixels.

Displacement from True Eye Location	0 to 3 pixels	3 to 5 pixels	> 5 pixels
Percentage of test patterns	96.4%	2.5%	1.1 %

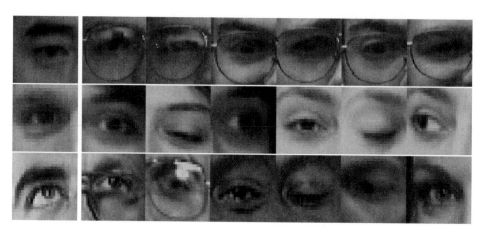

Figure 8.5. Eye images detected by the PDBNN eye localizer.

to form the training set (250 annotated images) and a higher resolution image patch for the input feature ($14 \times 14 \times 2 = 392$) than the face detector. The PDBNN eye localizer used one class subnet with four clusters to learn the distribution of the "eye" class. To simplify the training effort, only left eye images were used to form the training data set. Right eyes were detected by the mirrored image. Table 8.3 shows the experimental result on a test database of 323 images. The errors were normalized with the assumption that the eye-to-eye distance is 40 pixels. For a face recognition system, a misalignment of 5 pixels or less is tolerable. For this test database, 96.4% of the errors were within 3 pixels, and 98.9% were within 5 pixels.

Figure 8.5 shows several eye images detected by the PDBNN eye localizer, which is robust against variations in size, shape (including closed eyes), and orientation but somewhat vulnerable to eyeglass glare. Two failure cases are shown in Figure 8.6 in which the localizer mistook the glare to be a real eye and thus reported the wrong eye location.

8.4.3　Assisting Realtime Face Recognition

The application of face recognition technology can be categorized into two types: The first deals with controlled format photographs (e.g., photos in a police database).

Figure 8.6. Detection failures of the PDBNN eye localizer. The failures were caused by the specular reflection on the eyeglasses.

The number of images is usually small and additional images are not easy to obtain if more training images are needed. The second type of application receives realtime video streams (e.g., gateway security control), where the number of images can be very large. A video camera with a rate of 30 frames per second produces 1,800 images in one minute. The system developer therefore has the luxury of choosing many clear and distinguishable facial images to train the recognizer. Such an abundance of resources can greatly increase the chances of successful system training, but it also consumes a great deal of development time if human effort is required in the process of selecting "good" facial images. Thus, a complete realtime face recognition system demands not only a powerful face recognizer but also an effective, automatic scheme to acquire acceptable training samples.

The face detector and eye localizer play a crucial role in this acquisition scheme. The confidence score of the PDBNN provides a convenient and accurate criterion for selecting useful facial images from a large amount of image data. Because the face detector's confidence scores and eye localizer faithfully reflect the correctness of the pattern's detected position, an image with high confidence scores in both modules can almost always generate qualified facial patterns for face recognition.

The following description of an experiment shows more explicitly how to use the confidence scores of a face/eye detector to extract useful facial images. An in-house facial image database called the SCR 40×150 database was used for this study. This database was constructed by asking the tester to slowly move or rotate his or her head for 10 seconds in front of a video camera. Since the video frame rate was 15 per second, a total of 150 images were taken for each person. Forty people from different races, age groups, and genders participated in the construction of this database. While the 10-second video were taken, testers were asked to rotate their heads not only at a wide angle (up to 45 degrees) but also along various axes (i.e., left-right, up-down, and tilted rotations). All of the images were taken in front of a uniformly illuminated white background. The face detector and eye localizer worked correctly for 75% of the 6,000 images in this database, which was considered a *valid data set*. Although the face detector and eye localizer were trained only on frontal views, they nevertheless handled reasonably well the faces within 30 degrees. Most of the failures occurred in the case of faces with a large rotation/tilt angle (45 degrees or more). Moreover, the confidence scores of face detector and eye localizer

Table 8.4. Performance of a PDBNN face recognition system using a database with a large head orientation. *Note*: With the assistance of confidence scores from the PDBNN face detector and eye localizer, useful training patterns can be automatically selected (the valid set). The identification accuracy was highly improved by using a valid training set.

PDBNN	Trained by Original Set	Trained by Valid Set
Recognition	84.64%	98.34%
False rejection	10.03%	0.97%
Misclassification	5.33%	0.69%

were very low for all the failure cases. Therefore, images with a large head rotation angle or in poor condition due to thresholding were automatically screened out.

Twenty percent of the valid data set was selected as the training set for the PDBNN face recognizer. Table 8.4 shows its performance. The test image set was formed in two phases; first, 60 images were randomly selected from each individual's image data (excluding those used in the training set). Second, these 2,400 images were fed into the PDBNN face detector and eye localizer. By thresholding their confidence scores, 2,176 images were selected as the "valid test set."

Table 8.4 shows the performance of the PDBNN face recognizer on the valid test set. Here a false rejection pattern means that the confidence score generated by the face recognizer is below the threshold, and a misclassified pattern means that its confidence score is higher than the threshold but is classified to the wrong person. For the sake of comparison, another PDBNN was trained by a new set of training images formed by purely random selection from the original data set (the entire 6,000 images). Poor facial images might have been selected into this training set, which would account for a decrease in the recognition rate.

8.5   PDBNN Face Recognition System Case Study

A PDBNN-based face recognition system was developed through a collaboration between Siemens Corporate Research, Princeton, and Princeton University [188, 208–210]. The total system diagram is depicted in Figure 8.7. All four main modules—face detector, eye localizer, feature extractor, and face recognizer—were implemented on a Sun Sparc 10 workstation. An RS-170 format camera with 16mm, f1.6 lens was used to acquire image sequences. The S1V digitizer board digitized the incoming image stream into 640×480 8-bit grayscale images and stored them into the frame buffer. The image-acquisition rate was approximately 4 to 6 frames per second. The acquired images were downsized to 320×240 for the following processing [212].

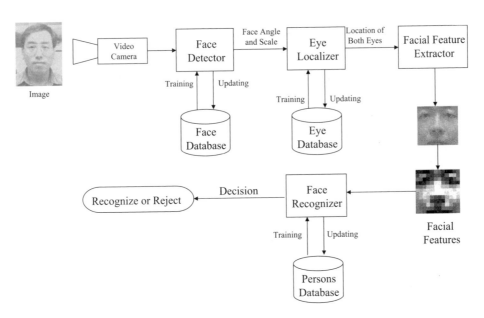

Figure 8.7. System configuration of the face recognition system, which acquires images from a video camera. The face detector determines whether there are faces inside the images, and the eye localizer indicates the exact coordinates of both eyes. The coordinates are passed to a facial feature extractor to extract low-resolution facial features for the face recognizer.

Figure 8.7 shows that the processing modules were executed sequentially. A module was activated only when the incoming pattern passed the preceding module (with an agreeable confidence level). After a scene was obtained by the image-acquisition system, a quick detection algorithm based on binary template matching was applied to detect the presence of a properly sized moving object. A PDBNN face detector was then activated to determine the presence of a human face. If positive, a PDBNN eye localizer was activated to locate both eyes. A subimage (approximately 140×100) corresponding to the face region was then extracted [212]. Finally, the feature vector was fed into a PDBNN face recognizer for recognition and subsequent verification.

This method is applicable under reasonable variations of pose and lighting and is very robust against large variations in facial features, eye shape, presence of eyeglasses, and cluttered backgrounds [207]. The algorithm took only 200ms to find human faces in an image with 320×240 pixels on the Sun workstation. For a 320×240 pixel facial image, the algorithm took 500ms to locate two eyes. In the face recognition stage, the computation time was linearly proportional to the number of persons in the database. For a 200-person database, it took less than 100ms to recognize a face. Because of the inherent parallel and distributed processing nature

Figure 8.8. The thin white box represents the location found by the face detector. Based on face location, the searching windows for locating eyes were assigned, as illustrated by the two thick white boxes.

of a DBNN, this technique can be easily implemented via specialized hardware for realtime performance.

8.5.1 Face Detection and Eye Localization

The training patterns of the PDBNN face detector and eye localizer were from a set of manually annotated facial images. Using these annotated coordinates, the 12×12 facial feature vector was extracted from the original facial image by the method described in Section 8.5.2. In operation mode, the face detector was activated when a moving object was detected from the video stream. A multiresolution window search was applied, and at each searching step the 12×12 feature vector was extracted and fed into the PDBNN face detector. A confidence score was produced by the PDBNN, indicating the system's confidence in the detection result. The presence of a face was declared if the score was above a predefined threshold. The network has consistently and reliably determined actual face positions, based on experiments performed on more than 1,000 testing patterns.

After the face detection phase, a PDBNN eye localizer was applied to a face image to locate the left and right eyes. Figure 8.8 shows an example of a detected face and the search windows for eye localization. Because the locations of both eyes were used to normalize face size and reorient the facial image, the detection result requires high precision. One prerequisite is that the pattern resolution used for the eyes is much higher than that used for faces. The proposed technique is insensitive to small changes in head size, face orientation (up to approximately 30%), and eyeglass style. It is also insensitive to partial occlusion by specular lens reflection.

8.5.2 Facial Region

Based on the given location of left and right eyes, a facial region—eyes, eyebrows, and nose—was extracted. An example is shown in Figure 8.9(a). Such a facial region—consisting of eyes and nose, but excluding mouth—provides distinctive fa-

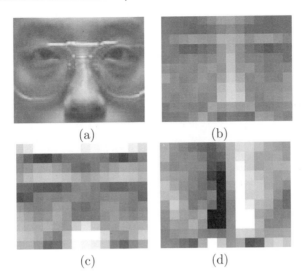

Figure 8.9. (a) Facial region used for face recognition, (b) intensity feature extracted from (a), (c) x-directional gradient feature, and (d) y-directional gradient feature.

cial features and also offers stability against different facial expressions, hair styles, and mouth movement. Hairline and mouth are suitable for secondary facial features; compare to Section 8.5.7.

Feature normalization and resolution reduction are important tasks for the creation of facial features. Two kinds of features were used for face recognition: intensity and edge. Pixel intensities were reconditioned by histogram modification techniques and edges were obtained by Sobel filtering. These two feature vectors were fed into two PDBNNs. The final recognition result was the fusion of the outputs of these PDBNNs. Intensities and edges in the facial region were normalized (to a range of 0 and 1) to compensate for changing illumination. The normalized and reconditioned (e.g., 140×100) images were then reduced to coarser (e.g., 14×10) feature vectors. The advantages of adopting lower-resolution facial features are (1) alleviating the *curse of dimensionality*, (2) reducing computational cost and storage space, and (3) increasing tolerance of location errors incurred by previous face detection and eye localization steps. Figure 8.9 shows an example of such extracted features.

8.5.3   Frontal View Faces

The experiment was conducted on three image databases: the SCR 80×20 database, the ARPA/ARL FERET database, and the ORL database.[3] The SCR 80×20 database consists of 80 people of different race, age, and gender and has 20 im-

---

[3] *http://www.uk.research.att.com/facedatabase.html.*

ages for each person. (If a person wears glasses, 10 of the images were taken with glasses and 10 without.) All images were taken under natural indoor conditions with uniform backgrounds. The facial orientations are roughly between -15 and 15 degrees. In many images, the person's head is tilted up to 15 degrees. A training data set was created by using four images per person (two with eyeglasses and two without, if the person wore eyeglasses). The testing image set included 16 images per person—1,280 images total. For all images, the face detector always correctly detected the center of the face (i.e., 100% success rate). Eye localization is a more difficult task than face detection, in particular when eyeglasses are present. The eye is typically 20 pixels wide and 10 pixels high. Among the 1,280 images, there were 5 images in which the eye localizer could not locate the eyes within the displacement tolerance of 5 pixels. For the remaining 1,275 images (in which the eyes were successfully located), the PDBNN face recognizer achieved a 100% recognition rate.

The ARPA/ARL FERET database contains 304 individuals, each having two frontal view images. The variation between the two images is much larger than those of SCR 80×20 and SCR 40×150 in terms of illumination, size, and facial expression. Phillips [274] achieved a 97% top-one identification rate on a preselected 172 faces from the FERET database. Moghaddam and Pentland [251] reported a 99% recognition rate on 155 preselected faces using a PCA-based eigenface method. In this experiment, because the confidence scores on the face detector and eye localizer reflect the accuracy of the detected pattern position, the images whose confidence scores were above the threshold were selected. Among the $304 \times 2 = 608$ images, 491 images passed both the face detector and eye localizer (success rate = 80.8%). Two hundred individuals passed both frontal view images; therefore, their images were used in the face recognition experiment; one image per person was used for training and the other for testing. The face recognition results are shown in Table 8.5.

Under reasonably high training accuracy (97%), PDBNN achieved a higher recognition rate (99%) than traditional DBNN (96%). MLPs were also used to implement the face recognizer, but their performance was inferior to both types of DBNN. The recognition results of this experiment do not imply that the PDBNN face recognizer has superior performance to the eigenspace approach [251] (99% for 155 people) because some of the images chosen may not be selected by the PDBNN face detector and eye localizer. Still, three conclusions can be drawn from these experimental results: (1) the PDBNN face recognizer can recognize up to 200 people with only one training pattern per person, (2) the face detector and eye localizer can help choose recognizable images automatically, and (3) PDBNN has better recognition performance than multi-layer perceptrons.

An experiment was also conducted on the face database from the Olivetti Research Laboratory in Cambridge, UK (the ORL database). The database contains 10 different images of 40 different people with variations in facial expression (open/closed eyes, smiling/nonsmiling), facial details (glasses/no glasses), scale (up to 10%), and orientation (up to 20 degrees). An HMM-based approach was applied

Table 8.5. Performance of different face recognizers on 200 persons in the FERET database.

	Training Accuracy	Testing Accuracy
Probabilistic DBNN	97.0%	99.0%
Traditional DBNN	100.0%	96.0%
Multi-layer perceptron	99.5%	87.5%

Table 8.6. Performance of different face recognizers on the ORL database. Adapted from Lawrence et al. [195].

System	Error Rate	Classification Time	Training Time
PDBNN	4%	< 0.1 seconds	20 minutes
SOM + CN	3.8%	< 0.5 seconds	4 hours
Pseudo 2D HMM	5%	240 seconds	n/a
Eigenface	10%	n/a	n/a
HMM	13%	n/a	n/a

to this database and a 13% error rate was achieved [329]. The eigenface algorithm of Turk and Pentland [356] reported an error rate around 10% [195, 329]. In Samaria [333], a pseudo two-dimensional HMM method was used and achieved 5% at the expense of long computation time (4 minutes/pattern on a Sun Sparc II). Lawrence et al. [195] used the same training and test set size as Samaria did and used a combined neural network (self-organizing map and convolutional neural network) to perform the recognition. This scheme spent 4 hours training the network and less than 1 second recognizing one facial image—the error rate was 3.8%. The PDBNN-based system reached similar performance (4%) but with a much faster training and recognition speed (20 minutes for training and less than 0.1 second for recognition). Both approaches were run on a SGI Indy. Table 8.6 summarizes the ORL database's performance.

8.5.4   Presence of Intruder Patterns

It is crucial for a recognition system to have the ability to reject intruders. An experiment was conducted to investigate the false acceptance and false rejection of the face recognition system. Among the 80 persons in the frontal-view database, 20 were chosen as "known persons" and 40 as "intruders." The remaining 20 individuals served as "negative examples" for network training. The training data sets were

Table 8.7. FARs of face patterns on various neural networks. *Note:* Detection threshold was set so that both false rejection rate and misclassification rate were equal to 0%.

	W/o Negative Examples	W/ Negative Examples
Probabilistic DBNN	13.75%	8.13%
Traditional DBNN	33.75%	22.5%
Multi-layer perceptron	33.75%	12.5%

formed by randomly choosing two images per person from the database and then generating 50 *virtual training patterns* (cf. Section 8.5.6) from each image. Both the known person group and negative training set have $20 \times 2 \times 50$ training patterns. The test sets were formed by selecting four images from each person. There were 20×4 test patterns from the known persons and 40×4 patterns from the intruders. If a test pattern is from a known person but is rejected by the network, the pattern is *falsely rejected*. If it's accepted but categorized to the wrong identity, the pattern is *misclassified*. If the pattern is from an intruder but is misrecognized to a known person, this pattern is *falsely accepted*.

The performance of probabilistic DBNNs, traditional DBNNs [192], and multi-layer perceptrons (MLPs) were compared. The number of clusters (or hidden neurons) that generated the best performance were chosen. There were two clusters for each subnet in either probabilistic or traditional DBNNs and 30 hidden neurons in the MLPs. The thresholds were adjusted so that all three networks had zero false rejection and misclassification rates [195]. The FARs are shown in Table 8.7. As mentioned in Section 7.5.2, likelihood density generates more locally conserved decision regions than posterior probability, so the probabilistic DBNN achieved a lower FAR than the traditional DBNN and MLP, whose outputs were proved to converge to posterior probabilities [311, 324]. Using negative training examples significantly reduced the false acceptance rate. This effect can be observed in all three networks.

8.5.5 Invariance Assurance

To understand the performance of the PDBNN face recognizer under the influence of different variation factors, an experiment was conducted on a face database of 55 known persons. The variation of this training set is similar to that in the IM66 database (five different head orientations, four different lighting conditions, and three expressions; see Section 8.5.7). Four PDBNNs were trained on different subsets of the training data, and were evaluated on a test set containing many variations that were not found in the training set (e.g., different head orientations, facial expressions, outdoor lighting, spotlight, heavy shadows, etc.). The variation

Recognition Rate (%)

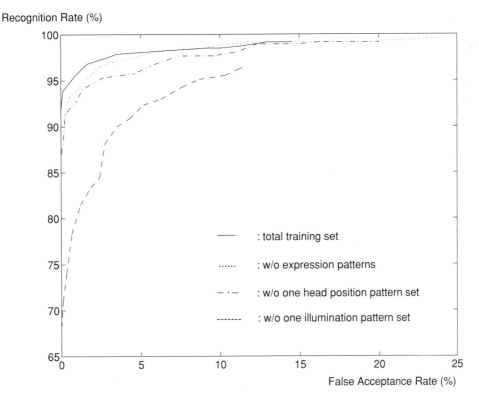

Figure 8.10. PDBNN face recognition experiment in which four PDBNNs were trained on the four different training data sets depicted here. The test set consisted of images of 55 known persons and 100 intruders, and the results show that the PDBNNs are insensitive to facial expression but sensitive to lighting variation.

is larger than that in the FERET database or the ORL database. Figure 8.10 depicts the ROC curves of the four PDBNNs—the PDBNN trained by the complete set achieved the best performance.

The performance of the PDBNN trained by the set without expression patterns was close to the best performance because the facial region chosen for feature extraction was not sensitive to facial expressions. The figure shows that performance deteriorates rapidly when the training patterns in one illumination condition are missing (dotted line), especially for those test images taken in an outdoor lighting environment. (All of the training patterns were taken under indoor lighting conditions.) The experimental results imply two directions of improvement for the PDBNN recognition system. First, although PDBNNs are immune to variations, such as facial expression, head size, and head orientation, performance varies under different lighting conditions. Second, current illumination normalization procedures (e.g., histogram normalization, edge preserving filtering) need to be improved.

8.5.6 Training Pattern Generation

The training pattern generation scheme for PDBNNs has the following three main aspects.

1. *Virtual training patterns.* To ensure sufficient diversity of real facial images in the training set, the algorithm took the acquired sensor image and transformed it in various ways to create additional training exemplars called *virtual training patterns.* As many as 200 virtual training patterns were generated from one original pattern by applying various affine transformations (e.g., rotation, scaling, shifting) and mirroring processes. The robustness of the trained network was consistently improved with the use of a virtual training set.

2. *Positive and negative training patterns.* Not all virtual training patterns were considered to be good face or eye patterns. If a virtual pattern was slightly perturbed from the original exemplar pattern, it was included in the positive training set. On the other hand, if the perturbation exceeded a certain threshold (predefined empirically), the virtual pattern was included in the negative training set. When training PDBNNs, positive patterns were used for reinforced learning, and negative patterns were used for antireinforced learning.

3. *Runtime negative pattern generation.* During the training phase, the PDBNN—while still under training—could be used to examine the whole image database every k epochs. If the network falsely detected a face (eye), that particular subimage was added to the negative training set.

8.5.7 Hierarchical Face Recognition System

The PDBNN face recognizer described in the preceding sections can be extended to a hierarchical recognition system in order to increase recognition accuracy. One example appears in Figure 8.11. A *face verifier* was cascaded with the *face recognizer.* Possible candidates of verifier input are facial regions with not as much discriminating information in the eye–nose region, such as hairline or mouth area. In this system, the forehead/hairline region was captured and downsampled to a 12×8 image before being fed to the face verifier, which was another PDBNN classifier. Its function was to verify/reject the decision of the primary recognizer. Because the hairline (forehead) region is smoother than the eye–nose region, it is easier to normalize lighting effects in this area. Therefore, the influence of lighting variation on final recognition results was reduced by the presence of the hairline verifier.

The verification scheme is as follows. After the PDBNN verifier was trained, an auxiliary data structure called the similarity list was constructed for each object class. The similarity list of class j records the IDs of all classes in the database that appear to be similar to class j from the face verifier's viewpoint. In other words,

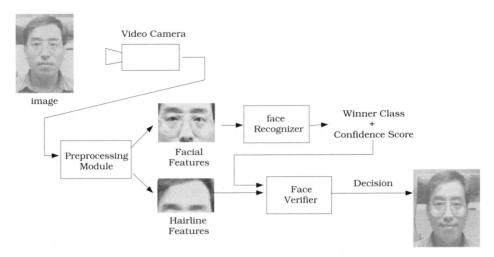

Figure 8.11. Hierarchical information processing system based on PDBNN. Primary features are from the facial region, and hairline features are used as a supporting verifier.

if the highest score of a class k training pattern comes from the j-th subnet, j is then recorded into class k's similarity list. The construction process of the lists is completed when all of the training patterns in the known-person database have been presented to the system. After the similarity lists were constructed, the following rules were used to verify the face recognizer's decisions.

- If the highest confidence score of the face recognizer is from subnet i but is below the recognizer threshold, the input pattern is recognized as class i if (a) the highest score of the face verifier is also from subnet i, *and* (b) the score exceeds the verifier's threshold. Otherwise, the pattern is rejected.

- If the highest confidence score of the face recognizer is from subnet i and is above the recognizer threshold, the input pattern is recognized as class i if (a) the highest score of the face verifier is from one of the classes on the similarity list of class i, *and* (b) the score exceeds the verifier's threshold. Otherwise, the pattern is rejected.

A face database (IM66) containing images of 66 individuals was used for the experiment. There were more variations in this database than in SCR 80×20 and SCR 40×150. For each person, the database has images of five different head orientations, four illumination conditions, and two images of three facial expressions (smile, frown, and surprised), thus yielding a total of $4 \times 5 + 3 \times 2 = 26$ images per person. A 38-class hierarchical PDBNN face recognition system was built. This system successfully recognized the 38 persons in the positive training data set and rejected the remaining 28 who were considered intruders.

Table 8.8. Performance of the hierarchical face recognition system

	Without Face Verifier	**With Face Verifier**
Recognition	92.65%	97.75%
False rejection	7.29%	2.25%
Misclassification	0.06%	0.00%
False acceptance	9.35%	0.00%

To handle the high feature dimension (13×9 for the face recognizer and 12×8 for the face verifier), EBF was applied for both face recognizer and face verifier as the discriminant function. The K-means algorithm was used for LU learning. For both face recognizer and face verifier, 10 images per person in the 38-person group formed the training set, and the remaining 16 images were used for evaluation. The images in the ORL face database were used as *negative training patterns* (for the face recognizer only). The images of the other 28 individuals in the IM66 database were used as *intruder test patterns*.

The face recognizer's recognition result is summarized in Table 8.8. The rates on the first three rows were calculated from the known-person data set, so they sum to 100% in each column. The FARs, which are in the fourth row, were obtained from the intruder data set. Because the recognition rate is the percentage of patterns in the known persons database correctly recognized by the system, the summation of recognition, false rejection and misclassification rates sum to 100%. This experiment shows that with face recognition alone the performance was worse than that of SCR 80×20. This is not surprising because the variation in IM66 was larger than that of SCR 80×20. With the help of the face verifier, the combined error rate was greatly reduced. In Table 8.8, the decision thresholds of the PDBNNs were set to lower the false acceptance and misclassification rates. Such a setting is favorable if the recognition system is for access-control applications.

8.6   Application Examples for Face Recognition Systems

The PDBNN face recognizer can be considered an extension of the PDBNN face detector. As mentioned in Section 7.5, PDBNN possesses the OCON structure—that is, there is one subnet in the PDBNN designated for a person to be recognized. For a K-person recognition problem, a PDBNN face recognizer consists of K different subnets. Analogous to the PDBNN detector, a subnet i in the PDBNN recognizer models the distribution of person i only and treats those patterns that do not belong to person i as "non-i" patterns.

The network architecture, learning rules, and performance evaluation of the multiple subnet PDBNNs were discussed in Chapter 7. This section focuses on their application to face recognition; the performance of the PDBNN face recognizer will be evaluated based on various experimental results. A hierarchical PDBNN face recognition system that exploits the discriminative power of PDBNNs is introduced in Section 8.5.7. This section also discusses several possible applications for the face recognition system.

The performance of PDBNN face recognition system is discussed from three aspects and corresponding experiments. The first experiment trained and tested PDBNN with three face databases consisting of frontal view face images (frontal view face databases are used in most face recognition research groups). The second experiment explored PDBNN's superior capability of rejecting intruders, which makes PDBNN attractive for applications such as access control. The third experiment compared PDBNN's performance under different variation factors.

8.6.1 Network Security and Access Control

The main purpose of designing the PDBNN face recognition system is access control. The PDBNN face recognition system diagram in Figure 8.7 fits right into the distributed biometric access-control system in Figure 2.3. The sensory device for the PDBNN system is a video camera. The reference database can either be stored locally in a central database or evenly distributed across a network. With the OCON structure of PDBNNs, it is possible to implement the face recognition network on a smart card system; the network parameters in the subnet representing the person can be stored in a smart card after proper encryption. When approaching the access point, a user should simply insert the card into the reader and then smile at the camera. The advantage of the smart card system is that the local matcher need not store the reference data.

The face recognition access-control system can also be adopted to user authentication over a network. With advances in multimedia infrastructure, more PCs are now equipped with video cameras. It is therefore possible to use a face image to assist, if not replace, the text-based password for the identity authentication process. The same technique can also be used in applications such as video conferencing or videophones. Figure 8.12 shows several frames from a live sequence, which demonstrates the recognition accuracy and processing speed of the PDBNN face recognition system.

8.6.2 Video Indexing and Retrieval

In many video applications, browsing through a large amount of video material to find relevant clips is an extremely important task [402]. The video database indexed by human faces allows users to efficiently acquire video clips about a person of interest. For example, a film study student can easily extract clips of his or her favorite actor from a movie archive, or a TV news reporter can quickly find from the news database clips containing images of the governor of California in 1980s.

A video indexing and retrieving scheme based on human face recognition was proposed by Lin [206]. The scheme contains three steps. First, the video sequence

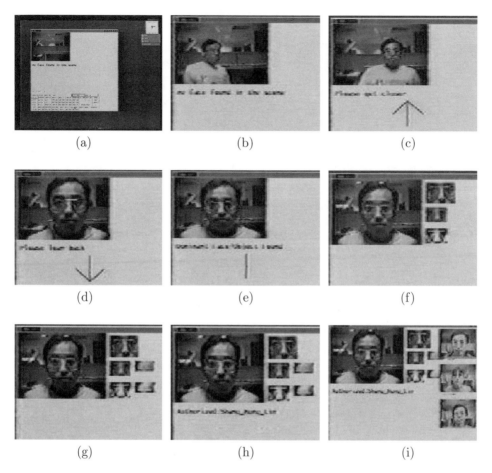

Figure 8.12. A live sequence of the hierarchical face recognition system. (a) X Window interface; the image in the upper-left corner was acquired from a video camera in real time. (b) A person entering the camera range (time index: 0 sec). (c) The system detected the presence of a face; since the face in the image was too small, the system asked the person to stand closer. Text on picture: "Please step closer"—time index: 0.33 sec. (d) The person stood too close to the camera; the system asked the person to step back. Text on picture: "Please lean back"–time index: 4.7 sec. (e) The system located the face. Text on picture: "Dominant Face/Object Found"—time index: 5 sec. (f) Facial features were extracted—time index: 6 sec. (g) Hairline features were extracted—time index: 6.33 sec. (h) The person was recognized—text on picture: "Authorized: Shang Hung Lin"—time index: 9.1 sec. *Note:* Special sound effects (a doorbell sound and a human voice calling the name of the authorized person) were generated between time 6.33 sec and time 9.1 sec. The actual recognition time took less than half a second. (i) The pictures of the three individuals most similar to the person under test—time index: 9.2 sec.

is segmented by applying a scene change detection algorithm [392]. Scene change detection indicates when a new shot starts and ends. Each segment created by scene change detection is considered as a story unit of the sequence. Second, a PDBNN face detector is invoked to find the segments most likely to contain human faces. From every video shot, the representative frame (Rframe) is selected and fed to the face detector. The representative frames for which the detector gives high face detection confidence scores are annotated and serve as the indices for browsing. Third, a PDBNN face recognizer finds all shots containing the persons of interest based on the user's request. The PDBNN recognizer is composed of several subnets, each representing a person to be recognized in the video sequence. If a representative frame has a high confidence score on one of the subnets, it is considered a frame containing that particular person. Figure 8.13 shows the system diagram of this face-based video browsing scheme; *not* all people who appear in the video sequence are assigned a PDBNN recognizer. This is similar to browsing, where a neural network needs not be built to recognize persons who are not in the cast.

A preliminary experiment was conducted using a 4-minute news report sequence. This sequence, which was sampled at 15 frames/sec with a spatial resolution of 320×240, contained three persons: an anchorman, a male interviewee, and a female interviewee. The goal was to find all of the frames that contained the anchorman. The sequence was divided into 32 segments; representative frames are illustrated in Figure 8.14(a). The anchorman's face appeared in 7 segments (4 in frontal view, 3 in side view). The PDBNN face detector and eye localizer mentioned before were again applied here to pinpoint the face location on each representative frame. It took about 10 seconds to process one frame. A PDBNN recognizer was then built to recognize the anchorman; approximately 2 seconds' length of frames (32 frames) were used as training examples. It took about 1 minute for the network to learn all of the training examples, and only seconds to process all of the index frames. The experimental result showed that the four segments containing the anchorman's frontal view images were successfully recognized with very high confidence scores—Figure 8.14(b). No other video segments were incorrectly identified as the anchorman.

Notice that since this approach requires the appearance of both eyes, the three side-view segments of the anchorman could not be recognized; overcoming this limitation is considered as future work. The face detection step needed much more processing time than the face recognition step did. For this news report sequence, the system took about 5 minutes (32 Rframes × 10 sec/frame) in the detection step. The long processing time is the result of a multiresolution search that locates faces with different sizes in the face detection step. Several methods can be applied to reduce total processing time. For example, the number of Rframes can be reduced by roughly merging similar Rframes based on their color histogram or shape similarity. In fact, since the face detector's job is to locate and annotate faces in the sequence, it need run only once. Thus, face detection time does not affect the system's online performance.

Video Sequence

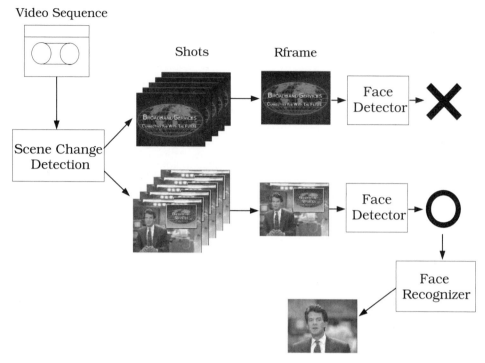

Figure 8.13. A PDBNN face-based video browsing system. The scene change algorithm divides the video sequences into a number of shots, and the face detector examines all representative frames to determine whether they contain human faces. If so, the detector passes the frame to a face recognizer to find out whose face it is.

8.6.3   Airport Security Application

Several field studies have suggested that it is still premature to claim reliable and robust performance by any face recognition product for airport security applications. On an optimistic note, Cognitec seems to have had some initial success. Its automatic passport checking system, SmartGate, had successfully processed 62,000 transactions of 4,200 enrolled Qantas Air crew members at Sydney's airport by November 2003. In contrast, disappointing results were reported at Boston's Logan International Airport after it tested both Identix and Viisage technologies in 2002 [145]; the system failed to identify positive matches 38% of the time. Although false positives based on an operator's decision didn't exceed 1%, machine-generated false positives exceeded 50%. The major difference between Sydney's success and Boston's failure is that face recognition technology was applied to different applications. Unlike the trial in Boston, where illumination at checkpoints can change drastically, SmartGate was installed under a controlled lighting environment and received testers' full cooperation.

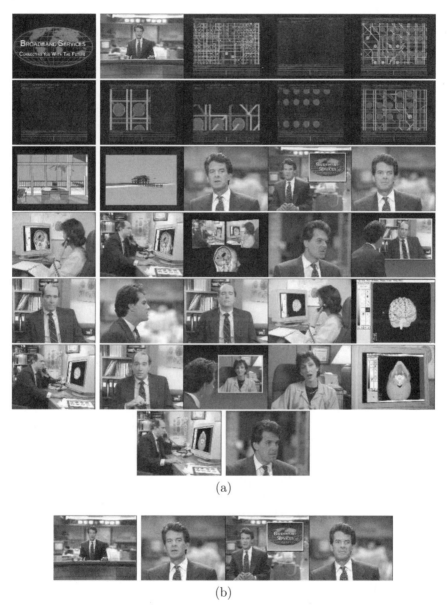

(a)

(b)

Figure 8.14. Representative (a) frames of the news report sequence and (b) frames of the anchorman found by the PDBNN face recognizer.

In general, when applications show vulnerability with respect to illumination and pose variation, poor performance and unacceptable results are often anticipated. A major R&D investment is needed to enable face recognition to become a mature and effective commercial product for airport security applications.

8.6.4 Face Recognition Based on a Three-Dimensional CG Model

In conventional approaches to face recognition, facial features are extracted from two-dimensional facial images. The main problem of such features is that a straight viewing angle and well-controlled lighting conditions are critically important for good performance. To overcome such difficulties, a novel three-dimensional computer graphics (CG) face model was proposed by Sakamoto et al. [328]. A range finder, as illustrated in Figure 8.15, was developed to acquire the three-dimensional face models. Facial depth (i.e., the distance between the face and camera) can be calibrated by projecting two point-source lights through sinusoidal grating films onto the facial surface. This allows the facial depth to be estimated in very fine resolution (on a scale of millimeters). This comes at a huge computation cost in the offline training phase. In fact, in terms of computational requirements, the range finder requires approximately 180 giga-operations (GOs) per person.

In the recognition phase, when an input photo query is presented, the registered CG faces are modified so that the adjusted face pose and illumination conditions can best match that of the input photo. The optimally adjusted CG face and query photo can thus be directly compared using the straightforward two-norm distance metrics. During this phase, the matching operation must be performed for all possible viewing angles (1 angle in each 5-degree interval). This amounts to a computational requirement of 10 GOs per person in realtime processing. This implies that a total of 1 tera-operation per second (TOPS) ($= 100 \times 10$ GOs) will be required for a database containing 100 registrants. This in turn implies a required realtime processing power of 1 TOPS if the allowed recognition latency is 1 second. Such a high speed far exceeds the capability offered by today's state-of-the-art digital signal processing technology. However, according to Moore's law on the rapid growth of integrated circuit technology, such a processing speed could become a reality in less than a decade.

8.6.5 Opportunities for Commercial Applications

Face recognition technology can find attractive commercial application opportunities when an application has a controlled environment or, even better, can expect user cooperation. In such cases, current face recognition technology usually works very well, as seen in Cognitec's passport checking system and Viisage's Illinois driver's license photo scan.

Another promising business opportunity can occur in situations where false decisions cause only inconvenience but are not really damaging. For instance, Fuji film's digital photo processing kiosk uses face recognition techniques to detect the presence of people in everyday photos. When human faces are detected, the kiosk

3 Cameras

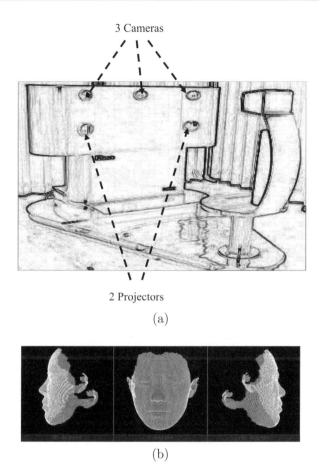

2 Projectors

(a)

(b)

Figure 8.15. Generation of three-dimensional face CG models. (a) The range finder consists of three cameras and two projectors. To calibrate the depth, the projector projects a point-source light through a sinusoidal grating film; this allows the facial depth to be estimated in very fine resolution (in mm scale). (b) The resulting face images of the three-dimensional CG face model with three different angles of approximately −45%, 0%, and +45%. Adapted from Sakamoto et al. [328].

balances the color based on the face region and makes the skin tone of the printed photo more vivid [140]. Automatic red-eye reduction and advanced autofocus are immediate extensions of this line of applications. Face recognition represents yet another promising add-on feature for video game products. Many face recognition companies (e.g., Viisage and Identix) have already provided authorization software (or screensaver) and data-mining tools to search the Internet or image databases for

PC and mobile devices. The technology will become more mature in the future and find a broader domain of applications including automatic teller machine (ATM) access and gate control for buildings.

8.7 Concluding Remarks

Face detection and eye localization are two very crucial preprocessing steps in the automatic face recognition system. The system needs the face detection module to find human face patterns from arbitrary scenes and the eye localizer to pinpoint the location of human eyes to normalize the detected face pattern. This chapter provided a case study of the automatic face recognition system, whose core technology is a modular neural network—the PDBNN. The configuration of the PDBNN face recognition system contains a face detector, eye localizer, and face recognizer. The system can be made automatic and requires no human operator. The processing speed (from the user appearing in front of the camera to the system permitting access) could easily meet realtime requirements. Performance data for the PDBNN face detector and eye localizer were also presented. Experimental results show that the PDBNN can perform face detection and eye localization at very high accuracy— 0%, 10-pixel displacement error for face detection, and 1.1%, 5-pixel displacement error for eye localization.

The face recognizer not only must deal with translational, rotational, and scaling variations, but also needs to distinguish different faces under various lighting conditions and possible facial deformity. The PDBNN face recognizer was evaluated based on experiments on three face databases—FERET, ORL, and SCR. Because of its high recognition rate, fast processing speed, and full modularity/portability, the PDBNN recognition scheme is suitable for access-control systems such as gateway surveillance or network security. In fact, the Cognitec FaceVACS system is very similar to the PDBNN face recognition system discussed in this chapter.

Many fundamental limitations exist in current face recognition systems. For example, a recent study suggests that for current face recognition technology to work well, several conditions must be met [139].

- The subject could not be wearing glasses: "Eyeglasses were problematic," according to a summary of the test findings. "Glare from ambient light and tinted lenses diminished the system's effectiveness."

- The angle of the facial image could not vary: "There was a substantial loss in matching if test subject had a pose 15 to 30 degrees (up/down, right/left) off of input camera focal point."

- The subject had to be perfectly still: "Motion of test subject head has a significant effect on the system ability to both capture and alarm on test subject."

- The subject had to be properly lit: "System required approximately 250 lux of directional lighting" to work.

- The airport had to have high-quality photographs. "Input photographs populating the database need to be of a good quality."

The technology must overcome these obstacles before it can be adopted as a truly automatic surveillance and security control method. Face recognition will require significant level of research and development before it can become a mature and effective commercial product.

Problems

1. Compare PCA-based and FLD-based facial feature extraction techniques. Under what conditions does FLD fail to gain the advantage of feature dimension reduction over PCA?

2. You are assigned to build a "mouth localizer" to assist with the functioning of the eye localizer of the face recognition system.

   (a) What kind of "mouth features" do you plan to use for detecting mouths? Describe the step-by-step procedure that you plan to use to implement the localizer.

   (b) What factors may affect your localizer's detection accuracy? How do you plan to make your localizer more robust against variation factors?

3. Changes of lighting condition can drastically affect the performance of face recognition systems.

   (a) What methods were applied to the PDBNN face recognition system to reduce the influence of lighting changes?

   (b) Research the literature to find more algorithms that could achieve illumination invariance.

4. Explain how you would use the following to build a face verification system.

   (a) Gaussian mixture models (GMMs)

   (b) Radial basis function networks (RBFNs)

   (c) Probabilistic decision-based neural networks (PDBNNs)

   (d) Support vector machines (SVMs)

   Draw a block diagram for each case. Also, explain how each of these models characterizes impostors' faces.

5. In designing a face detection system, we are sometimes provided with a collection of facial as well as nonfacial patterns. In training a face detection model, these patterns can be used, respectively, as positive and negative training samples. The network training strategy is dictated by its reaction to the positive or negative samples.

    (a) When a positive sample for a class is presented, it may be used to strengthen the statistical validity of the model. Show that a plausible positive-learning strategy is to adopt the reinforced learning rule introduced in Chapter 7. Explain why the learning rule can help reach a consensus between the newly observed positive sample and the old training samples.

    (b) When a negative sample of a class is presented, it can be used to cross-check and, if necessary, to further rectify the original model so that the chance of future error can be reduced. In other words, the main objective of a negative training strategy is to make the adjusted model react less favorably to samples resembling the negative sample observed. Design a negative-learning strategy based on the antireinforced learning rule introduced in Chapter 7.

6. A face recognition system usually contains these two distinct tasks:

    (a) Preprocessing in which a subset of potentially influential features is found

    (b) Postprocessing in which optimal and classification are performed

Discuss the two tasks in the context of a face recognition system. Suggest the types of features, optimization algorithms, and classification schemes for each system.

7. *Matlab Exercises.* Download the public "ORL face database" from AT&T Laboratories Cambridge (*http://www.uk.research.att.com/facedatabase.html*). The database contains 10 different images of 40 distinct subjects. The size of each image is 92×112 pixels, with 256 levels of gray per pixel.

    (a) Rearrange the images into $10,304 \times 1$ vector form ($10,304 = 92 \times 112$). Design a Matlab program to calculate the covariance matrix of all images in the database, and generate the corresponding eigenvectors and eigenvalues. (*Hint*: see Eq. 8.2.1.)

    (b) Sort the eigenvectors in order of descending eigenvalues. Plot the eigenvalues. How many eigenvectors are sufficient to reduce the reconstruction error to 10%? How many for 5%?

    (c) Discuss possible preprocessing techniques to reduce the sufficient number of eigenvectors in Problem 7b.

(d) Now equally divide the database into a training set and a test set. In other words, each set consists of 5 images of 40 persons. Repeat Problem 7a and Problem 7b, but use only the training set.

(e) Use the eigenvectors that reach 10% of reconstruction error to form an eigenspace. Project all the images in the test set onto this eigenspace. Implement a simple pattern classifier with Matlab as follows:

- For each subject class, calculate the mean vector of the 5 training images in the eigenspace.

- For each test image, calculate its distance to the mean vector of each of the 40 subject classes. Find the minimum distance.

- Define a thresholding parameter T. If the minimum distance is smaller than T, categorize the test image to that subject class; otherwise, categorize the test image as "unknown." Plot the ROC curve.

(f) Discuss possible classification techniques to improve the ROC performance.

Chapter 9

BIOMETRIC AUTHENTICATION BY VOICE RECOGNITION

9.1 Introduction

Chapter 3 outlined the basic principles of the EM algorithm. The key idea of expectation-maximization is to introduce a set of missing variables so that a density estimation problem without a closed-form solution can be decomposed into a number of iterative steps, from which closed-form solutions can be easily obtained. The EM algorithm has been shown to be a powerful tool for estimating the parameters of mixture densities.

This chapter demonstrates how the powerful EM technique can be applied to the modeling of speaker features, which is a key step in building speaker verification systems. Specifically, this chapter details and compares several kernel-based probabilistic neural networks that are particularly appropriate for this task. In addition to discussing density estimation, this chapter explains how the EM algorithm can be applied to transform speakers' features or speakers' models in a telephone-based speaker verification task. More specifically, the transformation is formulated as a maximum-likelihood problem where the unknown transformation parameters are estimated by the EM algorithm.

Experimental evaluations based on speech passing through 10 different phone handsets, including head-mounted, carbon-button, electret, and portable, and seven speech coders—G.711, G.726, GSM, G.729, G.723.1, MELP, LPC—are detailed. The evaluations demonstrate the effectiveness of kernel-based neural networks in capturing speaker characteristics. The evaluations also show that stochastic feature transformation and model transformation perform much better than conventional techniques in minimizing handset- and coder-distortion.

9.2   Speaker Recognition

The goal of automatic speaker recognition—Campbell [45] and Furui [109]—is to recognize a speaker from his or her voice. Speaker recognition can generally be divided into two categories: *speaker identification* and *speaker verification*. Speaker identification determines the identity of an unknown speaker from a group of known speakers; speaker verification authenticates the identity of a speaker based on his or her own voice. A speaker claiming an identity is called a *claimant*, and an unregistered speaker pretending to be a registered speaker is called an *impostor*. An ideal speaker recognition system should not reject registered speakers (*false rejections*) or accept impostors (*false acceptances*).

Speaker recognition can also be divided into text-dependent and text-independent. In *text-dependent* systems, the same set of keywords are used for enrollment and recognition; in *text-independent* systems, on the other hand, different phrases or sentences are used. Text-dependent systems, which require the cooperation of users and typically use hidden Markov models to represent speakers' speech, usually outperform text-independent systems because they can use phonetic information to align the unknown speech with reference templates. However, text-independent systems are more appropriate for forensic and surveillance applications, where predefined keywords are not available and users are usually uncooperative or unaware of the recognition task.

9.2.1   Components of Speaker Verification Systems

Typically, a speaker verification system is composed of a front-end feature extractor, a set of client speaker models, a set of background speaker models, and a recognition unit. The feature extractor derives speaker-specific information from the speech signals. It is well known from the source-filter theory of speech production [93] that spectral envelopes implicitly encode vocal-tract shape information (e.g., length and cross-section area) of a speaker and that pitch harmonics encode the glottal source information. Because it is commonly believed that vocal-tract shape varies from speaker to speaker, spectral features, such as the LPCCs or the MFCCs (see Section 2.6), are often used. A set of speaker models is trained from the spectral features extracted from client utterances. A background model is also trained using the speech of a large number of speakers to represent speaker-independent speech. Basically, the background models are used to normalize the scores of the speaker models to minimize nonspeaker related variability such as acoustic noise and channel effect. To verify a claimant, speaker scores are normalized by the background scores and the resulting normalized score is compared with a decision threshold. The claimant is accepted (rejected) if the score is larger (smaller) than the threshold.

9.2.2   Speaker-Specific Features

Speech signals are generated from quasi-stationary processes. Therefore, short-term spectral analysis can be applied to short speech segments, which results in a sequence of short-time spectra. In speech and speaker recognition, the short-time spectra are further transformed into feature vectors. In addition to spectral analysis, many speech and speaker recognition systems use linear prediction (LP) analysis [228] to extract the feature vectors (known as the LP coefficients) from short segments of speech waveforms. One advantage of LP coefficients is that they can be computed efficiently. More important, the LP coefficients represent the spectral envelopes of speech signals (i.e., information about the formant frequencies and their bandwidth). The spectral envelopes are characterized by the vocal-tract resonance frequencies, vocal-tract length, and spatially varied cross-section areas. Because all of these entities are known to be speaker-dependent, the LP coefficients are one of the candidate features for speaker recognition.

Several sets of features (e.g., LP coefficients, impulse responses, autocorrelation coefficients, cross-sectional areas, and cepstral coefficients) can be derived from LP analysis. Of particular interest is that a simple and unique relationship exists among these features. Despite this simple relationship, it has been shown that the cepstral coefficients are the most effective feature for speaker recognition [13] because the components of cepstral vectors are almost orthogonal.

In the field of psychoacoustics, it is well known that the human auditory system can be approximately described by a set of overlapped bandpass filters whose frequencies follow a scale known as the *critical band scale*. To capture the phonetically important characteristics of speech using auditory-based principles, Davies and Mermelstein [69] proposed using triangular *mel-scale filter banks* to extract spectral features from speech signals. These filters follow the mel-scale and space linearly at low frequencies and logarithmically at high frequencies. The resulting coefficients are therefore called mel-frequency cepstral coefficients (MFCCs). This chapter uses both LPCCs and MFCCs as speaker features.

9.2.3 Speaker Modeling

Over the years, a variety of speaker modeling techniques have been proposed [305]. This section describes four approaches to speaker modeling.

Template Matching

In this technique, reference templates are used as speaker models. Templates are composed of a sequence of feature vectors extracted from a set of fixed sentences uttered by a registered speaker. During recognition, an input utterance is dynamically aligned with the reference templates, and match scores are obtained by measuring the similarity between the aligned utterance and the templates [80]. The use of fixed templates, however, cannot model the wide variability present in the speech signals.

Vector Quantization

Vector quantization (VQ) is a coding technique typically used in transmitting signals at low bit rate. To use VQ in speaker recognition, a personalized codebook is created for each speaker. During recognition, an unknown speaker is identified by selecting the codebook whose code vectors are closest to the input vectors. Since its introduction by Soong et al. [342] in 1985, VQ has been a benchmark method for speaker recognition systems [234], and improvement in the standard VQ approach has also been made [32]. The advantage of VQ is that the problem of segmenting speech into phonetic units can be avoided. Additionally, VQ is more computationally efficient than template matching. The disadvantage of VQ, however, lies in the complexity of codebook search during recognition.

Hidden Markov Models

Hidden Markov models (HMMs) encode both the temporal structure of feature sequences and the statistical variation of the features. As a result, HMMs can be used as speaker models in text-dependent speaker recognition. The earliest attempt to use HMMs in speaker recognition was reported by Portiz [282]. After Portiz's work, several improved methods were proposed; that is, the mixture autoregressive HMMs [349], subword HMMs [123], and semicontinuous HMMs [102]. HMMs parameters can be estimated based on maximum-likelihood (ML) or maximum a posteriori (MAP) criteria. Criteria that use discriminative training, such as minimum classification error (MCE) [175, 238] or maximum mutual information (MMI) [264, 357], can also be adopted. Multistate, left-to-right HMMs can be used as speaker- and utterance-specific models.

The HMM approach is similar to the VQ approach in that the HMM states are found by a VQ-like procedure. However, unlike VQ, the probabilities of transition between states are encoded, and the order of presentation of the speech data is important. This may cause problems in text-independent speaker recognition where no temporal correlation exists between the training data and the test data. On the other hand, single-state HMMs—also known as Gaussian mixture models (GMMs) [300, 307]—can be applied to text-independent speaker recognition. Like VQ, the feature space of speakers is divided into a number of clusters. However, the probability density function is continuous rather than discrete, and the cluster membership is soft rather than hard. GMMs provide a probabilistic model for each speaker but unlike HMMs, there is no Markov constraint among the sound classes. As a result, the order of presentation of speech data will not affect recognition decisions.

Neural Networks

Neural networks, which have been used for speaker recognition, can be considered supervised classifiers that learn the complex mappings between data in the input and output space. This capability is particularly useful when the statistical distributions of the data are not known. The speaker models can have many different forms, including multi-layer perceptrons (MLP), radical basis functions (RBFs), hybrid MLP-RBF models [7], multi-expert connectionist models [24], and modified neural tree networks [94]. For MLP and RBF networks, each speaker has a personalized network trained to output a 1 (one) for the voices associated with that speaker and a 0 (zero) otherwise. One advantage of using neural networks for speaker recognition is that discriminative information can easily be incorporated by means of supervised learning. Although this information can usually improve recognition performance, it requires a longer training time.

9.2.4   Threshold Determination

The determination of decision thresholds is a very important problem in speaker verification. A large threshold could make the system annoying to users, but a small threshold could result in a vulnerable system. Conventional threshold determination methods [41, 107] typically compute the distribution of inter- and intraspeaker distances and then choose a threshold to equalize the overlapping area of the distributions—that is, to equalize the *false acceptance rate* (FAR) and *false rejection rate* (FRR). The success of this approach, however, depends on the estimated distributions matching the speaker- and impostor-class distributions. Another approach derives the threshold of a speaker solely from his or her own voice and speaker model [259]. Session-to-session speaker variability, however, contributes a great deal of bias to the threshold, rendering the verification system unusable.

Because of the difficulty in determining a reliable threshold, researchers often report the equal error rate (EER) of verification systems based on the assumption that an a posteriori threshold can be optimally adjusted during verification. Real-world applications, however, are only realistic with a priori thresholds that should be determined before verification.

In recent years, research effort has focused on the normalization of speaker scores both to minimize error rates and to determine a reliable threshold. This includes the likelihood ratio scoring proposed by Higgins et al. [136], where verification decisions are based on the ratio of the likelihood that the observed speech is uttered by the true speaker to the likelihood that it is spoken by an impostor. The a priori threshold is then set to 1.0, with the claimant being accepted (rejected) if the ratio is greater (less) than 1.0. Subsequent work based on likelihood normalization [214, 235], cohort normalized scoring [318], and minimum verification error training [320] also shows that including an impostor model during verification not only improves speaker separability but also allows decision thresholds to be set easily. Rosenberg and Parthasarathy [319] established some principles for constructing impostor models and showed that those with speech closest to the reference speaker's model performed the best. Their result, however, differs from that of Reynolds [308], who found that a gender-balanced, randomly selected impostor model performs better, which suggests that more work in this area is required.

Although these previous approaches help select an appropriate threshold, they may cause the system to favor rejecting true speakers, resulting in a high FRR. For example, Higgins et al. [136] reported the FRR is more than 10 times larger than the FAR. A report by Pierrot et al. [276], based on a similar normalization technique but with a different threshold setting procedure, also found the average of FAR and FRR is approximately three to five times larger than the EER, thereby suggesting that the EER could be an overly optimistic estimate of true system performance.

9.2.5   Performance Evaluation

Performance of speaker verification systems is usually specified by two types of errors:

1. *Miss rate* $(P_{\mathrm{miss|target}})$—the chance of misclassifying a true speaker as an impostor.

2. *False Alarm Rate* $(P_{\mathrm{fa|nontarget}})$—the chance of falsely identifying an impostor as a true speaker.

The miss rate and false alarm rate are also known as the FRR and the FAR. In addition to these two error rates, it is also common to report the equal error rate (EER)—the error rate at which $P_{\mathrm{miss|target}} = P_{\mathrm{fa|nontarget}}$.

Because the miss rate and false alarm rate depend on the decision threshold, a $\{P_{\mathrm{miss|target}}, P_{\mathrm{fa|nontarget}}\}$ pair represents one operating point of the system under evaluation. To provide more information about system performance, it is necessary to evaluate the system for a range of thresholds. This results in a *receiver operating characteristic* (ROC) curve, where the miss probability is plotted against the probability of a false alarm, similar to the one used by the face recognition community. However, the speaker verification community has chosen to use a variant of the ROC plots called *detection error tradeoff* (DET) plots [233]. In a DET plot, the axes' scales are normally deviated so that Gaussian distributed scores result in a straight line; the advantage is that systems with almost perfect performance can be compared easily.

In addition to DET curves, speaker verification systems are also compared based on the detection cost:

$$C_{\mathrm{det}} = C_{\mathrm{miss}} \times P_{\mathrm{miss|target}} \times P_{\mathrm{target}} + C_{\mathrm{fa}} \times P_{\mathrm{fa|nontarget}} \times P_{\mathrm{nontarget}},$$

where C_{miss} and C_{fa} are the cost of making a false rejection error and false acceptance error, respectively, and where P_{target} and $P_{\mathrm{nontarget}}$ are, respectively, the chance of having a true speaker and an impostor. Typical values of these figures are $C_{\mathrm{miss}} = 10$, $C_{\mathrm{fa}} = 1$, $P_{\mathrm{target}} = 0.01$, and $P_{\mathrm{nontarget}} = 0.99$ [288]. These values give an expected detection cost of approximately 1.0 for a system without any knowledge of the speakers. The operating point at which the detection cost C_{det} is at a minimum can be plotted on top of the DET curve.

Because the performance of speaker verification systems depends on the amount of training data, acoustic environment, and the length of test segments, it is very important to report this information in any performance evaluations so that performance of different systems and techniques can be compared. Thus, the NIST established a common set of evaluation data and protocols [152] in 1996. Although only focusing on conversational speech, the NIST speaker recognition evaluations are one of the most important benchmark tests for speaker verification techniques.

9.2.6 Speaker Recognition in Adverse Environments

It is well known that acoustic variation in the environment seriously degrades the performance of speaker recognition systems. In particular, the performance of most systems degrades rapidly under adverse conditions such as in the presence of background noise, channel interference, handset variation, intersession variability, and long-term variability of speakers' voices. Many sources of distortion can affect speech signals; additive noise and convolutive distortion are the most common.

Additive noise can be classified into different categories according to its properties. For example, stationary noise, such as electric fans and air conditioners, has a time-invariant power spectral density, but nonstationary noise created by passing cars, keyboard clicks, and slamming doors has time-varying properties. Additive noise can also be divided into continuous and short-lived, depending on the noise duration with respect to the speech duration.

In addition to additive noise, distortion can also be convolutional. For example, microphones, transmission channels, and speech codecs can be considered to be digital filters with which speech signals are convolved. In particular, typical phone channels exhibit a bandpass filtering effect on speech signals, where different degrees of attenuation are exerted on different spectral bands. Reverberation of speech signals is another source of convolutive distortion, which results in the addition of a noise component to the speech signals in the log-spectral domain.

Speakers may alter the way they speak under high levels of background noise or when under stress (Lombard effect); during articulation, speakers may also produce breathing noises and lip-smacks. All of these distortions will cause serious performance degradation in speaker recognition systems.

This chapter considers only additive noise and convolutive distortion, which can be combined into a composite source of distortion. Specifically, the acquired signal $y(t)$ is expressed as

$$ y(t) \; = \; s(t) * h(t) + n(t), \tag{9.2.1} $$

where $s(t)$ is the clean speech signal, and $h(t)$, $n(t)$, and $*$ (an asterisk) represent the channel's transfer function, the additive noise, and the convolution operators, respectively.

9.3 Kernel-Based Probabilistic Speaker Models

Because the amount of speaker-dependent data in a speaker recognition task is typically very large, it is almost impossible to store all data in the form of templates for recognition. Early techniques, such as vector quantization [342], attempt to reduce the amount of data by replacing similar data with their corresponding centroids. This is equivalent to partitioning the feature space into a number of clusters. This technique, however, assumes that data falling on one cluster do not influence the other clusters. In recent years, a number of kernel-based probabilistic neural networks have been proposed to address the deficiency of VQ; they include Gaussian mixture models (GMMs) [307], elliptical basis function networks (EBFNs) [225], and probabilistic decision-based neural networks (PDBNNs) [211]. These networks seek to reduce the amount of data to be stored in the speaker templates by estimating the data's probability density rather than simply computing the centroids as in VQ. Another key difference between these networks and VQ is that the data falling in one cluster influences the position and spread of all other clusters. In this sense, these networks can be regarded as performing soft (fuzzy) partitioning, whereas the VQ can be considered a kind of hard partitioning machine.

9.3.1 Gaussian Mixture Models

To apply GMMs (see Section 3.2.4) to speaker verification, each registered speaker in the system is represented by a GMM. To enhance the discrimination between the client speakers and impostors, it is common practice to compute the ratio between the client likelihood and impostor likelihood, where the former is the output of the client's GMM and the latter is the output of a background model [136]. The background model is a GMM trained from the speech of a large number of speakers, who should accurately represent the characteristics of all possible impostors. Alternatively, a set of background models is formed during verification by selecting the GMMs of a small set of client speakers (cohort) whose acoustic characteristics are close to those of the claimant [318].

During verification, a sequence of feature vectors X from the claimant is extracted and the following normalized score is computed:

$$S(X) = \log p(X|\omega_s) - \log p(X|\omega_b), \qquad (9.3.1)$$

where $p(X|\omega_s)$ and $p(X|\omega_b)$ are the GMMs' outputs (Eq. 3.1.5) corresponding to the speaker class ω_s and impostor class ω_b, respectively. The normalized score is then compared with a decision threshold to make a decision:

$$\text{If } S(X) \begin{cases} > \zeta & \text{accept the claimant} \\ \leq \zeta & \text{reject the claimant.} \end{cases} \qquad (9.3.2)$$

That is, to adopt the GMM-based classifier shown in Figure 6.2 to speaker verification, K is set to 2 and Maxnet is changed to compute the log-likelihood difference.

9.3.2 Elliptical Basis Function Networks

To apply EBFNs (see Sections 5.4.2 and 5.5.1) to speaker verification, one EBFN is trained for each registered speaker. Specifically, each network is trained to recognize speech patterns from two classes: speaker class and antispeaker class. To achieve this, the hidden nodes are divided into two groups: one corresponds to the speaker class and the other to the antispeaker class. The former is denoted as the speaker kernels and the latter as the antispeaker kernels. The EM algorithm is applied independently to the speaker data and antispeaker data to obtain the speaker kernels and antispeaker kernels, respectively. Then least-squares techniques (Eq. 5.5.5) are applied to determine the output weights. Each network contains two outputs (i.e., $K = 2$ in Figure 5.6), with $y_1(\mathbf{x})$ giving a desired output of 1.0 and $y_2(\mathbf{x})$ giving a desired output of 0.0 for speaker's data, and vice versa for antispeakers' data.

During verification, speech patterns $X = \{\mathbf{x}_t; t = 1, \ldots, T\}$ are extracted from the claimant's utterance, and the following score is computed:

$$S(X) = \frac{1}{T} \sum_{t=1}^{T} \frac{\exp\{y_1(\mathbf{x})/2P(\omega_s)\} - \exp\{y_2(\mathbf{x})/2P(\omega_b)\}}{\exp\{y_1(\mathbf{x})/2P(\omega_s)\} + \exp\{y_2(\mathbf{x})/2P(\omega_b)\}}, \qquad (9.3.3)$$

where $P(\omega_s)$ and $P(\omega_b)$ are the prior probabilities of the speaker class and antispeaker class, respectively, and $y_k(\mathbf{x})$ is the k-th output of the network. $P(\omega_s)$ and $P(\omega_b)$ can be easily computed by counting the number of speaker and antispeaker patterns in the training set. Note that dividing the network outputs by the prior probabilities is to rescale the network outputs so that the scaled averages (over the training set X' containing both speaker data and antispeaker data) are approximately equal to 0.5 (i.e., $\frac{1}{T'} \sum_{x \in X'} y_k(\mathbf{x})/2P(\omega) \approx 0.5$, $\omega = \omega_s$ or ω_b). The *softmax* function inside the summation of Eq. 9.3.3 is intended to prevent any extreme value of $y_k(\mathbf{x})/2P(\omega)$ from dominating the average outputs. Verification decisions are based on the criterion:

$$\text{If } S(X) \begin{cases} > \zeta & \text{accept the claimant} \\ \leq \zeta & \text{reject the claimant,} \end{cases} \qquad (9.3.4)$$

where $\zeta \in [-1, 1]$ is a speaker-dependent threshold.

The similarity between GMM-based and EBFN-based classifiers can be observed from their architecture (Figures 5.6 and 6.2). For example, both of them compute the Mahalanobis distance (Eq. 6.2.4 and Eq. 5.4.10) between the input vectors and the kernel centers in the hidden-layer; however, there are two important differences. First, a GMM computes the likelihood of observing the input vector $\mathbf{x}$, whereas an EBF network maps data from the input space to the output space. Second, the kernel parameters of a GMM must be estimated from data derived from its corresponding class. On the other hand, data derived from all known classes (K classes in the case of Figure 5.6) are applied to estimate the kernel parameters of an EBF network. Even if the EBF kernels are divided into K groups and K sets of kernel parameters are estimated independently using the data derived from the K classes, EBF networks are still different from GMMs because each EBF network's output depends on the kernel outputs of the corresponding class as well as those from other classes. The output of a GMM ($p(\mathbf{x}|\omega_i)$), on the other hand, depends on the kernel outputs of only its class. The consequence of this difference is that, for EBF networks, discrimination among all known classes is considered during the training phrase, but for GMMs class discrimination is introduced during the recognition phase.

9.3.3   Probabilistic Decision-Based Neural Networks

Three modifications were made to the PDBNN's training algorithm (see Section 7.4.2) to make PDBNNs appropriate for speaker verification. First, the original PDBNNs used one threshold per network. However, in this case, one network is used to model the speaker class and another one to model the antispeaker class (i.e., $i = 1$ or 2 in Eq. 7.5.1 and $L = 2$ in Figure 7.10). To make PDBNNs applicable to speaker verification, the likelihood computation was modified such that only one threshold is required. Specifically, instead of comparing the network's log-likelihood against its corresponding threshold, as in the original PDBNNs, a normalized score was compared against a single decision threshold, as in Eqs. 9.3.1 and 9.3.2.

In the second modification, the frequency at which the threshold is updated was changed. The original PDBNN adopts the so-called batch-mode supervised learning (see Eq. 7.5.2). Speaker verification procedure, however, is based on a partial-sequential mode. Specifically, the GS training was modified as follows: Let X_n be the n-th segment extracted from the speaker's speech patterns $X^{(s)}$ or from antispeakers' speech patterns $X^{(b)}$, the normalized segmental score is computed by evaluating

$$S(X_n) = S_s(X_n) - S_b(X_n)$$
$$= \frac{1}{T} \sum_{\mathbf{x} \in X_n} \{\phi_s(\mathbf{x}) - \phi_b(\mathbf{x})\}, \qquad (9.3.5)$$

where $\phi_s(\mathbf{x})$ and $\phi_b(\mathbf{x})$ are the log-likelihood function (Eq. 7.5.1) of the speaker class and antispeaker (background) class, respectively. For the n-th segment, the following criteria were used to determine whether to update the decision threshold $\zeta_n^{(j)}$:

$$\text{If } S(X_n) \begin{cases} > \zeta_{n-1}^{(j)} \text{ and } X_n \in X^{(s)} & X_n \text{ is correctly classified, no need to update} \\ \leq \zeta_{n-1}^{(j)} \text{ and } X_n \in X^{(b)} & X_n \text{ is correctly classified, no need to update} \\ > \zeta_{n-1}^{(j)} \text{ and } X_n \in X^{(b)} & \text{false acceptance, need to update} \\ \leq \zeta_{n-1}^{(j)} \text{ and } X_n \in X^{(s)} & \text{false rejection, need to update} \end{cases}$$
$$(9.3.6)$$

where $\zeta_{n-1}^{(j)}$ is the decision threshold of the PDBNN speaker model after learning from segment X_{n-1} at epoch j. Therefore, whenever misclassification occurs, the threshold $\zeta_{n-1}^{(j)}$ is updated according to

$$\zeta_n^{(j)} = \begin{cases} \zeta_{n-1}^{(j)} - \eta_r l'(\zeta_{n-1}^{(j)} - S(X_n)) & \text{if } X_n \in X^{(s)} \text{ and } S(X_n) < \zeta_{n-1}^{(j)} \\ \zeta_{n-1}^{(j)} + \eta_a l'(S(X_n) - \zeta_{n-1}^{(j)}) & \text{if } X_n \in X^{(b)} \text{ and } S(X_n) \geq \zeta_{n-1}^{(j)} \end{cases}$$
$$(9.3.7)$$

where η_r and η_a are, respectively, reinforced and antireinforced learning parameters (more on this in next paragraph), $l(d) = \frac{1}{1+e^{-d}}$ is a penalty function, and $l'(d)$ is the derivative of $l(\cdot)$.

In the third modification, a new method was introduced to compute learning rates. In the original PDBNNs, the learning rates for optimizing the thresholds are identical for both reinforced and antireinforced learning. However, in some situations, there may be many false acceptances and only a few false rejections (or vice versa), which means antireinforced learning will occur more frequently than reinforced learning (or vice versa). To reduce the imbalance in the learning frequency, the reinforced (antireinforced) learning rate η_r (η_a) is made proportional to the rate of false rejections (acceptance) weighted by the total number of impostor (speaker) segments:

$$\eta_r = \frac{\text{FRR}^{(j-1)}}{\text{FAR}^{(j-1)} + \text{FRR}^{(j-1)}} \frac{N_{\text{imp}}}{N_{\text{imp}} + N_{\text{spk}}} \eta \qquad (9.3.8)$$

$$\eta_a = \frac{\text{FAR}^{(j-1)}}{\text{FAR}^{(j-1)} + \text{FRR}^{(j-1)}} \frac{N_{\text{spk}}}{N_{\text{imp}} + N_{\text{spk}}} \eta, \qquad (9.3.9)$$

where $\text{FRR}^{(j-1)}$ and $\text{FAR}^{(j-1)}$ represent the error rate of false rejections and false acceptances at epoch $j-1$, N_{imp}, respectively; and N_{spk} represent the total number of training segments from the impostors and the registered speaker, respectively; and η is a positive learning parameter. The first term of Eqs. 9.3.8 and 9.3.9 increases the learning rate if the corresponding error rate is large, which has the effect of rapidly reducing the corresponding error rate. The second term weights the learning rate according to the proportion of training segments in the opposite class, which has the effect of reducing the learning rate of the frequent learner and increasing the learning rate of the nonfrequent learner. This arrangement can prevent reinforced learning or antireinforced learning from dominating the learning process and aims to increase the convergence speed of the decision threshold.

9.3.4   Comparison of Probabilistic Speaker Models

This subsection compares the performance of GMMs, EBFNs, and PDBNNs in a text-independent speaker verification task involving 138 speakers from the YOHO corpus [44]. The decision boundaries created by these networks are also illustrated and compared.

Speech Corpus and Feature Extraction

The YOHO corpus [44] was collected by the ITT Defense Communication Division under a contract with the U.S. Department of Defense. It is a large-scale, scientifically controlled speech corpus for testing speaker verification systems at a high confidence level. The corpus features "combination-lock" phrases, 138 speakers (108 male, 30 female), intersession variability, and high-quality phone speech sampled at 8kHz with 16 bits per sample. The YOHO recording system was set up in the corner of a large office, where low-level noise could be heard from adjoining offices. A handset containing an omnidirectional electret microphone without noise-canceling features was used for recordings. There were four enrollment sessions for each speakers and each session contained 24 utterances. Likewise, there were 10 verification sessions for each speaker, with each session containing four utterances, each of which was composed of three 2-digit numbers (e.g., 34-52-67). The combination-lock phrases together with intersession variability make YOHO ideal for speaker verification research.

In this work, all 138 YOHO corpus speakers were used for experimental evaluations. Gaussian white noise with different noise power was also added to the clean YOHO corpus. Both clean and noisy YOHO corpora were used in the evaluations.

The feature extraction procedure is as follows: For each utterance, the silent regions were removed by a silent detection algorithm based on the energy and zero crossing rate of the signals. The remaining signals were preemphasized by a filter with transfer function $1 - 0.95z^{-1}$. Twelfth-order LP-derived cepstral coefficients were computed using a 28ms Hamming window at a frame rate of 71Hz.

Enrollment Procedures

In the verification experiments, each registered speaker was represented using three different speaker models: GMM, EBFN, and PDBNN. A GMM-based speaker model consists of two GMMs, one representing the individual speaker and the other representing all other speakers (called antispeakers hereafter). (An EBFN- or PDBNN-based speaker model consists of a single EBFN or PDBNN representing the corresponding speaker as well as all antispeakers.) For each registered speaker, all utterances in the four enrollment sessions, corresponding to the speaker and a predefined set of antispeakers (each speaker has his or her own set of antispeakers), were used to train a speaker model. The speaker model was trained to recognize the speech derived from two classes—speaker class and antispeaker class. To this end, two groups of kernel functions[1] were assigned to each speaker model. Henceforth, the group corresponding to the speaker class is denoted as the speaker kernels and the one corresponding to the antispeaker class as the antispeaker kernels. For each registered speaker, a unique antispeaker set containing 16 antispeakers was created. Speech features derived from this set were subsequently used to estimate the antispeaker kernels by using the EM algorithm. The antispeaker kernels enable the integration of scoring normalization [215] into the speaker models, which enhances the models' capability in discriminating true speakers from impostors.

Each of the GMMs and PDBNNs is composed of 12 inputs (12th-order LP-derived cepstral coefficients were used as features), a predefined number of kernels, and one output. On the other hand, the EBFNs contain 12 inputs, a predefined number of kernels, and two outputs with each output representing one class (speaker class and antispeaker class).

---

[1]One group represented the individual speaker and the other represented the speakers in the antispeaker class.

The k-means algorithm was applied to initialize the starting position of the speaker kernels. Then, the kernels' covariance matrices were initialized by the K-nearest neighbor algorithm ($K = 2$). In other words, all off-diagonal elements were 0 and the diagonal elements (being equal) of each matrix were set to the average Euclidean distance between the corresponding center and its K-nearest centers. The EM algorithm was subsequently used to fine-tune the mean vectors, covariance matrices, and mixture coefficients (see Eqs. 6.2.2, 6.2.3, and 6.2.4). The same procedure was also applied to determine the mean vectors and covariance matrices of the antispeaker kernels, using the speech data derived from the antispeaker set. Initializing the mean vectors by k-means and the covariance matrices by K-NN reduces the number of EM iterations required to determine the maximum-likelihood solution. Because the k-means and K-NN algorithms run much faster than the EM algorithm, this approach can considerably reduce the training time.

The enrollment process for constructing a PDBNN-based speaker model involves two phases: locally unsupervised (LU) training and globally supervised (GS) training. The LU training phase is identical to the GMM training described before. In the GS training phase, the speaker's enrollment utterances and the utterances from all enrollment sessions of the antispeakers were used to determine a decision threshold.

For the EBFN-based speaker models, the speaker kernels and antispeaker kernels obtained from the GMM training were combined to form a hidden-layer. In this work, γ_j in Eq. 5.4.10 was determined heuristically by

$$\gamma_j = \frac{9}{5} \sum_{k=1}^{5} \|\mu_k - \mu_j\|, \qquad (9.3.10)$$

where μ_k denotes the k-th nearest neighbor of μ_j in the Euclidean sense. Empirically, it was found that using five nearest centers and multiplying the resulting average distance by 9.0 give reasonably good results; however, no attempts have been made to optimize these values. Finally, singular value decomposition was applied to determine the output weights. Details of the enrollment procedure for EBFNs can be found in Mak and Kung [225].

Verification Procedures

Verification was performed using each speaker in the YOHO corpus as a claimant, with 64 impostors being randomly selected from the remaining speakers (excluding the antispeakers and the claimant) and rotating through all the speakers. For each claimant, the feature vectors of the claimant's utterances from his or her 10 verification sessions in YOHO were concatenated to form a claimant sequence. Likewise, the feature vectors of the impostor's utterances were concatenated to form an impostor sequence.

For PDBNNs and GMMs, the following steps were performed during verification. The feature vectors from the claimant's speech $X^{(c)} = \{\mathbf{x}_1, \mathbf{x}_2, \ldots, \mathbf{x}_{T_c}\}$ was divided into a number of overlapping segments containing $T(< T_c)$ consecutive vectors as shown here[2]

$$\overbrace{\mathbf{x}_1, \mathbf{x}_2, \mathbf{x}_3, \mathbf{x}_4, \mathbf{x}_5, \mathbf{x}_6, \cdots\cdots, \mathbf{x}_T}^{\text{1st segment, } X_1}, \mathbf{x}_{T+1}, \mathbf{x}_{T+2}, \cdots, \mathbf{x}_{T_c}$$

$$\mathbf{x}_1, \mathbf{x}_2, \mathbf{x}_3, \mathbf{x}_4, \mathbf{x}_5, \overbrace{\mathbf{x}_6, \cdots\cdots, \mathbf{x}_{T+5}}^{\text{2nd segment, } X_2}, \mathbf{x}_{T+6}, \cdots, \mathbf{x}_{T_c}.$$

For the n-th segment $(X_n \subset X^{(c)})$, the normalized segmental log-likelihood $S(X_n)$ is computed as in Eq. 9.3.5. Verification decisions were based on the criterion

$$\text{If } S(X_n) \begin{cases} > \zeta & \text{accept the claimant} \\ \leq \zeta & \text{reject the claimant,} \end{cases} \qquad (9.3.11)$$

where ζ is a speaker-dependent decision threshold (see Determination of Decision Threshold section). A verification decision was made for each segment, with the error rate (either false acceptance or false rejection) being the proportion of incorrect verification decisions to the total number of decisions. In this work, T in Eq. 9.3.5 was set to 500 (i.e., 7 seconds of speech), and each segment was separated by five vector positions. More specifically, the t-th segment contains the vectors

$$X_n = \{\mathbf{x}_{5(n-1)+1}, \mathbf{x}_{5(n-1)+2}, \ldots, \mathbf{x}_{5(n-1)+500}\}, \qquad (9.3.12)$$

where $5(n-1) + 500 < T_c$. Note that dividing the vector sequence into a number of segments has also been used successfully for increasing the number of decisions by Mak and Kung [225] and Reynolds and Rose [307].

For the EBF-based speaker models, verification decisions were based on the difference between the scaled network outputs [225]. As with the PDBNN-based and GMM-based speaker models, the claimant's utterance $X^{(c)}$ was divided into a number of overlapping segments. For each segment X_n (with segment length T), the score $z_{t,k}$

$$z_{t,k} = \frac{1}{T} \sum_{\mathbf{x} \in X_n} \frac{\exp\{y_k(\mathbf{x})/2P(C_k)\}}{\sum_{r=1}^{2} \exp\{y_r(\mathbf{x})/2P(C_r)\}} \quad k = 1, 2 \qquad (9.3.13)$$

corresponding to the speaker and antispeaker classes were computed, where $P(C_r)$ is the prior probability of class C_r. Note that the *softmax* function in Eq. 9.3.13 prevents any extreme value of $y_k(\mathbf{x})/2P(C_k)$ from dominating the average outputs.

Verification decisions were based on the criterion

$$\text{If } z_{t,1} - z_{t,2} \begin{cases} > \zeta & \text{accept the claimant} \\ \leq \zeta & \text{reject the claimant} \end{cases} \qquad (9.3.14)$$

where $\zeta \in [-1, 1]$ is a speaker-dependent threshold (see next page) controlling the FRR and FAR. Again, a verification decision was made for each segment (as defined

---

[2]The claimant can be either the true speaker (in which case $X^{(c)}$ represents a claimant sequence) or an impostor (in which case $X^{(c)}$ represents an impostor sequence).

in Eq. 9.3.12) at a rate of one decision per five feature vectors. Computing the difference between the two outputs is equivalent to normalizing the score in GMMs. Thus, this verification procedure integrates scoring normalization into the network architecture.

In this work, equal error rate (EER) was used as a performance index to compare the verification performance among different speaker models. Because the speaker models remain fixed once they have been trained, EER can be used to compare the models' ability in discriminating speaker features from impostor features.

Determination of Decision Thresholds

As mentioned before, it is necessary to determine a decision threshold for each speaker model during enrollment. These thresholds will be used during verification.

The procedures for determining the decision thresholds of PDBNNs, GMMs, and EBFNs are different. For the GMM and EBFN speaker models, the utterances from all enrollment sessions of 16 randomly selected antispeakers were used for threshold determination. Specifically, these utterances were concatenated and the verification procedures were applied. The thresholds ζ's in Eqs. 9.3.11 and 9.3.14 were adjusted until the corresponding FAR fell below a predefined level; in this work, the level was set to 0.5%. Antispeakers' utterances, rather than a speaker's utterances, were used because it is easier to collect the speech of a large number of antispeakers. Hence, the thresholds obtained are more reliable than those that would have been obtained from the speaker's speech. In addition, using a predefined FAR to determine the decision thresholds makes it easier to predict the robustness of the verification system against impostor attacks [403].

The modified threshold determination procedure described in Section 9.3.3 was used to determine PDBNNs' decision thresholds. To keep the mean vectors and covariance matrices the same as the maximum-likelihood estimates, only the decision thresholds in the globally supervised training were modified, with the mean vectors and covariance matrices remaining unchanged.

Pilot Experiments

The architecture of GMMs, EBFNs, and PDBNNs depends on several free parameters, including the number of speaker kernels, the number of antispeaker kernels, and the number of antispeakers for finding the antispeaker kernels. To determine these parameters, a series of pilot experiments involving 30 speakers from the YOHO corpus were performed. Equal error rates were used as the performance indicator. To determine an appropriate number of speaker kernels, speaker models with different numbers of speaker kernels were constructed, and the numbers of antispeaker kernels and antispeakers were fixed to 160 and 16, respectively.

Table 9.1(a) summarizes the average EERs obtained by the GMM-based speaker models. Evidently, the EER decreases as the number of kernels increases, although the decrease becomes less significant after the number of speaker kernels reaches 40. To figure out an appropriate number of antispeakers for determining the anti-

Table 9.1.   Average EERs based on 30 GMM-based speaker models. *Note:* This shows that there were different numbers of (a) speaker kernels, where the number of antispeakers and the number of antispeaker kernels were set to 16 and 160, respectively; (b) antispeakers, where the number of speaker kernels and antispeaker kernels were set to 40 and 160, respectively; and (c) antispeaker kernels, where the number of speaker kernels and antispeakers were set to 40 and 16, respectively.

Number of Speaker's Kernels	EER (%)
10	2.78
20	1.51
40	0.77
80	0.57
160	0.48

(a)

Number of Antispeakers	EER (%)
4	2.02
8	1.30
16	0.77
32	0.48
64	0.81

(b)

Number of Antispeaker Kernels	EER (%)
40	0.83
80	0.83
160	0.77
320	0.75
640	0.79

(c)

kernels, the number of antispeakers were varied while fixing the number of speaker kernels and antispeaker kernels to 40 and 160, respectively. Table 9.1(b) summarizes the average EER obtained by the GMM-based speaker models. Optimal performance is obtained when the number of antispeakers is 32. To reduce processing time, 16 antispeakers were used in the rest of the experiments.

The number of antispeaker kernels was varied while the number of speaker kernels and the number of antispeakers were fixed to 40 and 16, respectively. Table 9.1(c) shows the average EER obtained by the GMM-based speaker models. The results show that no significant reduction in error rate can be achieved when the number of antispeaker kernels reaches 160. Hence, 160 antispeaker kernels were used in subsequent experiments.

Table 9.2. Average error rates achieved by the GMMs, EBFNs, and PDBNNs based on 138 speakers in the YOHO corpus. *Note*: The decision thresholds for GMMs and PDBNNs were determined by setting the predefined FAR to 0.5%; whereas, the thresholds for PDBNNs were determined by reinforced learning (see Determination of Decision Thresholds section).

Speaker Model	FAR (%)	FRR (%)	EER (%)
GMMs	8.01	0.08	0.33
EBFs	15.24	0.50	0.48
PDBNNs	1.10	1.87	0.33

Because the EBFNs, GMMs, and PDBNNs use the same set of kernels, it is not necessary to repeat the preceding experiments for EBFNs and PDBNNs.

Large-Scale Experiments

Based on the results in the pilot experiments (see previous page), the number of speaker kernels and the number of antispeaker kernels was set to 40 and 160, respectively; and for each speaker model, 16 antispeakers were used for determining the parameters in the antispeaker kernels. A suboptimal number of antispeakers was selected to reduce the computation time in creating the speaker models.

All speakers (108 male, 30 female) in the YOHO corpus were used to evaluate the performance of EBFNs, GMMs, and PDBNNs in closed-set text-independent speaker verification. The aim was to evaluate the robustness of different pattern classifiers (speaker models) for speaker verification. To demonstrate the robustness of these classifiers, speech from the enrollment sessions of the YOHO corpus was used for training, and speech from the verification sessions was used for testing.

Table 9.2 summarizes the average FAR, FRR, and EER obtained by the PDBNNs, GMMs, and EBFNs. All results are based on the average of 138 speakers in the YOHO corpus. The results, in particular the EERs, demonstrate the superiority of the GMMs and PDBNNs over the EBFNs. The EER of GMMs and PDBNNs are the same because their kernel parameters are identical. Table 9.2 shows that the smallest EER obtained by the EBFN models is 0.48%, which is greater than 0.33%—the EER of GMMs and PDBNNs.

In terms of FAR and FRR, Table 9.2 demonstrates the superiority of the threshold determination procedure of PDBNNs. In particular, the table clearly shows that the GS learning of PDBNNs can maintain the average FAR at a very low level during verification, whereas the ad hoc approach used by the EBFNs and GMMs produces a much larger average FAR. Recall from the previous discussion that the predefined FAR for determining the decision thresholds of EBFNs and GMMs was set to 0.5%. The average FAR of EBFNs and GMMs are, however, very different

Table 9.3. Performance of the PDBNN, GMM, and EBFN in the two-dimensional speaker verification problem.

	PDBNN and GMM		EBFN	
	Train	Test	Train	Test
EER (%)	4.12	24.61	6.86	27.17

from this value. This suggests that it may be difficult to predict the performance of EBFNs and GMMs in detecting impostor attacks.

Figure 9.1 depicts the FAR and FRR of individual speakers in the GMM-, EBFN- and PDBNN-based speaker verification systems. Evidently, most of the speakers in the PDBNN-based system exhibit a low FAR. On the other hand, the GMMs and EBFNs exhibit a much larger variation in FAR. The globally supervised learning in PDBNNs probably are able to find decision thresholds that minimize the variation in FAR.

Figure 9.2 shows the DET curves (see Martin et al. [233]) corresponding to speaker 164 for different types of speaker models. The DET plots use a nonlinear scale for both axes so that systems producing Gaussian distributed scores will be represented by straight lines. This property helps spread out the receiver operating characteristics (ROCs), making comparison of high-performing systems much easier. Note that the DET curves for the GMM and PDBNN are identical in this experiment because the GS training updates the thresholds of PDBNNs only. It is evident from Figure 9.2 that the GMM- and PDBNN-based speaker models outperform the EBFN one.

Decision Boundaries

To further illustrate the difference among the speaker models, the first and second cepstral coefficients of speaker 162 and those of his antispeakers and impostors were extracted to create a set of two-dimensional speech data. Similar to the enrollment procedure in the speaker verification experiments, a PDBNN, a GMM, and an EBFN (all with 2 inputs and 6 centers) were trained to classify the patterns into two classes. Therefore, except for the reduction in feature dimension, the training methods, learning rate, and verification methods are identical to the speaker verification experiments described previously.

Table 9.3 compares the performance of the three speaker models, and Figure 9.3 shows the test data, decision boundaries, function centers, and contours of basis function outputs formed by these models. The decision boundaries are based on the equal error thresholds obtained from the data set. Figure 9.3(a) shows that the decision boundaries formed by the EBFN enclose two regions, which belong to the speaker class, with a large amount of test data, whereas the complement region, which belongs to the impostor class, extends to infinity.

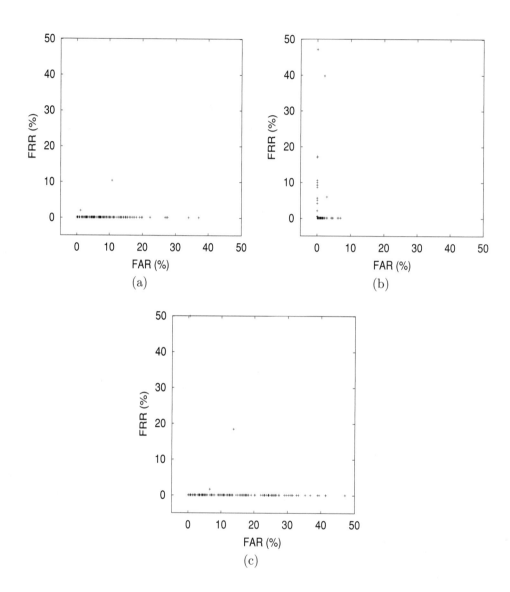

Figure 9.1. FRRs versus FARs (during verification) of 138 speakers using (a) GMMs, (b) PDBNNs, and (c) EBFNs as speaker models.

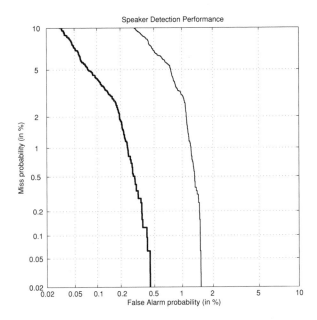

Figure 9.2. DET curves corresponding to speaker 164. *Thin curve*: EBFN-based speaker model. *Thick curve*: GMM-based and PDBNN-based speaker models.

On the other hand, the decision boundaries created by the GMM and PDBNN extend to infinity in the feature space for both speaker class and impostor class. Both the decision boundaries (Figure 9.3) and the EER (Table 9.3) suggest that GMM and PDBNN provide better generalization than EBFN. These results also agree with what was shown in Table 9.2. The poor performance in EBFNs may be caused by the least-squares approach to finding the output weights. Because the EBFNs formulate the classification problem as a function interpolation problem (mapping from the feature space to 0.0 or 1.0), overfitting can occur easily if there are too many hidden nodes but too few training samples.

Robustness against Noise

To test the robustness of different speaker models against noise, zero-mean Gaussian noise was added to the YOHO speech so that the resulting corrupted speech has an SNR of 10dB, 6dB, 3dB, and 0dB. Segmentation files[3] derived from the clean YOHO corpus were used to specify the silence and speech regions of the corrupted YOHO speech. In practice, the same speech detector that was applied to YOHO speech should also be used to segment the corrupted YOHO speech. The objective,

---

[3]These files specify only the silence and speech regions of the utterances. A speech detection algorithm based on the zero crossing rate and average amplitude was used to determine the speech regions.

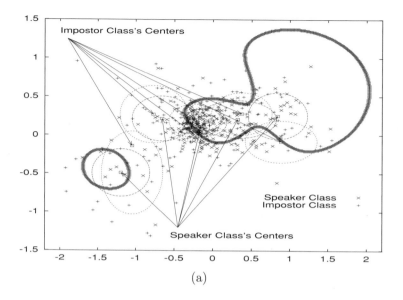

(a)

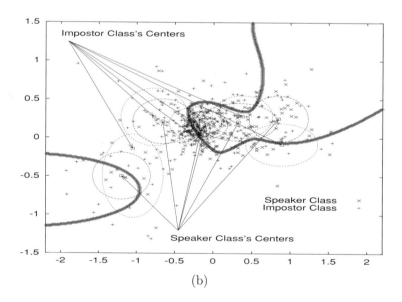

(b)

Figure 9.3. Speaker verification problem using two-dimensional speech features. The figures depict the decision boundaries, function centers, and contours of constant basis function outputs (*thin* ellipses) produced by (a) EBFNs and (b) GMMs and PDBNNs. The × and + signs represent the speakers' data and impostors' data, respectively.

however, is to compare the robustness of different speaker models against additive noise. Therefore, using the same segmentation files to define the speech regions of both clean and corrupted speech prevents the error introduced by the speech detector from interfering with the comparison.

Table 9.4 summarizes the average FAR, FRR, and EER obtained by the GMMs, PDBNNs, and EBFNs under different SNRs. The results show that the error rates of all models increase as the noise power increases. Such performance degradation is caused mainly by mismatches in the training and testing environments. Additive white noise contaminates the speech signals and therefore changes their acoustic characteristics. The speaker models, which were trained with clean speech, produced a reasonably low error rate for speech with a high SNR; however, their performance degraded rapidly when they were applied to noisy speech. Evidently, the EERs of PDBNNs and GMMs are smaller than those of the EBFNs under different SNR. Although PDBNNs and GMMs provide better generalization, the performance of PDBNNs and GMMs is still unacceptable at low SNR. In addition to additive noise, phone speech may be distorted by handsets and phone channels. These issues are addressed in Section 9.4.

Compared with Related Work

In the literature, several speaker verification evaluations are based on the YOHO corpus. For example, Reynolds [302] obtained a 0.51% EER and Higgins et al. [136] achieved a 1.7% EER. Both systems are based on GMMs. The performance (0.33% EER) of the system discussed here is better than that of Reynolds [302] and Higgins et al. [136]. However, these error rates can be only loosely compared because the evaluations were not performed under identical conditions (i.e., different training/testing paradigms and background speaker sets were used).

To compare with a more classical approach, the same experiment was repeated using vector quantization (VQ) speaker models [342]. Using 64-center VQ speaker codebooks,[4] an EER of 1.29% was obtained. When the number of code vectors per VQ codebook was reduced to 32, the EER increased to 1.61%. These EERs are much higher than those of the PDBNNs and GMMs.

It is also important to point out that EER is not the only criterion for judging speaker verification performance. In practical situations, the tradeoff between false acceptance and false rejection must be considered. The key finding of this work is that the PDBNNs can effectively control this tradeoff through their threshold determination mechanism.

9.4   Handset and Channel Distortion

Because of the proliferation of e-banking and e-commerce, recent research on speaker verification has focused on verifying speakers' identity over the phone. A challenge of phone-based speaker verification is that transducer variability could result in

---

[4]The number of centers in VQ codebooks must be a power of 2.

Table 9.4. Average error rates. *Note*: Results obtained by (a) GMM speaker models, (b) PDBNN speaker models, and (c) EBFN speaker models at different signal-to-noise ratios.

SNR	FAR (%)	FRR (%)	EER (%)
0dB	43.98	55.47	34.00
3dB	43.52	54.91	27.30
6dB	42.51	53.59	20.32
10dB	41.20	50.70	12.79
clean	8.01	0.08	0.33

(a)

SNR	FAR (%)	FRR (%)	EER (%)
0dB	21.63	76.34	34.00
3dB	19.52	77.53	27.30
6dB	17.03	77.53	20.32
10dB	13.67	76.38	12.79
clean	1.10	1.87	0.33

(b)

SNR	FAR (%)	FRR (%)	EER (%)
0dB	30.95	66.57	37.51
3dB	30.48	65.91	30.32
6dB	29.97	65.16	22.45
10dB	29.22	61.06	14.58
clean	15.24	0.50	0.48

(c)

acoustic mismatches of the speech data gathered from different handsets. The recent popularity of mobile and Internet phones further complicates the problem, because speech coders in these phones also introduce acoustic distortion to the speech signals. The sensitivity to handset variations and speech coding algorithms means that handset compensation techniques are essential for practical speaker verification systems.

This section describes several compensation techniques to resolve the channel mismatch problems. These techniques combine handset selectors [352–354] with stochastic feature and model transformation [226, 227, 394] to reduce the acoustic mismatch between different handsets and different speech coders. Coder-dependent GMM-based handset selectors [397, 398] are trained to identify the most likely handset used by the claimants. Stochastic feature and model transformations are then applied to remove the acoustic distortion introduced by the coder and the handset. Experimental results show that the proposed technique outperforms conventional approaches and significantly reduces error rates under seven different coders (G.711, G.726, GSM, G.729, G.723.1, MELP, and LPC) with bit rates ranging from 2.4Kbps to 64Kbps. Strong correlation between speech quality and verification performance has also been observed.

9.4.1   Handset and Channel Compensation Techniques

Current approaches to handset and channel compensation can be divided into feature transformation and model transformation. The former transforms the distorted speech features to fit the clean speaker models, whereas the latter adapts or transforms the parameters of the clean models to fit the distorted speech. These two types of compensation approaches are discussed next.

Feature Transformation

As mentioned before, one of the key problems in phone-based speaker verification is the acoustic mismatch between speech gathered from different handsets. One possible approach to resolving the mismatch problem is *feature transformation*. Feature-based approaches attempt to modify the distorted features so that the resulting features fit the clean speech models better. These approaches include cepstral mean subtraction (CMS) [13] and signal bias removal [294], which approximate a linear channel by the long-term average of distorted cepstral vectors. These approaches, however, do not consider the effect of background noise. A more general approach, in which additive noise and convolutive distortion are modeled as codeword-dependent cepstral biases, is *codeword-dependent cepstral normalization* (CDCN) [3]. The CDCN, however, works well only when the background noise level is low.

When stereo corpora are available, channel distortion can be estimated directly by comparing the clean feature vectors against their distorted counterparts. For example, in SNR-dependent cepstral normalization (SDCN) [3], cepstral biases for different signal-to-noise ratios are estimated in a maximum-likelihood framework. In probabilistic optimum filtering [261], the transformation is a set of multidimensional least-squares filters whose outputs are probabilistically combined. These methods, however, rely on the availability of stereo corpora. The requirement of stereo corpora can be avoided by making use of the information embedded in the clean speech models. For example, in stochastic matching [332], the transformation parameters are determined by maximizing the likelihood of the clean models given the transformed features.

Model Transformation

Instead of transforming the distorted features to fit the clean speech model, the clean speech models can be modified such that the density functions of the resulting models fit the distorted data better. This is known as *model-based transformation* in the literature. Influential model-based approaches include (1) stochastic matching [332] and stochastic additive transformation [317], where the models' means and variances are adjusted by stochastic biases; (2) maximum-likelihood linear regression (MLLR) [200], where the mean vectors of clean speech models are linearly transformed; and (3) constrained reestimation of Gaussian mixtures [78], where both mean vectors and covariance matrices are transformed. Recently, MLLR has been extended to maximum-likelihood linear transformation [110], in which the transformation matrices for the variances can be different from those for the mean vectors. Meanwhile, the constrained transformation noted in Digalakis et al. [78] has been extended to piecewise-linear stochastic transformation [76], where a collection of linear transformations is shared by all the Gaussians in each mixture. The random bias shown by Sankar and Lee [332] has also been replaced by a neural network to compensate for nonlinear distortion [344]. All of these extensions show improvement in recognition accuracy.

Because the preceding methods "indirectly" adjust the model parameters via a small number of transformations, they may not be able to capture the fine structure of the distortion. Although this limitation can be overcome by the Bayesian techniques [157, 197], where model parameters are adjusted "directly," the Bayesian approach requires a large amount of adaptation data to be effective. Because both direct and indirect adaptations have their own strengths and weaknesses, a natural extension is to combine them so that the two approaches can complement each other [252, 341].

Limitations of Current Approaches

Although the preceding methods have been successful in reducing channel mismatches, most of them operate on the assumption that the channel effect can be approximated by a linear filter. Most phone handsets, in fact, exhibit energy-dependent frequency responses [308] for which a linear filter may be a poor approximation. Recently, this problem has been addressed by considering the distortion as a nonlinear mapping [205, 291]. However, these methods rely on the availability of stereo corpora with accurate time alignment.

To address these limitations, the Mak and Kung [226] and Mak et al. [227] studies proposed a method in which nonlinear transformations can be estimated under a maximum-likelihood framework, thus eliminating the need for accurately aligned stereo corpora. The only requirement is to record a few utterances generated by a few speakers using different handsets. These speakers need not utter the same set of sentences during the recording sessions, although this may improve the system's performance. The nonlinear transformation technique is detailed in the next section.

9.4.2   Stochastic Feature Transformation

The stochastic feature transformation (SFT) technique is inspired by the stochastic matching method of Sankar and Lee [332].   Stochastic matching was originally proposed for speaker adaptation and channel compensation. Its main goal is to transform the distorted data to fit the clean speech models or to transform the clean speech models to better fit the distorted data. In the case of feature transformation, the channel is represented by either a single cepstral bias ($\mathbf{b} = [b_1, b_2, \ldots, b_D]^T$) or a bias together with an affine transformation matrix ($A = \mathrm{diag}\ \{a_1, a_2, \ldots, a_D\}$). In the latter case, the componentwise form of the transformed vectors is given by

$$\hat{x}_{t,i} = f_\nu(\mathbf{y}_t)_i = a_i y_{t,i} + b_i, \qquad (9.4.1)$$

where $\mathbf{y}_t$ is a D-dimensional distorted vector, $\nu = \{a_i, b_i\}_{i=1}^D$ is the set of transformation parameters, and $f_\nu(\cdot)$ denotes the transformation function. Intuitively, the bias $\mathbf{b}$ compensates the convolutive distortion and the matrix A compensates the effects of noise. Figure 9.4(a) illustrates the concept of stochastic feature transformation with a single set of linear transformation parameters ν per handset.

The first-order transformation in Eq. 9.4.1, however, has two limitations. First, it assumes that all speech signals are subject to the same degree of distortion, which may be incorrect for nonlinear channels where signals with higher amplitude are subject to a higher degree of distortion, because of the saturation effect in transducers. Second, the use of a single transformation matrix is inadequate for an acoustic environment with varying noise levels. A new approach to overcome these limitations is proposed next.

Nonlinear Feature Transformation

The proposal is based on the notion that different transformation matrices and bias vectors can be applied to transform the vectors in different regions of the feature space. This can be achieved by extending Eq. 9.4.1 to

$$\hat{x}_{t,i} = f_\nu(\mathbf{y}_t)_i = \sum_{k=1}^K g_k(\mathbf{y}_t)(c_{ki} y_{t,i}^2 + a_{ki} y_{t,i} + b_{ki}), \qquad (9.4.2)$$

where $\nu = \{a_{ki}, b_{ki}, c_{ki}; k = 1, \ldots, K; i = 1, \ldots, D\}$ is the set of transformation parameters and

$$g_k(\mathbf{y}_t) = P(k|\mathbf{y}_t, \Lambda_Y) = \frac{\omega_k^Y\, p(\mathbf{y}_t|\mu_k^Y, \Sigma_k^Y)}{\sum_{l=1}^K \omega_l^Y\, p(\mathbf{y}_t|\mu_l^Y, \Sigma_l^Y)} \qquad (9.4.3)$$

is the posterior probability of selecting the k-th transformation given the distorted speech $\mathbf{y}_t$. Note that the selection of transformation is probabilistic and data-driven. In Eq. 9.4.3, $\Lambda_Y = \{\omega_k^Y, \mu_k^Y, \Sigma_k^Y\}_{k=1}^K$ is the speech model that characterizes the distorted speech and

$$p(\mathbf{y}_t|\mu_k^Y, \Sigma_k^Y) = (2\pi)^{-\frac{D}{2}} |\Sigma_k^Y|^{-\frac{1}{2}} \exp\left\{ -\frac{1}{2}(\mathbf{y}_t - \mu_k^Y)^T (\Sigma_k^Y)^{-1} (\mathbf{y}_t - \mu_k^Y) \right\} \qquad (9.4.4)$$

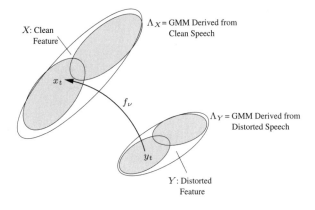

$$\nu' = \arg\max_{\nu} p(f_\nu(Y)|\Lambda_X, \nu)$$

$$\hat{x}_{t,i} = f_\nu(\mathbf{y}_t)_i = a_i y_{t,i} + b_i$$

(a)

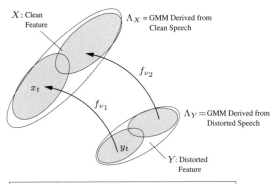

$$\nu = (\nu_1, \nu_2)$$

$$\nu' = \arg\max_{\nu} p(f_\nu(Y)|\Lambda_X, \nu)$$

$$\hat{x}_{t,i} = \sum_{k=1}^{2} g_k(\mathbf{y}_t)(c_{ki} y_{t,i}^2 + a_{ki} y_{t,i} + b_{ki})$$

(b)

Figure 9.4. The idea of stochastic feature transformation is illustrated here. (a) Linear transformation with a single set of transformation parameters per handset and (b) nonlinear transformation with two sets of transformation parameters (ν_1 and ν_2) per handset.

is the density of the k-th distorted cluster. Figure 9.4(b) illustrates the idea of nonlinear feature transformation with different transformations for different regions of the feature space. Note that when $K = 1$ and $c_{ki} = 0$, Eq. 9.4.2 is reduced to Eq. 9.4.1 (i.e., the standard stochastic matching is a special case of the proposed approach).

Given a clean speech model $\Lambda_X = \{\omega_j^X, \mu_j^X, \Sigma_j^X\}_{j=1}^M$ derived from the clean speech of several speakers (10 speakers in this work), the maximum-likelihood estimates of ν can be obtained by maximizing an auxiliary function

$$
\begin{aligned}
Q(\nu'|\nu) &= \sum_{t=1}^T \sum_{j=1}^M \sum_{k=1}^K h_j(f_\nu(\mathbf{y}_t)) g_k(\mathbf{y}_t) \log \left\{ \omega_j^X \omega_k^Y p(\mathbf{y}_t|\mu_j^X, \Sigma_j^X, \nu_k') \right\} \\
&= \sum_{t=1}^T \sum_{j=1}^M \sum_{k=1}^K h_j(f_\nu(\mathbf{y}_t)) g_k(\mathbf{y}_t) \log \left\{ \omega_j^X \omega_k^Y p(f_{\nu_k'}(\mathbf{y}_t)|\mu_j^X, \Sigma_j^X)|J_{\nu_k'}(\mathbf{y}_t)| \right\}
\end{aligned}
$$
(9.4.5)

with respect to ν'. In Eq. 9.4.5, ν' and ν represent the new and current estimates of the transformation parameters, respectively. T is the number of distorted vectors; $\nu_k' = \{a_{ki}', b_{ki}', c_{ki}'\}_{i=1}^D$ denotes the k-th transformation; $|J_{\nu_k'}(\mathbf{y}_t)|$ is the determinant of the Jacobian matrix, the (r, s)-th entry of which is given by $J_{\nu_k'}(\mathbf{y}_t)_{rs} = \partial f_{\nu_k'}(\mathbf{y}_t)_s / \partial y_{t,r}$; and $h_j(f_\nu(\mathbf{y}_t))$ is the posterior probability given by

$$
h_j(f_\nu(\mathbf{y}_t)) = P(j|\mathbf{y}_t, \Lambda_X, \nu) = \frac{\omega_j^X p(f_\nu(\mathbf{y}_t)|\mu_j^X, \Sigma_j^X)}{\sum_{l=1}^M \omega_l^X p(f_\nu(\mathbf{y}_t)|\mu_l^X, \Sigma_l^X)},
$$
(9.4.6)

where

$$
\begin{aligned}
p(f_\nu(\mathbf{y}_t)|\mu_j^X, \Sigma_j^X) &= (2\pi)^{-\frac{D}{2}} |\Sigma_j^X|^{-\frac{1}{2}} \\
&\quad \exp\left\{ -\frac{1}{2}(f_\nu(\mathbf{y}_t) - \mu_j^X)^T (\Sigma_j^X)^{-1}(f_\nu(\mathbf{y}_t) - \mu_j^X) \right\}.
\end{aligned}
$$
(9.4.7)

Ignoring the terms independent of ν' and assuming diagonal covariance (i.e., $\Sigma_j^X = \text{diag}\{(\sigma_{j1}^X)^2, \ldots, (\sigma_{jD}^X)^2\}$, and likewise for Σ_k^Y), Eq. 9.4.5 can be written as

$$
\begin{aligned}
Q(\nu'|\nu) &= \sum_{t=1}^T \sum_{j=1}^M \sum_{k=1}^K h_j(f_\nu(\mathbf{y}_t)) g_k(\mathbf{y}_t) \cdot \\
&\quad \left\{ -\frac{1}{2} \sum_{i=1}^D \frac{(c_{ki}' y_{t,i}^2 + a_{ki}' y_{t,i} + b_{ki}' - \mu_{ji}^X)^2}{(\sigma_{ji}^X)^2} + \sum_{i=1}^D \log(2c_{ki}' y_{t,i} + a_{ki}') \right\}
\end{aligned}
$$
(9.4.8)

The generalized EM algorithm can be applied to find the maximum-likelihood estimates of ν. Specifically, in the E-step, Eqs. 9.4.6 and 9.4.7 are used to compute

$h_j(f_\nu(\mathbf{y}_t))$ and Eqs. 9.4.3 and 9.4.4 are used to compute $g_k(\mathbf{y}_t)$; then in the M-step, ν' is updated according to

$$\nu' \leftarrow \nu' + \eta \frac{\partial Q(\nu'|\nu)}{\partial \nu'}, \tag{9.4.9}$$

where η ($= 0.001$ in this work) is a positive learning factor and $\partial Q(\nu'|\nu)/\partial \nu'$ is the derivative of $Q(\nu'|\nu)$ with respect to a'_{ki}, b'_{ki} and c'_{ki}, that is,

$$\frac{\partial Q(\nu'|\nu)}{\partial a'_{ki}} = \sum_{t=1}^{T} \sum_{j=1}^{M} h_j(f_\nu(\mathbf{y}_t)) g_k(\mathbf{y}_t) \cdot$$
$$\left\{ -\frac{y_{t,i}(c'_{ki} y_{t,i}^2 + a'_{ki} y_{t,i} + b'_{ki} - \mu_{ji}^X)}{(\sigma_{ji}^X)^2} + \frac{1}{2c'_{ki} y_{t,i} + a'_{ki}} \right\} \tag{9.4.10}$$

$$\frac{\partial Q(\nu'|\nu)}{\partial b'_{ki}} = \sum_{t=1}^{T} \sum_{j=1}^{M} h_j(f_\nu(\mathbf{y}_t)) g_k(\mathbf{y}_t) \cdot$$
$$\left\{ -\frac{(c'_{ki} y_{t,i}^2 + a'_{ki} y_{t,i} + b'_{ki} - \mu_{ji}^X)}{(\sigma_{ji}^X)^2} \right\} \tag{9.4.11}$$

$$\frac{\partial Q(\nu'|\nu)}{\partial c'_{ki}} = \sum_{t=1}^{T} \sum_{j=1}^{M} h_j(f_\nu(\mathbf{y}_t)) g_k(\mathbf{y}_t) \cdot$$
$$\left\{ -\frac{y_{t,i}^2(c'_{ki} y_{t,i}^2 + a'_{ki} y_{t,i} + b'_{ki} - \mu_{ji}^X)}{(\sigma_{ji}^X)^2} + \frac{2y_{t,i}}{2c'_{ki} y_{t,i} + a'_{ki}} \right\}. \tag{9.4.12}$$

These E- and M-steps are repeated until $Q(\nu'|\nu)$ ceases to increase. In this work, Eq. 9.4.9 was repeated 20 times in each M-step.

The posterior probabilities $g_k(\mathbf{y}_t)$ and $h_j(f_\nu(\mathbf{y}_t))$ suggest that there are K regions in the distorted feature space and M regions in the clean feature space. As a result, there are KM possible transformations; however, this number can be reduced to K by arranging the indexes j and k such that the symmetric divergence

$$D(\Lambda_{X,j}||\Lambda_{Y,k}) = \frac{1}{2} tr \left\{ (\Sigma_j^X)^{-1}\Sigma_k^Y + (\Sigma_k^Y)^{-1}\Sigma_j^X - 2I \right\}$$
$$+ \frac{1}{2}(\mu_j^X - \mu_k^Y)^T \left[(\Sigma_k^Y)^{-1} + (\Sigma_j^X)^{-1} \right] (\mu_j^X - \mu_k^Y)$$

between the j-th mixture of Λ_X and the k-th mixture of Λ_Y is minimal.

Piecewise Linear Feature Transformation

When $c_{ji} = 0$, Eq. 9.4.2 becomes a piecewise linear version of the standard stochastic matching in Eq. 9.4.1. The maximum-likelihood estimate of ν can be obtained by the EM algorithm. Specifically, in the M-step, the derivative of Eq. 9.4.8 with respect to ν' is set to 0, which results in

$$s_{ki}{a'_{ki}}^2 + (u_{ki} - q_{ki}b'_{ki})a'_{ki} - v_k = 0 \quad \text{and} \quad q_{ki}a'_{ki} + r_{ki}b'_{ki} - p_{ki} = 0, \quad (9.4.13)$$

where

$$p_{ki} = \sum_{t=1}^{T}\sum_{j=1}^{M} h_{tj}g_{tk}\mu_{ji}^{X}/(\sigma_{ji}^{X})^2,$$

$$q_{ki} = \sum_{t=1}^{T}\sum_{j=1}^{M} h_{tj}g_{tk}y_{t,i}/(\sigma_{ji}^{X})^2,$$

$$r_{ki} = \sum_{t=1}^{T}\sum_{j=1}^{M} h_{tj}g_{tk}/(\sigma_{ji}^{X})^2,$$

$$s_{ki} = \sum_{t=1}^{T}\sum_{j=1}^{M} h_{tj}g_{tk}y_{t,i}^2/(\sigma_{ji}^{X})^2,$$

$$u_{ki} = \sum_{t=1}^{T}\sum_{j=1}^{M} h_{tj}g_{tk}\mu_{ji}^{X}y_{t,i}/(\sigma_{ji}^{X})^2, \quad \text{and}$$

$$v_k = \sum_{t=1}^{T}\sum_{j=1}^{M} h_{tj}g_{tk},$$

where $h_{tj} \equiv h_j(f_\nu(\mathbf{y}_t))$ and $g_{tk} \equiv g_k(\mathbf{y}_t)$ are estimated during the E-step.

Justification for Using Second-Order Transformation

Research has shown that handset distortion is the major source of recognition errors [131], that different handsets cause different degrees of distortion of speech signals [253], and that distortion can be nonlinear [308, 393]. As a result, the probability density functions of the distorted cepstra caused by different handsets are different. A set of GMMs can be used to estimate the probability that the observed speech comes from a particular handset.

To compare the characteristics of Eqs. 9.4.1 and 9.4.2, Gaussian noise was added to a section of switchboard conversations $x(n)$ and the resulting signals were passed through a bandpass filter with cutoff frequencies at 625Hz and 1,875Hz. The resulting signals were subsequently passed through a nonlinear function $f(x) = \tanh(7x)$, where x was normalized to $(-0.25, +0.25)$ to obtain a distorted signal $y(n)$. Figure 9.5 depicts the projection of the cepstral coefficients obtained at different stages of processing on the first two components $c1 - c2$. Evidently, the Gaussian noise shrinks the clean cluster ($\square$), suggesting that at least one affine transformation matrix is needed. The bandpass filter shifts the noise-corrupted cluster ($*$) to another region of the feature space, which explains the inclusion of the bias terms b_i in Eqs. 9.4.1 and 9.4.2. The nonlinear filter twists the bandpass filtered cluster ($\circ$)

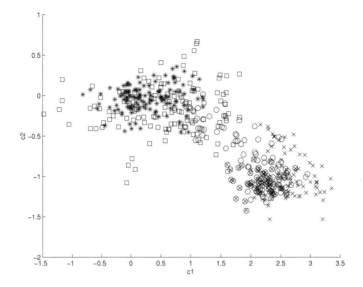

Figure 9.5. Plot of $c2$ against $c1$ at different stages of processing. $\square$ = clean (no processing); $*$ = after adding Gaussian noise; $\times$ = after bandpass filtering; $\circ$ = after nonlinear filtering

and moves it closer to the clean cluster. This suggests the necessity of using more than one affine transformation matrix.

Figure 9.6 depicts the cluster recovered by Eq. 9.4.1, and Figure 9.7 illustrates how the linear regressor in Eq. 9.4.1 solves the regression problem. A closer look at Figure 9.7 reveals that the amount of componentwise variation in the clean and distorted speech is almost identical. This suggests that a single regressor is inadequate. Figure 9.8 plots the clean cluster, the distorted cluster, and the cluster recovered by the second-order transformation in Eq. 9.4.2. Figure 9.9 illustrates how the nonlinear probabilistic regressors solve the problem. These figures show that the nonlinear regressors are better than the linear ones in fitting the data. In particular, Figure 9.8 shows that almost none of the recovered data falls outside the region of the clean cluster. Evidently, having more than one nonlinear regressor for each component helps reduce transformation error.

Handset Selector

Unlike speaker adaptation in which the adapted system is used by the same "adapted" speaker in all subsequent sessions, speaker verification allows the claimant to be a different person during each verification session. As a result, the claimant's speech cannot be used for adaptation because doing so would transform the claimant's speech to fit the client model in the case of feature transformation (or transform the client model to fit the claimant's speech in the case of model adaptation). This

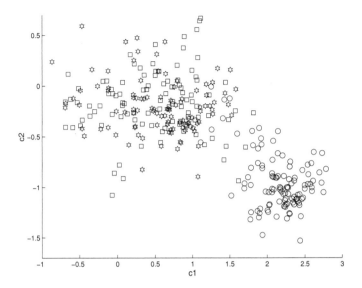

Figure 9.6. Plot showing the clean cluster ($\square$), nonlinearly distorted cluster ($\circ$), and recovered cluster ($\star$). Eq. 9.4.1 was used to produce the recovered cluster.

results in verification error if the claimant turns out to be an impostor. Therefore, instead of using the claimant speech for determining the transformation parameters or adapting the client model directly, it can be used indirectly as follows.

Before verification takes place, one set of transformation parameters (or adaptation parameters) is obtained for each type of handset that the clients are likely to use. During verification, the most likely handset to be used by the claimant is identified and the best set of transformation parameters is selected accordingly. Specifically, during verification an utterance of the claimant's speech is fed to H GMMs—denoted as $\{\Gamma_k\}_{k=1}^{H}$. The most likely handset is selected according to

$$k^* = \arg\max_{k=1}^{H} \sum_{t=1}^{T} \log p(\mathbf{y}_t|\Gamma_k), \qquad (9.4.14)$$

where $p(\mathbf{y}_t|\Gamma_k)$ is the likelihood of the k-th handset. Finally, the transformation parameters corresponding to the k^*-th handset are used to transform the distorted vectors. Figure 9.10 illustrates the idea of combining a handset selector with feature transformation in a speaker verification system (see p. 314).

Experimental Evaluations

Evaluation of HTIMIT Speech. The HTIMIT corpus [304] was used to evaluate the feature transformation approach. HTIMIT was constructed by playing a gender-balanced subset of the TIMIT corpus through nine phone handsets and a

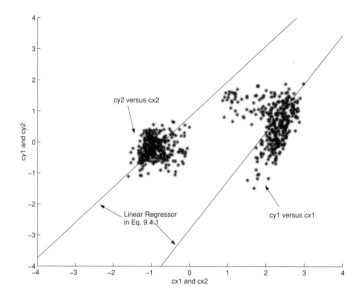

Figure 9.7. Plot showing the linear regressor in Eq. 9.4.1 and the componentwise variation of clean ($cx1$ and $cx2$) and distorted ($cy1$ and $cy2$) vectors.

Sennheizer head-mounted microphone. Unlike other phone speech databases, where no handset labels are given, every utterance in HTIMIT is labeled with a handset name (cb1–cb4, el1–el4, pt1, or senh). This feature makes HTIMIT amenable to the study of transducer effects in speech and speaker recognition. Because of this special characteristic, HTIMIT has been used in several speaker recognition studies, including Mak et al. [227] and Quartieri et al. [289].

Speakers in the corpus were divided into a speaker set (50 male and 50 female) and an impostor set (25 male and 25 female). Each speaker was assigned a personalized 32-center GMM (with diagonal covariance) that models the characteristics of his or her own voice. For each GMM, the feature vectors derived from the SA and SX sentence sets of the corresponding speaker were used for training. A collection of all SA and SX sentences uttered by all speakers in the speaker set was used to train a 64-center GMM background model ($\mathcal{M}_b$). The feature vectors were 12th order LP-derived cepstral coefficients computed at a frame rate of 71Hz using a Hamming window of 28ms.

For each handset in the corpus, the SA and SX sentences of 10 speakers were used to create two 2-center GMMs (Λ_X and Λ_Y in Figure 9.4).[5] For each handset, a set of feature transformation parameters ν were computed based on the estimation algorithms described in Section 9.4.2. Specifically, the utterances from the handset

---

[5]Only a few speakers will be sufficient for creating these models; however, no attempt was made to determine the optimum number.

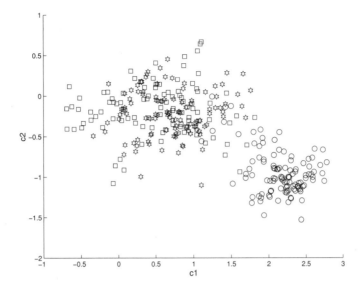

Figure 9.8. Plot showing the clean cluster ($\square$), nonlinearly distorted cluster ($\circ$), and recovered cluster ($\star$). Eq. 9.4.2 (with $K = 2$) was used to produce the recovered cluster.

"senh" were used to create Λ_X, and those from the other 9 handsets were used to create $\Lambda_{Y_1}, \ldots, \Lambda_{Y_9}$. The number of transformations for all handsets was set to 2 (i.e., $K = 2$ in Eq. 9.4.2).

During verification, a vector sequence Y derived from a claimant's utterance (SI sentence) was fed to a GMM-based handset selector $\{\Gamma_i\}_{i=1}^{10}$ described in the Handset Selector section. A set of transformation parameters was selected according to the handset selector's outputs (Eq. 9.4.14). The features were transformed and then fed to a 32-center GMM speaker model ($\mathcal{M}_s$) to obtain a score ($\log p(Y|\mathcal{M}_s)$), which was then normalized according to

$$S(Y) = \log p(Y|\mathcal{M}_s) - \log p(Y|\mathcal{M}_b), \qquad (9.4.15)$$

where $\mathcal{M}_b$ is a 64-center GMM background model. $S(Y)$ was compared against a threshold to make a verification decision. In this work, the threshold for each speaker was adjusted to determine an EER (i.e., speaker-dependent thresholds were used). Similar to Mak and Kung [225] and Reynolds and Rose [307] studies, the vector sequence was divided into overlapping segments to increase the resolution of the error rates.

Table 9.5 compares different stochastic feature transformation approaches against CMS and the baseline (without any compensation). All error rates were based on the average of 100 genuine speakers and 50 impostors. Evidently, all cases of stochastic feature transformation show significant reduction in error rates. In particular,

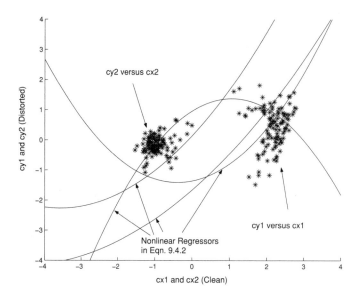

Figure 9.9. Plot showing the nonlinear regressors in Eq. 9.4.2 and the componentwise variation of clean and distorted vectors.

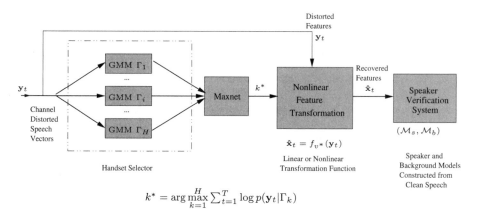

$$k^* = \arg\max_{k=1}^{H} \sum_{t=1}^{T} \log p(\mathbf{y}_t | \Gamma_k)$$

Figure 9.10. Combining a handset selector with feature transformation in speaker verification.

second-order stochastic transformation achieves the highest reduction. The second last column of Table 9.5 shows that when the enrollment and verification sessions use the same handset (senh), CMS can degrade the performance. On the other hand, in the case of feature transformation (rows 3 to 8), the handset selector can detect the fact that the claimant is using the enrollment handset. As a result, the error rates become very close to the baseline. This suggests that the combination of

Table 9.5. EERs achieved by the baseline, CMS, and different transformation approaches. *Note*: ST0, ST1, and ST2 stand for stochastic transformation with zero- first-, and second-order, respectively; the enrollment handset is senh. The second last column represents the case where enrollment and verification use the same handset. The average handset identification accuracy is 98.29%. Note that the baseline and CMS do not require the handset selector. The value of K in column 2 represents the range of summation in Eq. 9.4.2. Ave. stands for the average EER of the ten handsets.

	Transformation	Equal Error Rate (%)										
Row	Method	cb1	cb2	cb3	cb4	el1	el2	el3	el4	pt1	senh	Ave.
1	Baseline	7.89	6.93	26.96	18.53	5.79	14.09	7.80	13.85	9.51	2.98	11.43
2	CMS	5.81	5.02	12.07	9.41	5.26	8.88	8.44	6.90	6.97	3.58	7.23
3	ST0, $K = 1$	4.06	3.63	8.86	6.05	3.57	6.78	6.66	4.79	5.43	2.99	5.28
4	ST0, $K = 2$	4.27	3.74	9.19	6.74	3.68	6.95	7.06	5.00	5.38	3.09	5.51
5	ST1, $K = 1$	4.33	4.06	8.92	6.26	4.30	7.44	6.39	4.83	5.74	3.47	5.63
6	ST1, $K = 2$	4.27	3.84	9.14	6.73	3.83	7.01	6.98	5.04	5.74	3.16	5.57
7	ST2, $K = 1$	4.10	3.65	8.98	6.06	3.63	6.94	7.23	4.87	5.41	3.03	5.93
8	ST2, $K = 2$	4.04	3.57	8.85	6.82	3.53	6.43	6.41	4.76	5.02	2.98	5.24

handset selector and stochastic transformation can maintain the performance under matched conditions.

Evaluation of Transcoded HTIMIT Speech.   To evaluate the performance of the feature transformation technique on the coded HTIMIT corpora, seven codecs were employed in this work: G.711 at 64kpss, G.726 at 32Kbps, GSM at 13Kbps, G.729 at 8Kbps, G.723.1 at 6.3Kbps, MELP at 2.4Kbps, and LPC at 2.4Kbps. Six sets of coded corpora were obtained by coding the speech in HTIMIT using these coders. The encoded utterances were then decoded to produce resynthesized speech. Feature vectors were extracted from each of the utterances in the uncoded and coded corpora. The feature vectors were twelfth-order mel-frequency cepstral coefficients (MFCC) [69]. These vectors were computed at a frame rate of 14ms using a Hamming window of 28ms.

Six handset selectors, each consisting of 10 GMMs $\{\Gamma_k^{(i)}; i = 1, \ldots, 6 \text{ and } k = 1, \ldots, 10\}$, were constructed from the SA and SX sentence sets of the coded corpora. For example, GMM $\Gamma_k^{(i)}$ represents the characteristics of speech derived from the k-th handset of the i-th coded corpus. Assuming that in most practical situations the receiver will know the type of coders being used (otherwise it will not be able to decode the speech), there will not be any error in choosing the handset selector. The only error that will be introduced is the incorrect decisions made by the chosen handset selector. This error, however, is very small, as demonstrated in the latter part of this section.

During verification, a vector sequence Y derived from a claimant's utterance

(SI sentence) was fed to a coder-dependent handset selector corresponding to the coder being used by the claimant. According to the outputs of the handset selector (Eq. 9.4.14), a set of coder-dependent transformation parameters was selected. The experimental results are summarized in Tables 9.6, 9.7, and 9.8. A baseline experiment (without using the handset selectors and feature transformations) and an experiment using CMS as channel compensation were also conducted for comparison. All error rates are based on the average of 100 genuine speakers, with each speaker impersonated by 50 impostors.

Average EERs of the uncoded and coded corpora are plotted in Figure 9.11. The average EER of a corpus is computed by taking the average of all of the EERs corresponding to the different handsets of the corpus. The results show that the transformation technique achieves significant error reduction for both uncoded and coded corpora. In general, the transformation approach outperforms the CMS approach except for the LPC-coded corpus.

The results in Table 9.6 show that the error rates of LPC-coded corpus are relatively high before channel compensation is applied. An informal listening test reveals that the perceptual quality of LPC-coded speech is very poor, which means that most of the speaker's characteristics have been removed by the coding process. This may degrade the performance of the transformation technique.

Table 9.6. EERs achieved by the baseline approach. *Note:* Results are without handset selectors and feature transformation on speech corpora coded by different coders.

Codec	Equal Error Rate (%)										
	cb1	cb2	cb3	cb4	el1	el2	el3	el4	pt1	senh	Ave.
Uncoded (128Kbps)	4.85	5.67	21.19	16.49	3.60	11.11	5.14	11.56	11.74	1.26	9.26
G.711 (64Kbps)	4.88	5.86	21.20	16.73	3.67	11.08	5.21	12.28	12.04	1.34	9.43
G.726 (32Kbps)	6.36	8.71	22.67	19.61	6.83	14.98	6.68	16.81	16.42	2.66	12.17
GSM (13Kbps)	6.37	6.10	19.90	15.93	6.21	17.93	9.86	15.29	16.42	2.35	11.64
G.729 (8Kbps)	6.65	4.59	20.15	15.08	6.18	14.28	6.71	9.63	11.93	2.67	9.79
G.723.1 (6.3Kbps)	7.33	5.49	20.83	15.59	6.56	14.71	6.58	10.63	14.03	3.30	10.51
MELP (2.4Kbps)	8.56	7.36	24.99	19.71	7.48	16.13	7.17	16.97	13.31	3.23	12.49
LPC (2.4Kbps)	10.81	10.30	29.68	24.21	8.56	19.29	10.56	19.19	14.97	3.43	15.10

Currently, G.711 and GSM coders are widely used in fixed-line and mobile communication networks, respectively; and G.729 and G.723.1 have become standard coders in teleconferencing systems. These are the areas where speaker verification is useful. LPC coders, on the other hand, are mainly used for applications where speaker verification is not very important (e.g., toys). Because the feature transformation technique outperforms CMS in areas where speaker verification is more important, it is a better candidate for compensating coder- and channel-distortion in speaker verification systems.

Table 9.7. EERs achieved by the CMS approach on speech corpora coded by different coders.

Codec	Equal Error Rate (%)										Ave.
	cb1	cb2	cb3	cb4	el1	el2	el3	el4	pt1	senh	
Uncoded (128Kbps)	4.00	3.02	10.69	6.62	3.36	5.16	5.67	4.05	5.67	3.67	5.19
G.711 (64Kbps)	4.06	3.07	10.73	6.70	3.43	5.26	5.74	4.23	5.84	3.75	5.28
G.726 (32Kbps)	5.65	4.42	11.78	8.00	5.61	7.95	6.97	6.47	9.07	5.12	7.10
GSM (13Kbps)	5.25	4.10	11.32	8.00	4.95	7.04	7.47	6.05	7.58	4.73	6.65
G.729 (8Kbps)	5.43	4.37	11.81	7.98	5.16	7.38	7.32	5.42	7.21	4.69	6.68
G.723.1 (6.3Kbps)	6.40	4.60	12.36	8.53	6.11	8.50	7.31	6.11	8.28	5.62	7.38
MELP (2.4Kbps)	5.79	4.81	13.72	8.75	5.13	8.18	7.31	6.40	7.93	5.86	7.39
LPC (2.4Kbps)	6.34	5.51	14.10	9.22	6.35	8.95	8.95	6.34	9.55	4.57	7.99

Table 9.8. EERs achieved by combining zero-order feature transformation with coder-dependent handset selectors on speech corpora coded by different coders. The accuracy achieved by the handset selectors is also shown. "HS Acc." stands for handset identification accuracy.

Codec	Equal Error Rate (%)										Ave.	HS Acc. (%)
	cb1	cb2	cb3	cb4	el1	el2	el3	el4	pt1	senh		
Uncoded (128Kbps)	1.63	1.27	9.65	4.47	1.41	3.58	3.37	1.66	3.08	1.09	3.12	98.13
G.711 (64Kbps)	1.52	1.26	9.57	4.53	1.41	3.53	3.33	1.72	3.21	1.17	3.13	98.22
G.726 (32Kbps)	2.55	2.55	11.66	6.05	2.74	6.19	4.17	3.19	5.82	2.29	4.72	97.95
GSM (13Kbps)	3.13	2.44	11.13	7.10	3.10	6.34	6.29	3.38	5.58	2.67	5.12	97.19
G.729 (8Kbps)	3.94	3.27	9.99	6.63	4.18	6.17	6.20	3.08	4.70	2.89	5.11	96.73
G.723.1 (6.3Kbps)	3.94	3.42	10.74	6.83	4.49	6.70	5.80	3.44	5.71	3.41	5.45	96.51
MELP (2.4Kbps)	4.42	3.58	13.07	7.31	3.51	5.75	5.40	3.62	5.04	3.74	5.54	96.13
LPC (2.4Kbps)	5.68	5.93	17.33	11.05	7.14	10.50	9.34	4.94	8.89	3.95	8.48	94.86

It is obvious from the senh columns of Tables 9.6 and 9.7 that CMS degrades the performance of the system when the enrollment and verification sessions use the same handset (senh). When the transformation technique is employed under this matched condition, the handset selectors can detect the most likely handset (i.e., senh) and facilitate the subsequent transformation of the distorted features. As a result, the error rates become very close to the baseline.

As the experimental results show, verification based on uncoded phone speech performs better than that based on coded phone speech. However, since the distortion introduced by G.711 is very small, the error rates of uncoded and G.711 coded corpora are similar.

In general, the verification performance of the coded corpora becomes poor when the bit rate of the corresponding codec decreases (Figure 9.11). However, the per-

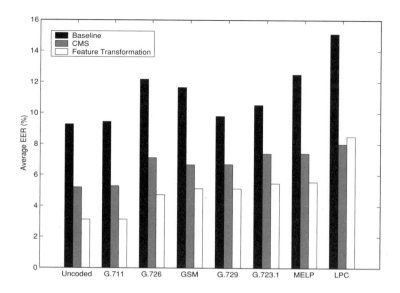

Figure 9.11. Average EERs achieved by the baseline, CMS, and transformation approaches. Note that the bit rate of coders decreases from left to right, with "uncoded" being the highest (128Kbps) and LPC the lowest (2.4Kbps).

formance among the GSM, G.729, and G.723.1 coded speech occasionally does not obey this rule even for the same handsets. After CMS was applied, the error rates were reduced for all of the uncoded and coded corpora, while a stronger correlation between bit rates and verification performance can be observed among GSM, G.729, and G.723.1 coded speech. Using the transformation technique, error rates are further reduced, and correlation between bit rates and verification performance becomes very obvious among the coded speech at various bit rates. Because the perceptual quality of the coded speech is usually poorer for lower rate codecs, it appears that a strong correlation between coded speech quality and verification performance exists.

Compared with the results in the Mak and Kung study in 2002 [226], it is obvious that using MFCC is more desirable than using LPCC. For example, when MFCC is used, the average error rate for the uncoded speech is 9.01%, whereas the error rate increases to 11.16% when LPCC is used [226].

9.4.3 Stochastic Model Transformation

In feature transformation, speech features are transformed so that the resulting features fit the clean speaker models better. On the other hand, model transformation and adaptation modifies the parameters of the statistical models so that the modified models characterize the distorted speech features better. Maximum a posteriori (MAP) adaptation [197] and maximum-likelihood linear regression (MLLR)

[200] are the two fundamental techniques for model adaptation. Although these techniques were originally designed for speaker adaptation, with some modification they can also be applied to compensating channel distortion. One of the positive properties of MAP is that its performance approaches that of maximum-likelihood-based methods, provided sufficient adaptation data are available. However, MAP is an unconstrained method in that adaptation of model parameters is performed only for those who have *seen* the adaptation data. MLLR, on the other hand, applies a transformation matrix to a group of acoustic centers so that all the centers are transformed. As a result, MLLR provides a quick improvement, but its performance quickly saturates as the amount of adaptation data increases.

This section investigates two model adaptation and transformation techniques: probabilistic decision-based neural networks (PDBNNs) [211] and maximum-likelihood linear regression (MLLR) [200] in the context of phone-based speaker verification. These techniques adapt or transform the model parameters to compensate for the *mismatch* between the training and testing conditions. Some preliminary results on PDBNN adaptation and MLLR adaptation were reported in Yiu et al. [394]. Here, the results of the 2003 study are extended by combining these two techniques with stochastic feature transformation (SFT) as described in Section 9.4.2.

Specifically, precomputed MLLR transformation matrices are used to transform the clean models to handset-dependent MLLR-adapted models. Then, PDBNN adaptation is performed on the MLLR-adapted models using handset-dependent, stochastically transformed patterns to obtain the final adapted models. The results are also compared with state-of-the-art normalization techniques, including CMS [13], handset normalization (Hnorm) [303] and test normalization (Tnorm) [14]. Unlike feature transformation and model adaptation, Hnorm and Tnorm are score normalization techniques that work on the score space to minimize the effect introduced by handset variability. The idea is to remove the handset-dependent bias by normalizing the distributions of speaker scores using the scores of nontarget speakers. The resulting score distribution should have zero mean and unit standard deviation. Experimental results based on 150 speakers of the HTIMIT corpus show that the proposed model transformation techniques outperform CMS, Hnorm and Tnorm.

Probabilistic Decision-Based Neural Networks

PDBNNs were discussed in Chapter 7. This section makes use of their globally supervised (GS) learning to adapt the speaker and background models to fit different acoustic environments.

A training strategy was adopted to make PDBNNs appropriate for environment adaptation. It begins with the training of a clean speaker model and a clean background model using the LU training of PDBNNs. This step aims to maximize the likelihood of the training data. The clean models were then adapted to a handset-dependent speaker model and a handset-dependent background model using the

GS training. Although clean speech data were used in the LU training, distorted training data derived from the target handset were used in the GS training. The GS training uses gradient descent and reinforced learning to update the models' parameters so that the classification error of the adapted models on the distorted data is minimized. Hence, the resulting models will be speaker- and handset-specific.

By using the distorted data derived from H handsets, the preceding training strategy produces H handset-dependent speaker models for each speaker. Likewise, H handset-dependent background models are also created, and they are shared among all of the registered speakers in the system. Figure 9.12 (left portion) illustrates the idea of PDBNN-based adaptation for robust speaker verification. For some handsets, however, speaker-specific training data could be sparse or might not even exist. In such cases, unsupervised adaptation, such as MLLR, may be more appropriate.

Maximum-Likelihood Linear Regression

MLLR was originally developed for speaker adaptation [200]; however, it can also be applied to environment adaptation. Specifically, a set of adaptation matrices $W^k, k = 1, \ldots, H$, are estimated to model the mismatches between the enrollment and verification conditions; during recognition, the most appropriate transformation W^{k^*}, where $k^* \in \{1, 2, \ldots, H\}$, is applied to the Gaussian means of the speaker and background models. More precisely, if $\mu_{s,j}$ is the j-th mean vector of the clean speaker model, the adapted mean vector $\mu_{ad\_s,j}$ will be

$$\mu_{ad\_s,j} = W^{k^*} \hat{\mu}_{s,j} = A^{k^*} \mu_{s,j} + \mathbf{b}^{k^*},$$

where $\hat{\mu}_{s,j} = [\mu_{s,j}^T, 1]^T$ is the extended mean vector of $\mu_{s,j}$. The adaptation matrices are estimated by maximizing the likelihood of the adaptation data using the EM algorithm. Each adaptation matrix is composed of a translation vector $\mathbf{b}^k \in \Re^D$ and a transformation matrix $A^k \in \Re^D \times \Re^D$, where D is the dimensionality of the feature vectors and $k = 1, \ldots, H$ (i.e., $W^k = [A^k, \mathbf{b}^k]$).

The transformation matrices are estimated by maximizing the likelihood of the transformed models given the adaptation data. The maximization can be achieved by using the EM algorithm. The transformation matrices can be full or diagonal. Although full transformation matrices can model the correlations among feature components, they require a lot more data to estimate robustly. The number of transformation matrices used depends on the amount of adaptation data. When only a small amount of observation data are available, it may be better to use a single matrix to transform all of the Gaussian centers of the clean speaker models and background models. When more adapted data become available, more specific transformation matrices can be computed.

Figure 9.12 (right portion with the bypass) illustrates the idea of MLLR-based model transformation for robust speaker verification.

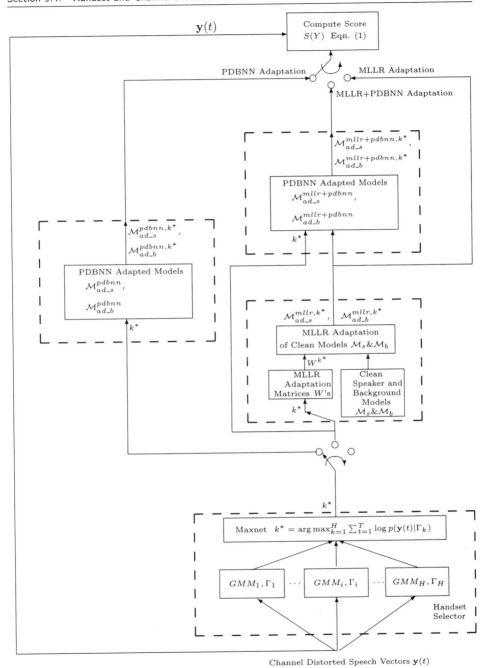

Figure 9.12. The combination of handset identification and model adaptation for robust speaker verification. *Note*: Adaptation was applied to both the speaker and background models.

Cascading MLLR Transformation and PDBNN Adaptation

Although PDBNN's reinforced learning uses handset-specific and speaker-specific patterns to adapt the model parameters, a previous study [394] showed that its performance is not as good as that of the MLLR adaptation. A possible reason is that PDBNN's discriminative training is sensitive to the centers' locations of the "initial model." Because the centers of the clean speaker models and the handset-dependent speaker models may be far apart, PDBNN's reinforced learning can only adapt some of the model centers in the clean speaker models. To overcome this obstacle, MLLR is proposed to transform the clean speaker models to a region close to the handset-distorted speech. These MLLR-transformed models are then used as the initial models for PDBNN's reinforced learning.

Figure 9.12 (right portion without the bypass) depicts the cascade of MLLR transformation and PDBNN adaptation.

Cascading MLLR Transformation and PDBNN Adaptation Using Transformed Features

Although cascading MLLR transformation and PDBNN adaptation should be better than using MLLR transformation or PDBNN adaptation alone, this approach is not very practical. This is because PDBNN's reinforced training requires client-dependent speech data to be collected from different environments that the client may encounter during verification. A possible solution is to use stochastic feature transformation (SFT) (Section 9.4.2) to transform the clean speech data to environment-specific speech data. Figure 9.13 illustrates the idea of the proposed method.

For each known handset, a precomputed MLLR adaptation matrix is used to transform the clean models to handset-dependent MLLR-adapted models. Then, PDBNN adaptation is performed on the MLLR-adapted models using handset-dependent, stochastically-transformed patterns to obtain the final adapted models. Because handset-specific, client-dependent speech patterns are difficult to obtain, SFT is applied to transform the clean speaker patterns, as illustrated in Figure 9.13, to handset-dependent client features. The key idea is that the SFT-transformed speaker patterns are artificially generated to provide the PDBNN adaptation with the required data in the handset-distorted space for fine-tuning the MLLR-adapted models. For the background speakers' data, transformed features are not necessary because plenty of environment-specific data are available.

A Two-Dimensional Example

Figure 9.14 illustrates the idea of environment adaptation in a two-class problem. Figure 9.14(a) plots the clean and distorted patterns of Class 1 and Class 2. The upper right (respectively lower left) clusters represent the clean (respectively distorted) patterns. The ellipses show the corresponding equal density contours. The decision boundary in Figure 9.14(a) was derived from the clean GMMs, which were

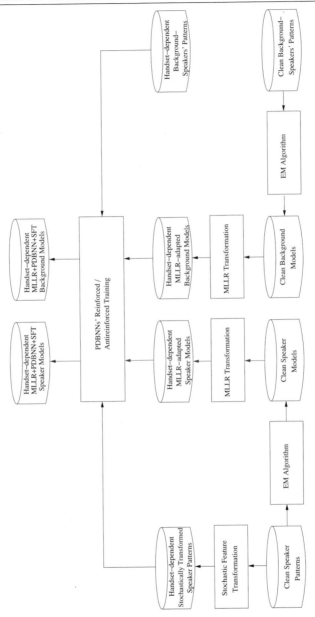

Figure 9.13. The process of fine-tuning MLLR-adapted models using transformed features. For each known handset, a precomputed MLLR adaptation matrix is used to transform the clean models into handset-dependent MLLR-adapted models. Then, PDBNN adaptation is performed on the MLLR-adapted models using handset-dependent, stochastically transformed patterns to obtain the final adapted models.

trained using the clean patterns. As the lower left portion of Figure 9.14(a) shows, this decision boundary is not appropriate for separating the distorted patterns.

The clean models were adapted using the methods explained before, and the results are shown in Figures 9.14(b) through 9.14(d). In Figures 9.14(b), 9.14(c), and 9.14(d), the markers ♦ and ■ represent the centers of the clean models and ● and ▲ represent the centers of the adapted models. The arrows indicate the adaptation of model centers, and the thick curves show the equal-error decision boundaries derived from the adapted models. For PDBNN-adaptation (Figure 9.14(b)), two clean GMMs were trained independently using the clean patterns of each class. The distorted patterns of both classes were used to adapt the clean models using PDBNN's reinforced learning. For MLLR-adaptation (Figure 9.14(c)), a clean GMM was trained using the clean patterns of both classes, which was followed by the estimation of MLLR parameters using the clean GMM and the distorted patterns of both classes. The two clean GMMs corresponding to the two classes were then transformed using the estimated MLLR parameters. For the cascaded MLLR–PDBNN adaptation (Figure 9.14(d)), the clean models were first transformed by MLLR, then PDBNN adaptation was performed on the adapted models to obtain the final model.

A comparison between Figure 9.14(b) and Figure 9.14(c) reveals that while PDBNN-adaptation can transform only some of the model centers to the region of the distorted clusters, MLLR-based adaptation can transform all of the centers of the clean GMMs to a region around the distorted clusters. By cascading MLLR transformation and PDBNN adaptation, all of the centers of the clean GMMs can be transformed to a region within the distorted clusters before the decision boundary is fine-tuned. The adaptation capability of cascading MLLR and PDBNN is also demonstrated in the speaker verification evaluation described in Section 9.4.3.

Applications to Channel Compensation

A recently proposed handset selector has been adopted (see Handset Selector section and references [226, 227, 352, 353]) to identify the most likely handset given an utterance. Specifically, H GMMs, $\{\Gamma_k\}_{k=1}^{H}$, as shown in Figure 9.12, were independently trained using the distorted speech recorded from the corresponding phone handsets. During recognition, the claimant's features, $\mathbf{y}(t), t = 1, \ldots, T$, were fed to all GMMs. The most likely handset k^* is selected by the Maxnet, as illustrated in Figure 9.12.

For PDBNN adaptation, the precomputed PDBNN-adapted speaker model ($\mathcal{M}_{ad\_s}^{pdbnn,k^*}$) and background model ($\mathcal{M}_{ad\_b}^{pdbnn,k^*}$) corresponding to the k^*-th handset were used for verification. For MLLR adaptation, the precomputed MLLR adaptation matrix (W^{k^*}) for the k^*-th handset was used to transform the clean speaker model ($\mathcal{M}_s$) to the MLLR-adapted speaker model ($\mathcal{M}_{ad\_s}^{mllr,k^*}$). The same matrix was also used to transform the clean background model ($\mathcal{M}_b$) to the MLLR-adapted background model ($\mathcal{M}_{ad\_b}^{mllr,k^*}$). For MLLR+PDBNN adaptation, the precomputed MLLR adaptation matrix (W^{k^*}) for the k^*-th handset was first used

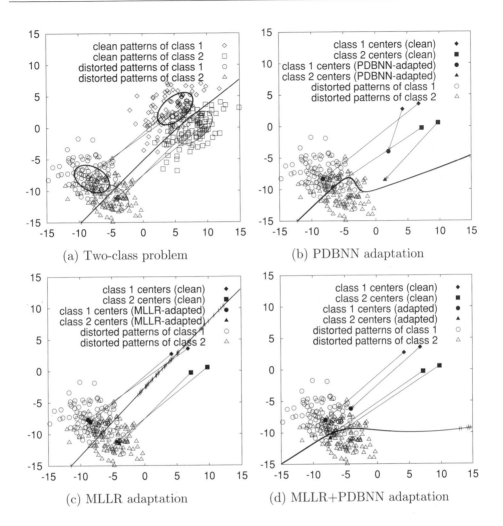

(a) Two-class problem   (b) PDBNN adaptation

(c) MLLR adaptation   (d) MLLR+PDBNN adaptation

Figure 9.14. (a) Scatter plots of the clean and distorted patterns corresponding to Class 1 and Class 2 in a two-class problem. The thick and thin ellipses represent the equal density contour of Class 1 and Class 2, respectively. The upper right (respectively lower left) clusters contain the clean (respectively distorted) patterns. The decision boundary (*thick curve*) was derived from the clean GMMs that were trained using the clean patterns. (b) Centers of the clean models and the PDBNN-adapted models. The thick curve is the decision boundary created by the PDBNN-adapted models. (c) Centers of the clean models and the MLLR-adapted models. The thick curve is the decision boundary created by the MLLR-adapted models. (d) Centers of the clean models and the MLLR+PDBNN adapted models. (The clean models were firstly transformed by MLLR, then PDBNN adaptation was performed to obtain the final adapted model.) In (b), (c) and (d), only the distorted patterns are plotted for clarity. The arrows indicate the displacement of the original centers and the adapted centers.

to transform the clean models ($\mathcal{M}_s$, $\mathcal{M}_b$) to the MLLR-adapted model ($\mathcal{M}_{ad\_s}^{mllr,k^*}$, $\mathcal{M}_{ad\_b}^{mllr,k^*}$). Then, PDBNN adaptation was performed on the MLLR-adapted models to obtain the MLLR+PDBNN adapted models ($\mathcal{M}_{ad\_s}^{mllr+pdbnn,k^*}$, $\mathcal{M}_{ad\_b}^{mllr+pdbnn,k^*}$). These models are to be used for verifying the claimant.

Experimental Evaluation

Enrollment Procedures. Similar to SFT, the HTIMIT corpus was used in the evaluation. Speakers in the HTIMIT corpus were divided into a speaker set (100 speakers) and an impostor set (50 speakers). Each speaker in the speaker set was assigned a 32-center GMM ($\mathcal{M}_s$) characterizing his or her own voice. For each speaker model, training feature vectors were derived from the SA and SX utterances of the corresponding speaker. A 64-center GMM universal background model ($\mathcal{M}_b$), which was trained using all of the SA and SX utterances from all speakers in the speaker set, was used to normalize the speaker scores (see Eq. 9.4.16). Utterances from the head-mounted microphone (senh) were considered to be clean and were used for creating the speaker models and background models. The feature vectors were twelfth-order MFCCs computed every 14ms using a Hamming window of 28ms.

Environment Adaptation. For PDBNN-based adaptation, the clean speaker model ($\mathcal{M}_s$) and clean background model ($\mathcal{M}_b$) described before were used as the initial models for GS training. The SA and SX utterances of the target speaker and the background speakers (excluding the target speaker) from a phone handset were used as positive and negative training patterns, respectively. Hence, for each target speaker, a handset-dependent speaker model and a handset-dependent background model were created for each handset (including the head-mounted microphone used for enrollment).

For MLLR-based adaptation, a single, full transformation matrix is used to compensate for the "mismatch" between two environments. Specifically, a clean background model ($\mathcal{M}_b$) was trained using the clean speech of all speakers in the speaker set. Then, the speech data from another handset were used to estimate a transformation matrix corresponding to that handset using MLLR. This procedure was repeated for all handsets in the HTIMIT corpus.

For MLLR+PDBNN adaptation, both MLLR transformation and PDBNN globally supervised training were applied to compensate for the mismatched conditions. First, MLLR was used to transform the clean speaker model ($\mathcal{M}_s$) and the clean background model ($\mathcal{M}_b$) to handset-specific models. PDBNN globally supervised training were then applied to the handset-specific models. The handset-specific SA and SX utterances of the target speaker and the background speakers (excluding the target speaker) were used as the training patterns in the PDBNN supervised training.

The MLLR+PDBNN+SFT adaptation is identical to MLLR+PDBNN adaptation except that the former uses stochastically transformed speaker features in the PDBNN globally supervised training. Specifically, the clean speaker patterns from the senh handset were transformed using SFT to obtain a set of artificial, handset-specific training patterns. With SFT, the necessity of handset-specific speaker patterns for PDBNN training is eliminated. Because handset-specific speech patterns for the background speakers are easily obtainable, performing SFT on the background speaker's speech is unnecessary.

A preliminary evaluation was performed to compare the performance of MLLR transformation using 5, 20, and 100 speakers to estimate the adaptation matrices Ws. Although the performance improves with the number of speakers, the computation time also increases with the total number of training patterns. The MLLR transformation matrices used in the following experiments were estimated using 20 and 100 speakers. Experiments have also been conducted using SFT [226], Hnorm [303], and Tnorm [14] for comparison. The parameters for SFT were estimated using the same 20 and 100 speakers, as in PDBNN model adaptation and MLLR model transformation. For Hnorm, the speech patterns derived from the handset-specific utterances of 49 same-gender (same as the client speaker), nontarget speakers were used to compute the handset-dependent means and standard deviations. As a result, each client speaker model is associated with 10 handset-dependent score means and variances. These means and variances were used during verification to normalize the claimant's scores. For Tnorm [14], verification utterances were fed to all of the 99 nontarget speaker models to calculate the mean and variance parameters. These parameters were then used to normalize the speaker scores.

Verification Procedures. During verification, a pattern sequence Y derived from each of the SI sentences of the claimant was fed to the GMM-based handset selector $\{\Gamma_i\}_{i=1}^{10}$. Handset-dependent speaker and background models/adaptation matrix were selected according to the handset selector's output (see Figure 9.12). The test utterance was then fed to an adapted speaker model $\mathcal{M}_{ad\_s}^{\Psi,k^*}$ and an adapted background model $\mathcal{M}_{ad\_b}^{\Psi,k^*}$ to obtain the score

$$S(Y) = \log p(Y|\mathcal{M}_{ad\_s}^{\Psi,k^*}) - \log p(Y|\mathcal{M}_{ad\_b}^{\Psi,k^*}), \qquad (9.4.16)$$

where $\Psi \in \{$ PDBNN, MLLR, PDBNN+MLLR, PDBNN+MLLR+SFT $\}$ represents the type of adaptation used to obtain the adapted models. $S(Y)$ was compared with a global, speaker-independent threshold to make a verification decision. In this work, the threshold was adjusted to determine the equal error rates (EERs).

Results and Discussions. Figure 9.15 and Table 9.9 show the results of different environment adaptation approaches, including CMS, Tnorm [14], Hnorm [303], PDBNN adaptation, SFT [226], MLLR, MLLR+PDBNN, and MLLR+PDBNN+ SFT adaptation. All error rates were based on the average of 100 target speakers and 50 impostors.

Evidently, all cases of environment adaptation show significant reduction in error rates when compared to CMS. In particular, MLLR+PDBNN and MLLR+PDBNN+SFT adaptation achieve the largest error reduction. Table 9.9 also demonstrates that model-based adaptation and feature-based transformation (SFT) are comparable in terms of error rate reduction (ERR).

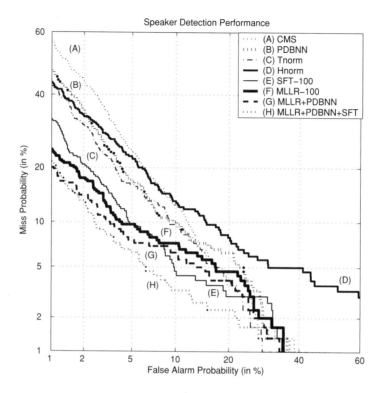

Figure 9.15. DET curves comparing speaker verification performance using different environment adaptation approaches: CMS, PDBNN, Tnorm, Hnorm, SFT, MLLR, MLLR+PDBNN, and MLLR+PDBNN+SFT. All the DET curves were obtained using the testing utterances from handset el2. For ease of comparison, methods labeled from (A) to (H) in the legend are arranged in descending order of EER.

Although PDBNN adaptation uses discriminative training to adapt the model parameters to fit the new environment, the results show that using PDBNN adaptation alone is not desirable because its performance is not as good as that of MLLR adaptation. This may be due to the intrinsic design of PDBNNs (i.e., the supervised reinforced/antireinforced learning of PDBNNs are designed for fine-tuning the decision boundaries by slightly adjusting the model centers responsible for the misclassifications). As a result, some of the centers will not be adapted at all. Because only misclassified data are used for adaptation and their amount is usually small after LU training, moving all of the centers from the clean space to the environmentally distorted space may be difficult. Alternatively, MLLR adaptation finds a transformation matrix to maximize the likelihood of the adaptation data. Thus, moving all of the model centers to the environmentally distorted space is much easier.

Table 9.9. EER achieved by CMS, Tnorm, Hnorm, PDBNN, SFT, MLLR, MLLR+PDBNN, and MLLR+PDBNN+SFT adaptations. *Note*: CMS and Tnorm do not require the handset selector. All results were based on 100 target speakers and 50 impostors. The MLLR and SFT transformation matrices were estimated using 20 and 100 speakers (denoted as -20 and -100, respectively) in the speaker set. For MLLR+PDBNN and MLLR+PDBNN+SFT, 100 speakers were used to created the transformation matrices.

Adaptation	Equal Error Rate (%)										
Method	cb1	cb2	cb3	cb4	el1	el2	el3	el4	pt1	senh	Ave.
CMS	8.21	8.50	21.20	15.40	8.15	11.20	11.49	8.85	10.56	6.79	**11.03**
Tnorm	8.88	8.94	22.58	14.94	9.30	9.78	10.40	8.64	8.51	5.54	**10.74**
Hnorm	7.30	6.98	13.81	10.42	7.42	9.40	10.32	7.62	9.34	7.06	**8.96**
PDBNN-100	7.72	8.48	10.02	9.66	6.72	11.59	8.64	9.59	8.99	3.01	**8.44**
SFT-20	4.18	3.61	17.64	11.81	4.93	7.24	7.82	3.85	6.64	3.60	**7.13**
SFT-100	4.28	3.61	17.55	11.06	4.81	7.60	7.34	3.87	6.12	3.65	**6.98**
MLLR-20	4.69	3.34	17.23	10.21	5.52	7.35	9.66	4.76	8.84	3.54	**7.51**
MLLR-100	4.52	3.14	15.17	9.58	4.79	7.60	6.46	4.84	6.94	3.69	**6.67**
SFT-20+Hnorm	4.18	3.61	17.64	11.81	4.93	7.24	7.87	3.85	6.64	3.60	**7.13**
MLLR+PDBNN (100)	4.30	3.12	10.16	6.89	5.01	6.84	6.36	4.87	6.53	3.43	**5.75**
MLLR+PDBNN+SFT (100)	4.21	3.01	12.86	7.65	4.97	5.65	6.05	4.26	5.46	3.43	**5.75**

PDBNN adaptation requires speaker-specific training data from all possible handsets that the users can use. MLLR adaptation, on the other hand, only requires some environment-specific utterances to estimate the global transformation matrices, which requires much less training data; better still, the utterances do not necessarily need to be produced by the same client speaker. PDBNN adaptation also requires additional training for the inclusion of new speakers because the new speaker models and the background models should be adapted using gradient descent to environment-specific models. On the other hand, for MLLR adaptation, transformation is applied to new speaker models and background models once the MLLR transformation matrices have been estimated.

The results also demonstrate that SFT and MLLR achieve a comparable amount of error reduction. A comparison between SFT-20 and MLLR-20 (where the training utterances of 20 speakers were used to estimate the transformation parameters) reveals that SFT performs slightly better when the amount of training data is small. This is because the number of free parameters in feature transformation is much less than that of MLLR. However, the performance of SFT quickly saturates when the total number of training patterns increases, as indicated in SFT-100 and MLLR-100. While MLLR requires much more data to estimate the global transformation matrices robustly, its performance is better than that of SFT when sufficient training data are available.

The performances of the proposed environment adaptation approaches were also compared with that of Hnorm [303] and Tnorm [14]. As shown in Table 9.9, both methods outperform the classical CMS; however, the improvement is not as significant as SFT and MLLR. Recall that both Hnorm and Tnorm work on the likelihood-ratio score space using two normalization parameters only (bias and scale). SFT and MLLR, on the other hand, aim to compensate the mismatch effect at the feature space and model space. Both methods can translate and scale the components of feature vectors or models' centers. Because the number of free parameters in SFT and MLLR is much larger than that of Hnorm and Tnorm, SFT and MLLR are more effective provided that the free parameters can be estimated correctly.

Because SFT and Hnorm work on different spaces (SFT transforms the speech features in the feature space while Hnorm performs score normalization in the score space), these two techniques were also combined to see whether further improvement could be obtained. The results (third row—"SFT-20+Hnorm"—from bottom in Table 9.9) show that no improvement was obtained when compared with SFT-20. A possible reason is that the score distributions of clients and impostors after SFT are already well separated; as a result, handset normalization cannot improve the performance further in the score space.

Although it may not be desirable to use PDBNN's reinforced learning alone, it is amenable to the fine-tuning of the MLLR-transformed centers. This is evident from the second row from the bottom of Table 9.9, where error reductions of 13.8% with respect to MLLR-100 and 31.9% with respect to PDBNN-100 were achieved. The main reason behind these error reductions is that MLLR can transform the clean model centers to the region of distorted clusters; these initial centers give a better *seed* for reinforced learning.

For MLLR+PDBNN+SFT adaptation, an average EER of 5.75% is achieved (bottom row of Table 9.9), which is identical to that of the MLLR+PDBNN adaptation. Bear in mind that MLLR+PDBNN adaptation requires speaker-specific and handset-specific training data, whereas MLLR+PDBNN+SFT adaptation requires handset-specific training data only. Having the same EER means that MLLR+PDBNN+SFT adaptation is a better method.

To compare the computation complexity, the training and verification time of different adaptation approaches were measured. The simulations were performed on a Pentium IV 2.66GHz CPU. The measurements were based on 30 speakers and 50 impostors in the HTIMIT corpus and the total CPU time running the whole task was recorded. The training time is composed of two parts: systemwise computation time and enrollment computation time. The systemwise computation time represents the time to carry out the tasks that only need to be performed once, including computation of the MLLR transformation matrices and background model training. The enrollment computation time represents the time to enroll a new speaker and to adapt his or her models. It was determined by taking the average enrollment time of 30 client speakers using seven utterances per speaker. The verification time is the time taken to verify a claimant based on a single utterance. It was determined by taking the average verification time of 80 claimants (30 client speakers and 50 impostors) using three utterances per claimant.

Table 9.10. Total training and verification time used by different adaptation approaches. The training time is composed of two parts: systemwise computation time, which represents the time to carry out tasks that only need to be performed once, and enrollment computation time. The systemwise tasks include the computation of the MLLR transformation matrices and background model training. The enrollment tasks are the time required to enroll a new speaker and adapt his or her models.

Adaptation Method	Training Time		Verification Time (seconds)
	Systemwise Computation (seconds)	Enrollment Computation (seconds)	
Hnorm	324.30	240.78	2.48
Tnorm	324.30	0.96	37.54
PDBNN-100	324.30	8489.54	1.20
MLLR-100	130713.30	0.96	0.38
MLLR+PDBNN-100	130713.30	2561.74	1.59
MLLR+PDBNN+SFT-100	130713.30	5157.77	1.27

Table 9.10 shows the training and verification time for Hnorm, Tnorm, PDBNN-100, MLLR-100, MLLR+PDBNN-100, and MLLR+PDBNN+SFT-100. Evidently, PDBNN adaptation requires extensive computational resources during enrollment because a handset-specific speaker model and a handset-specific background model were created for each speaker for each acoustic environment. The results also show that computing the MLLR transformation matrices requires an extensive amount of computational resources. Hnorm requires only a small amount of resources for enrollment because it needs to estimate only the score mean and score variance. Tnorm also requires a small amount of computation during enrollment. However, because all of the normalization parameters are estimated from the scores of all speaker models during verification, Tnorm is very computation intense during verification. All other adaptations can achieve near realtime performance during verification.

Although both Hnorm and Tnorm work on the likelihood-ratio score space, they have different complexities when it comes to training and testing. The main computation for Hnorm lies in the training phase and is proportional to the total number of nontarget utterances because these utterances are fed to the speaker and background models to calculate the nontarget scores, which in turn are used to calculate the normalization parameters. Once the Hnorm parameters have been calculated, rapid verification can be achieved. For Tnorm, on the other hand, verification is computationally intense and the amount of computation is proportional to the total number of speaker models used for deriving the normalization scores. For a better estimation of Tnorm parameters, a large number of scores are required, which in turn require a large number of speaker models. This can greatly lengthen verification time.

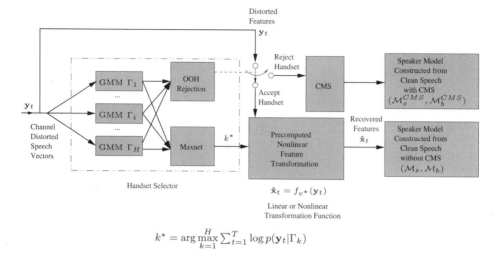

$$k^* = \arg\max_{k=1}^{H} \sum_{t=1}^{T} \log p(\mathbf{y}_t | \Gamma_k)$$

Figure 9.16. Speaker verification system with handset identification, OOH rejection, and handset-dependent feature transformation.

9.4.4 Out-of-Handset Rejection

The handset selector defined by Eq. 9.4.14 simply selects the most likely handset from a set of known handsets even for speech coming from an *unseen* handset. In addition, the approach requires a handset database containing all types of handsets that users are likely to use; at least two reasons make this requirement difficult to fulfill in practical situations. First, because there are so many types of phone handsets on the market, it is almost impossible to include all of them in the handset database. Second, new handset models are released every few months, which means the handset database must be updated frequently to keep it current. If a claimant uses a handset that has not been included in the handset database, the verification system may identify the handset incorrectly, resulting in verification error.

To address the preceding problems, combining SFT and a GMM-based handset selector with out-of-handset (OOH) rejection capability for assisting speaker verification has been proposed [352, 354]. Specifically, each handset in the database of the handset selector is assigned a set of transformation parameters. During verification, the handset selector determines whether the handset used by the claimant is one of the handsets in the database. If so, the selector identifies the most likely handset and transforms the distorted vectors according to the transformation parameters of the identified handset. Otherwise, the selector identifies the handset as "unseen" and processes the distorted vectors by cepstral mean subtraction (CMS). Figure 9.16 illustrates the idea of incorporating OOH rejection into the handset selector.

The Jensen Difference

For each utterance, the selector either identifies or rejects the handset. The decision is based on the following rule:

$$\text{If} \begin{cases} J(\vec{\alpha}, \vec{r}) \geq \varphi & \text{identify the handset} \\ J(\vec{\alpha}, \vec{r}) < \varphi & \text{reject the handset,} \end{cases} \qquad (9.4.17)$$

where $J(\vec{\alpha}, \vec{r})$ is the *Jensen difference* [39, 360] between $\vec{\alpha}$ and $\vec{r}$—whose values will be discussed next—and φ is a decision threshold. $J(\vec{\alpha}, \vec{r})$ can be computed as

$$J(\vec{\alpha}, \vec{r}) = S\left(\frac{\vec{\alpha} + \vec{r}}{2}\right) - \frac{1}{2}\left[S(\vec{\alpha}) + S(\vec{r})\right], \qquad (9.4.18)$$

where $S(\vec{z})$, called the *Shannon entropy*, is defined by

$$S(\vec{z}) = -\sum_{i=1}^{H} z_i \log z_i, \qquad (9.4.19)$$

where z_i is the i-th component of $\vec{z}$.

The Jensen difference has a nonnegative value and can be used to measure the divergence between two vectors. If all elements of $\vec{\alpha}$ and $\vec{r}$ are similar, $J(\vec{\alpha}, \vec{r})$ will have a small value. On the other hand, if the elements of $\vec{\alpha}$ and $\vec{r}$ are different, the value of $J(\vec{\alpha}, \vec{r})$ will be large. For the case where $\vec{\alpha}$ is identical to $\vec{r}$, $J(\vec{\alpha}, \vec{r})$ becomes 0. Therefore, *Jensen difference* is an ideal candidate for measuring the divergence between two n-dimensional vectors.

The handset selector here uses the Jensen difference to compare the probabilities of a test utterance produced by the known handsets. Let $Y = \{\mathbf{y}_t : t = 1, \ldots, T\}$ be a sequence of feature vectors extracted from an utterance recorded by an unknown handset, and $l_h(\mathbf{y}_t)$ be the log-likelihood of observing the pattern $\mathbf{y}_t$, given it is generated by the h-th handset (i.e., $l_h(\mathbf{y}_t) = \log p(\mathbf{y}_t | \Gamma_h)$). Hence, the average log-likelihood of observing the sequence Y, given it is generated by the h-th handset, is

$$L_h(Y) = \frac{1}{T}\sum_{t=1}^{T} l_h(\mathbf{y}_t). \qquad (9.4.20)$$

For each vector sequence X, $\vec{\alpha} = [\alpha_1 \, \alpha_2 \, \cdots \, \alpha_H]^T$ is created with each of its element

$$\alpha_h = \frac{\exp\{L_h(Y)\}}{\sum_{i=1}^{H} \exp\{L_i(Y)\}} \qquad 1 \leq h \leq H \qquad (9.4.21)$$

representing the probability that the test utterance is produced by the h-th handset such that $\sum_{h=1}^{H} \alpha_h = 1$ and $\alpha_h > 0$ for all h. If all the elements of $\vec{\alpha}$ are similar, the probabilities of the test utterance produced by each handset are close, and it is difficult to identify which handset the utterance comes from. On the other hand, if the elements of $\vec{\alpha}$ are not similar, some handsets will be more likely than others

Table 9.11. Confusion matrix containing the normalized log-likelihoods of 10 different handsets. The normalized log-likelihoods were found using Eq. 9.4.23, and have a range between 0.00 and 1.00. Entries with a large value implies that the handset corresponding to that row and the handset corresponding to that column have similar characteristics; a small value implies that the characteristics between the two handsets are different. Elements in bold on the diagonal are the largest because the i-th GMM should give the largest log-likelihood for utterances from the i-th handset.

Normalized Log-Likelihood of Handset (P'_{ij})										
Utterances from	**Handset Model (Γ_j)**									
Handset (i)	cb1	cb2	cb3	cb4	el1	el2	el3	el4	pt1	senh
cb1	**0.80**	0.66	0.38	0.41	0.64	0.51	0.63	0.47	0.52	0.53
cb2	0.61	**0.76**	0.22	0.36	0.45	0.33	0.56	0.55	0.39	0.54
cb3	0.72	0.62	**1.00**	0.86	0.70	0.55	0.65	0.64	0.60	0.58
cb4	0.66	0.62	0.76	**0.94**	0.65	0.43	0.59	0.56	0.51	0.56
el1	0.56	0.45	0.13	0.21	**0.73**	0.49	0.54	0.35	0.52	0.48
el2	0.57	0.47	0.01	0.02	0.61	**0.81**	0.69	0.46	0.64	0.43
el3	0.60	0.57	0.20	0.27	0.61	0.63	**0.77**	0.53	0.57	0.59
el4	0.40	0.54	0.25	0.28	0.40	0.37	0.50	**0.75**	0.28	0.40
pt1	0.60	0.53	0.20	0.27	0.64	0.66	0.69	0.47	**0.84**	0.51
senh	0.43	0.49	0.00	0.11	0.46	0.24	0.50	0.30	0.29	**0.71**

to produce the test utterance. In this case, being able to identify the handset from which the utterance is recorded is likely.

The similarity among the elements of $\vec{\alpha}$ is determined by the Jensen difference $J(\vec{\alpha}, \vec{r})$ between $\vec{\alpha}$ (with the elements of vector $\vec{\alpha}$ defined in Eq. 9.4.21) and a reference vector $\vec{r} = [r_1 \, r_2 \cdots r_H]^T$, where $r_h = \frac{1}{H}$, $h = 1, \ldots, H$. A small Jensen difference indicates that all elements of $\vec{\alpha}$ are similar, while a large value means that the elements of $\vec{\alpha}$ are different.

During verification, when the selector finds that the Jensen difference $J(\vec{\alpha}, \vec{r})$ is greater than or equal to the threshold φ, the selector identifies the most likely handset according to Eq. 9.4.14, and the transformation parameters corresponding to the selected handset are used to transform the distorted vectors. On the other hand, when $J(\vec{\alpha}, \vec{r})$ is less than φ, the selector considers the sequence Y to be coming from an unseen handset. In the latter case, the distorted vectors will be processed differently, as shown in Figure 9.16.

Similarity and Dissimilarity among Handsets

Because the divergence-based handset classifier is designed to reject dissimilar, unseen handsets, handsets that are either similar to one of the seen handsets or dissimilar to all seen handsets must be used for evaluation. The similarity and dissimilarity among the handsets can be observed from a confusion matrix. Given N utterances (denoted as $Y^{(i,n)}, n = 1, \dots, N$) from the i-th handset and the GMM of the j-th handset (denoted as Γ_j), the average log-likelihood of the j-th handset is

$$P_{ij} = \frac{1}{N} \sum_{n=1}^{N} \log p(Y^{(i,n)}|\Gamma_j) = \frac{1}{N} \sum_{n=1}^{N} \frac{1}{T_n} \sum_{t=1}^{T_n} \log p(\mathbf{y}_t^{(i,n)}|\Gamma_j), \qquad (9.4.22)$$

where $p(\mathbf{y}_t^{(i,n)}|\Gamma_j)$ is the likelihood of the model Γ_j given the t-th frame of the n-th utterance recorded from the i-th handset, and T_n is the number of frames in $Y^{(i,n)}$.

To facilitate comparison among the handsets, the average log-likelihood is normalized according to

$$P'_{ij} = \frac{P_{ij} - P_{min}}{P_{max} - P_{min}}, \qquad (9.4.23)$$

where P_{max} and P_{min} are respectively the maximum and minimum log-likelihoods found in the matrix (i.e., $P_{max} = \max_{i,j} P_{ij}$ and $P_{min} = \min_{i,j} P_{ij}$). As a result, the largest normalized log-likelihood has a value of 1.00, and the smallest has a value of 0.00. Each of the P'_{ij}'s, $0 \leq i, j \leq H$, corresponds to an element in the confusion matrix. This matrix enables the identification of those handsets with similar (or dissimilar) characteristics. For example, a large value of P'_{ij} suggests that the i- and j-th handsets are similar; on the other hand, a small value of P_{ij} implies that they are different.

A confusion matrix containing the normalized log-likelihoods of the 10 handsets is shown in Table 9.11. The utterances of these handsets were obtained from 100 speakers in the HTIMIT corpus. Note that the diagonal elements are the largest among all other elements in their respective rows. This is because the log-likelihood of a handset given the utterances from the same handset should be the largest.

The confusion matrix in Table 9.11 demonstrates the similarity and dissimilarity among all handsets. For example, a value of 1.00 means very similar and a value of 0.00 implies very dissimilar. However, the matrix does not show how similar a particular handset is to another handset. The normalized log-likelihood differences ($\tilde{P}_{ij}$) provide this information:

$$\tilde{P}_{ij} = \left\{ \max_{k=1}^{H} P'_{ik} \right\} - P'_{ij} \qquad 1 \leq i, j \leq H. \qquad (9.4.24)$$

Table 9.12 depicts a matrix containing the values of $\tilde{P}_{ij}$'s. The table clearly shows that the cb1 handset is similar to cb2, el1, and el3 because their normalized log-likelihood differences with respect to handset cb1 are small (≤ 0.17). On the other hand, it is likely that cb1 has characteristics different from those of handsets cb3 and cb4 because their normalized log-likelihood differences are large (≥ 0.39).

The confusion matrix in Table 9.12 labels some handsets as unseen, while the remaining are considered seen handsets. These two categories of handsets—seen and unseen—will be used to test the OOH rejection capability of the proposed handset selector.

Table 9.12. Normalized log-likelihood differences of 10 handsets (see Eq. 9.4.24). Entries with small (large) value means that the corresponding handsets are similar (different).

Normalized Log-Likelihood Difference ($\tilde{P}_{ij}$)										
Utterances from	Handset Model (Γ_j)									
Handset (i)	cb1	cb2	cb3	cb4	el1	el2	el3	el4	pt1	senh
cb1	**0.00**	0.14	0.42	0.39	0.16	0.29	0.17	0.33	0.28	0.27
cb2	0.15	**0.00**	0.54	0.40	0.31	0.43	0.20	0.21	0.37	0.22
cb3	0.28	0.38	**0.00**	0.14	0.30	0.45	0.35	0.36	0.40	0.42
cb4	0.28	0.32	0.18	**0.00**	0.29	0.51	0.35	0.38	0.43	0.38
el1	0.17	0.28	0.60	0.52	**0.00**	0.24	0.19	0.38	0.21	0.25
el2	0.24	0.34	0.80	0.79	0.20	**0.00**	0.12	0.35	0.17	0.38
el3	0.17	0.20	0.57	0.50	0.16	0.14	**0.00**	0.24	0.20	0.18
el4	0.35	0.21	0.50	0.47	0.35	0.38	0.25	**0.00**	0.47	0.35
pt1	0.24	0.31	0.64	0.57	0.20	0.18	0.15	0.37	**0.00**	0.33
senh	0.28	0.22	0.71	0.60	0.25	0.47	0.21	0.41	0.42	**0.00**

Experimental Evaluations

In this experiment, the OOH rejection capability of the handset selector was studied. Different approaches were applied to integrate the OOH rejection into a speaker verification system, and utterances from "seen" and "unseen" handsets were used to test the resulting system. The following section discusses how the seen and unseen handsets were chosen for testing the OOH rejection capability of the handset selector. The discussion is followed by a detailed description of the approaches to integrating the OOH rejection into a speaker verification system.

Selection of Seen and Unseen Handsets. When a claimant uses a handset that has not been included in the handset database, the characteristics of this unseen handset may be different from all of those in the database, or its characteristics may be similar to one or a few handsets in the database. Therefore, it is important to test the handset selector under two scenarios: (1) unseen handsets with characteristics different from those of the seen handsets, and (2) unseen ones with characteristics similar to seen handsets.

Seen and unseen handsets with different characteristics. Table 9.12 shows that the cb3 and cb4 handsets are similar. In the table, the normalized log-likelihood difference in row cb3, column cb4, has a value of 0.14 and the normalized log-likelihood difference in row cb4, column cb3, is 0.18; both of these entries have small values. On the other hand, these two handsets (cb3 and cb4) are not similar to all other handsets because the log-likelihood differences in the remaining entries of row cb3 and row cb4 are large. Therefore, in the first part of the experiment, the cb3 and cb4 handsets were used as the unseen ones, and the other eight handsets were used as the seen handsets.

Table 9.13. Three different approaches to integrating OOH rejection into a speaker verification system.

Approach	OOH Rejection Method	Rejection Handling
I	None	N/A
II	Euclidean distance-based	Use CMS-based speaker models to verify the rejected utterances
III	Divergence-based	Use CMS-based speaker models to verify the rejected utterances

Seen and unseen handsets with similar characteristics. The confusion matrix in Table 9.12 shows that the el2 handset is similar to el3 and pt1 since their normalized log-likelihood differences with respect to el2 are small (i.e., 0.12 and 0.17, respectively, in row el2 of the table). It is also likely that cb3 and cb4 have similar characteristics as stated in the previous paragraph. Therefore, if cb3 and el2 are used as the unseen handsets, and the remaining are left as the seen handsets, it will be possible to find some seen handsets (e.g., cb4, el3, and pt1) that are similar to the two unseen ones. In the second part of the experiment, cb3 and el2 are used as the unseen handsets and the other eight handsets are used as the seen ones.

Approaches to Incorporating OOH Rejection into Speaker Verification. Three different approaches to integrating the handset selector into a speaker verification system were investigated; they are denoted as Approach I, Approach II, and Approach III, and are detailed in Table 9.13. In the experiment, nine handsets (cb1–cb4, el1–el4, and pt1) and one Sennheizer head-mounted microphone (*senh*) from HTIMIT [304] were used as the testing handsets. These handsets were divided into the seen and unseen categories, as described before. Speech from the senh handset was used for enrolling speakers, and speech from the other nine handsets was used for verifying speakers. The enrollment and verification procedures were identical to those in Section 9.4.2.

Approach I: Handset selector without OOH rejection. In this approach, if test utterances from an unseen handset are fed to the handset selector, the selector is forced to choose an incorrect handset and a wrong set of transformation parameters will be used to transform the distorted vectors. The handset selector consists of eight 64-center Gaussian mixture models (GMMs) $\{\Gamma_k\}_{k=1}^8$ corresponding to the eight seen handsets. Each GMM was trained with the distorted speech recorded from the corresponding handset. Also, for each handset, a set of feature transformation parameters ν that transform speech from the corresponding handset to the enrolled handset (senh) were computed (see Section 9.4.2). Note that utterances from the unseen handsets were not used to create any GMM models.

During verification, a test utterance was fed to the GMM-based handset selector, which then chose the most likely one from the eight handsets according to Eq. 9.4.14, with $H = 8$. Then, the transformation parameters corresponding to the k^*-th handset were used to transform the distorted speech vectors for speaker verification.

Approach II: Handset selector with Euclidean distance-based OOH rejection and CMS. In this approach, OOH rejection was implemented based on the Euclidean distance between two vectors: a vector $\vec{\alpha}$ (with the elements of vector $\vec{\alpha}$ defined in Eq. 9.4.21) and a reference vector $\vec{r} = [r_1\, r_2 \cdots r_H]^T$, where $r_i = \frac{1}{H}$, $i = 1, \ldots, H$. The vector distance $D(\vec{\alpha}, \vec{r})$ between $\vec{\alpha}$ and $\vec{r}$ is

$$D(\vec{\alpha}, \vec{r}) = \|\vec{\alpha} - \vec{r}\| = \sqrt{\sum_{i=1}^{H} (\alpha_i - r_i)^2}. \qquad (9.4.25)$$

The selector then identifies the most likely handset or rejects the handset using the decision rule:

$$\text{if } \begin{cases} D(\vec{\alpha}, \vec{r}) \geq \zeta & \text{identify the handset} \\ D(\vec{\alpha}, \vec{r}) < \zeta & \text{reject the handset,} \end{cases} \qquad (9.4.26)$$

where ζ is a decision threshold. Specifically, for each utterance, the handset selector determines whether the utterance is recorded from one of the eight known handsets according to Eq. 9.4.26. If so, the corresponding transformation is used to transform the distorted speech vectors; otherwise, cepstral mean subtraction (CMS) is used to compensate for the channel distortion.

Approach III: Handset selector with divergence-based OOH rejection and CMS. This approach uses a handset selector with divergence-based OOH rejection capability. Specifically, for each utterance, the handset selector determines whether it is recorded from one of the eight known handsets by making an accept or a reject decision according to Eq. 9.4.17. For an accept decision, the handset selector selects the most likely one from the eight handsets and uses the corresponding transformation parameters to transform the distorted speech vectors. For a reject decision, CMS was applied to the utterance rejected by the handset selector to recover the clean vectors from the distorted ones.

Scoring Normalization.   Recovered vectors were fed to a 32-center GMM speaker model. Depending on the handset selector's decision, the recovered vectors were either fed to a GMM-based speaker model without CMS ($\mathcal{M}_s$) to obtain the score ($\log p(Y|\mathcal{M}_s)$) or fed to a GMM-based speaker model with CMS ($\mathcal{M}_s^{CMS}$) to obtain the CMS-based score ($\log p(Y|\mathcal{M}_s^{CMS})$). In either case, the score was normalized according to

$$S(Y) = \begin{cases} \log p(Y|\mathcal{M}_s) - \log p(Y|\mathcal{M}_b) & \text{if feature transformation is used} \\ \log p(Y|\mathcal{M}_s^{CMS}) - \log p(Y|\mathcal{M}_b^{CMS}) & \text{if CMS is used,} \end{cases}$$
$$(9.4.27)$$

where $\mathcal{M}_b$ and $\mathcal{M}_b^{CMS}$ are the 64-center GMM background model without CMS and with CMS, respectively. $S(Y)$ was compared with a threshold to make a verification

Table 9.14. Results for seen and unseen handsets with different characteristics. *Note*: EERs (in %) achieved by the baseline, CMS, and the three handset selector integration approaches are shown here (see Table 9.13 for integration methods). Handsets cb3 and cb4 were used as the unseen handsets. The enrollment handset is senh. The average handset identification accuracy is 98.25%. Note that the baseline and CMS do not require the handset selector.

Compensation Method	Integration Method	Equal Error Rate (%)										
		cb1	cb2	*cb3*	*cb4*	el1	el2	el3	el4	pt1	**Ave.**	senh
Baseline	N/A	8.15	7.01	*25.78*	*18.08*	5.99	15.06	7.86	14.02	9.75	**12.41**	2.99
CMS	N/A	6.42	5.71	*13.33*	*10.17*	6.15	9.29	9.59	7.18	6.81	**8.29**	4.66
2nd-order SFT	Approach I	4.14	3.56	*19.02*	*18.41*	3.54	6.78	6.38	4.72	4.69	**7.92**	2.98
2nd-order SFT	Approach II	4.39	3.99	*13.37*	*12.34*	4.29	6.57	8.77	4.74	5.06	**7.05**	2.98
2nd-order SFT	Approach III	4.17	3.91	*13.35*	*12.30*	4.54	6.46	7.60	4.69	5.23	**6.92**	2.98

decision. In this work, the threshold for each speaker was adjusted to determine an equal error rate (EER).

Results and Discussions.   The experimental results using the cb3 and cb4 handsets as the unseen handsets are summarized in Table 9.14.[6] All of the stochastic feature transformations used in this experiment were of second-order. For Approach II, the threshold ζ (Eq. 9.4.26) for the decision rule used in the handset selector was set to 0.25, while for Approach III, the threshold φ (Eq. 9.4.17) for the handset selector was set to 0.06. These threshold values were found to produce the best result empirically.

In addition, Table 9.14 shows that Approach I substantially reduces the average EER. The average goes down to 7.92%, as compared to 12.41% for the baseline and 8.29% for CMS. However, no reductions in EERs for the "unseen" handsets (i.e., cb3 and cb4) were found. The EER of cb3 using this approach is even higher than the one obtained by the CMS method. Handset cb4's EER is even higher than that of the baseline. Therefore, it can be concluded that using an incorrect set of transformation parameters can degrade the verification performance when the characteristics of the unseen handset are different from the seen handsets.

Table 9.14 also shows that Approach II is able to achieve a satisfactory performance. With the Euclidean distance-based OOH rejection, there were 365 and 316 rejections out of 450 test utterances for the two unseen handsets (cb3 and cb4), respectively. As a result of these rejections, the EERs of cb3 and cb4 were reduced to 13.37% and 12.34%, respectively. These errors are significantly lower than those achievable by Approach I. Nevertheless, some utterances from the seen ones were rejected by the handset selector, causing a higher EER for other "seen" handsets. Therefore, OOH rejection based on Euclidean distance has limitations.

---

[6]Table 9.12 and the discussion in the Similarity and Dissimilarity among Handsets section demonstrate that cb3 and cb4 are different from all other handsets.

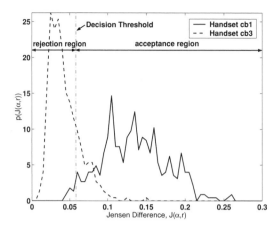

Figure 9.17. The distribution of the Jensen Difference $J(\vec{\alpha}, \vec{r})$ corresponding to the cb1 seen handset and the cb3 unseen handset.

As shown in the last row of Table 9.14, Approach III achieves the lowest average EER. The reduction in EERs is also the most significant for the two unseen handsets. In an ideal situation with this approach, all utterances of the unseen handsets will be rejected by the selector and processed by CMS, and the EERs of the unseen handsets are reduced to those achievable by the CMS method. In the experiment, 369 and 284 rejections were obtained out of 450 test utterances for cb3 and cb4, respectively. As a result of these rejections, the EERs corresponding to handsets cb3 and cb4 decrease to 13.35% and 12.30%, respectively; neither are significantly different from the EERs achieved by the CMS method. Although this approach may cause the EERs of the seen handsets (except for el2 and el4) to be slightly higher than those achieved by Approach I, it is a worthwhile tradeoff because its average EER is still lower than that of Approach I. Approach III also reduces the EERs of the two seen handsets (el2 and el4) because some of the incorrectly identified utterances in Approach I were rejected by the handset selector in Approach III. Using CMS to recover the distorted vectors of these utterances allows the verification system to recognize speakers correctly.

Figure 9.17 shows the distribution of the Jensen difference $J(\vec{\alpha}, \vec{r})$ for the cb1 seen handset and the cb3 unseen handset. The vertical dash-dot line defines the decision threshold used in the experiment (i.e., $\varphi = 0.06$). According to Eq. 9.4.17, the handset selector accepts the handsets for Jensen differences greater than or equal to the decision threshold (i.e., the region to the right of the dash-dot line) and rejects the handset for Jensen differences less than the decision threshold (i.e., the region to the left of the dash-dot line). For cb1, only a small area under its Jensen difference distribution is inside the rejection region, which means that few utterances from this handset were rejected by the selector—out of 450 test utterances in the experiment,

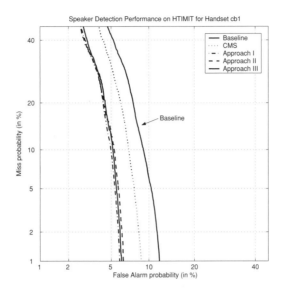

Figure 9.18. DET curves obtained by using cb1 seen handset in the verification sessions. Handsets cb3 and cb4 were used as the unseen ones.

only 14 were rejected. On the other hand, a large portion of cb3's distribution is inside the rejection region. As a result, most of the utterances from this unseen handset are rejected by the selector—out of 450 utterances, 369 were rejected.

To better illustrate the detection performance of the verification system, the DET curves are plotted, as introduced in Martin et al. [233], for the three approaches. The speaker detection performance, using the cb1 seen handset and the cb3 unseen handset in verification sessions, are shown in Figures 9.18 and 9.19, respectively. The five DET curves in each figure represent five different methods of processing the speech, and each curve was obtained by averaging the DET curves of 100 speakers. Note that the curves are almost straight because each DET curve is constructed by averaging the DET curves of 100 speakers, resulting in a normal distribution. Appendix B illustrates how the average DET curves were obtained.

The EERs obtained from the curves in Figure 9.18 correspond to the values in column cb1 of Table 9.14, and the EERs in Figure 9.19 correspond to the values in column cb3. Due to interpolation errors, there are slight discrepancies between the EERs obtained from the figures and those shown in Table 9.14.

In addition, Figures 9.18 and 9.19 show that Approach III achieves satisfactory performance for both seen and unseen handsets. In Figure 9.18, using Approach III, the DET curve for the cb1 seen handset is close to the curve achieved by Approach I. In Figure 9.19, using Approach III, the DET curve for the cb3 unseen handset is close to the curve achieved by the CMS method. Therefore, by applying Approach III (with divergence-based OOH rejection) to the speaker verification system, the error rates of a seen handset can be reduced to values close to that achievable

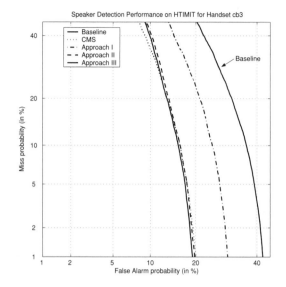

Figure 9.19. DET curves obtained by using the cb3 unseen handset in the verification sessions. Handsets cb3 and cb4 were used as the unseen ones.

Table 9.15. Results for seen and unseen handsets with similar characteristics. *Note*: EER (in %) achieved by the baseline, CMS, and the three handset selector integration approaches are shown here (see Table 9.13 for integration methods). Handsets cb3 and el2 were used as the unseen handsets. The enrollment handset is senh. The average handset identification accuracy is 98.38%. Notice that the baseline and CMS do not require the handset selector.

Compensation Method	Integration Method	Equal Error Rate (%)										
		cb1	cb2	*cb3*	cb4	el1	*el2*	el3	el4	pt1	**Ave.**	senh
Baseline	N/A	8.15	7.01	*25.78*	18.08	5.99	*15.06*	7.86	14.02	9.75	**12.41**	2.99
CMS	N/A	6.42	5.71	*13.33*	10.17	6.15	*9.29*	9.59	7.18	6.81	**8.29**	4.66
2nd-order ST	Approach I	4.14	3.56	*13.35*	6.75	3.53	*9.82*	6.37	4.72	4.69	**6.33**	2.98
2nd-order ST	Approach II	4.14	3.56	*13.30*	6.75	4.08	*9.46*	6.59	4.70	4.73	**6.37**	2.98
2nd-order ST	Approach III	4.14	3.56	*13.10*	6.75	3.48	*9.63*	6.20	4.72	4.69	**6.25**	2.98

by Approach I (without OOH rejection); the error rates of an unseen handset, the characteristics of which are different from all the seen ones, can be reduced to values close to that achievable by the CMS method.

The experimental results using cb3 and el2 as the unseen handsets are summarized in Table 9.15.[7] Again, all the stochastic transformations (STs) used in this

---

[7]According to Table 9.12 and the arguments in the Similarity and Dissimilarity among Handsets section, cb3 is similar to cb4, and handset el2 is similar to el3 and pt1.

experiment were second-order. For Approach II, the threshold ζ (Eq. 9.4.26) for the decision rule used in the handset selector was set to 0.25. For Approach III, the threshold φ used by the handset selector was set to 0.05. These threshold values were found to produce the best results empirically.

Table 9.15 shows that Approach I can achieve a satisfactory performance. Its average equal error rate is significantly smaller than that of the baseline and ceptral mean subtraction method. In addition, the EERs of the two unseen handsets, cb3 and el2, have values close to those of the CMS method even without out-of-handset rejection. This occurs because the characteristics of cb3 are similar to those of the cb4 seen handset, and those of el2 are similar to those of the el3 and pt1 seen handsets.

Therefore, when utterances from cb3 are fed to the handset selector, it chooses cb4 as the most likely handset in most cases (out of 450 test utterances from cb3, 446 were identified as coming from cb4). Because the transformation parameters of cb3 and cb4 are close, the recovered vectors (despite using an incorrect set of transformation parameters) can still be correctly recognized by the verification system. A similar situation occurs when utterances from handset cb2 are fed to the selector. In this case, the transformation parameters of either el3 or pt1 are used to recover the distorted vectors (out of 450 test utterances from el2, 330 were identified as coming from el3, and 73 utterances were identified as being from pt1).

In addition, Table 9.15 shows that the performance of Approach II is not satisfactory. Although this approach can bring further reduction in EERs for the two unseen handsets (as a result of 21 rejections for cb3 and 11 rejections for el2), the cost is a higher average EER over Approach I.

Results in Table 9.15 also show that Approach III, once again, achieves the best performance. Its average EER is the lowest, and further reduction in EERs of the two unseen handsets (cb3 and el2) was obtained. For the el2 handset, there were only two rejections out of 450 test utterances because most of the utterances were considered to be from the el3 or pt1 seen handset. With such a small number of rejections, the EER of el2 is reduced to 9.63%, which is close to the 9.29% of the CMS method. The EER of the cb3 handset is even lower than that obtained by the CMS method. Out of the 450 utterances from cb3, 428 were identified as being from cb4, 20 were rejected, and only two were incorrectly identified by the handset selector. Because most of the utterances were either transformed by the transformation parameters of the cb4 handset or recovered using CMS, its EER is reduced to 13.10%.

Figure 9.20 shows the distribution of the Jensen difference $J(\vec{\alpha}, \vec{r})$ for the cb1 seen handset and the cb3 unseen handset. The vertical dash-dot line defines the decision threshold used in the experiment (i.e., $\varphi = 0.05$). For cb1, all of the area under its probability density curve of the Jensen difference is in the handset acceptance region, which means that no rejection was made by the handset selector (In the experiment, all utterances from the cb1 handset were accepted by the handset selector.) For cb3, a large portion of the Jensen difference distribution is also in the handset acceptance region. This occurs because the characteristics of the cb3

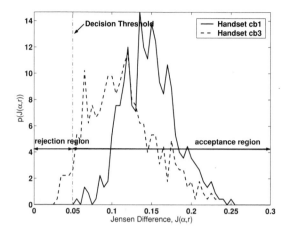

Figure 9.20. The distribution of the Jensen Difference $J(\vec{\alpha}, \vec{r})$ corresponding to the cb1 seen handset and the cb3 unseen handset.

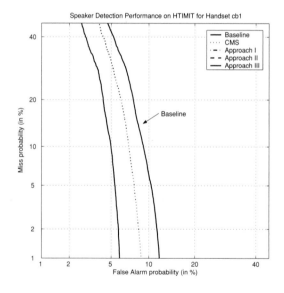

Figure 9.21. DET curves obtained by using the cb1 seen handset in the verification sessions. cb3 and el2 were used as the unseen handsets. Note that the DET curves corresponding to Approaches I, II and III are overlapped.

handset are similar to those of cb4; as a result, few rejections were made by the selector (only 20 out of 450 utterances were rejected in the experiment).

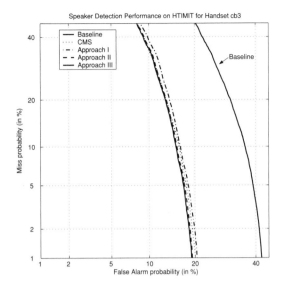

Figure 9.22. DET curves obtained by using the cb3 unseen handset in the verification sessions. cb3 and el2 were used as the unseen handsets.

The speaker detection performance for the seen handset cb1 and the unseen handset cb3 is shown in Figures 9.21 and 9.22, respectively. The EERs measured from the DET curves in Figure 9.21 correspond to the values in column cb1 of Table 9.15, and the EERs from Figure 9.22 correspond to the values in column cb3. Again, the slight discrepancy between the measured EERs and the EERs in Table 9.15 is due to interpolation error.

Figures 9.21 and 9.22 show that Approach III can achieve satisfactory performance for both seen and unseen handsets. In particular, Figure 9.21 shows that when Approach III was used, the DET curve of the cb1 seen handset overlaps the curve obtained by Approach I. This means that Approach III is able to keep the EERs of the seen handsets at low values. In Figure 9.22, using Approach III, the DET curve of the cb3 unseen handset is slightly on the left of the curve obtained by the CMS method, resulting in slightly lower error rates. Therefore, by applying Approach III to the speaker verification system, the error rates of a seen handset can be reduced to values close to those achievable by Approach I. On the other hand, the error rates of an unseen handset, with characteristics similar to some of the seen handsets, can be reduced to values close to or even lower than the values achievable by the CMS method.

9.5   Blind Handset-Distortion Compensation

As discussed in Section 9.4.2, one popular approach to compensating for handset distortion is to divide handsets into several broad categories according to the type of transducer (e.g., carbon button and electret). During operations, a handset selector is used to identify the most likely handset type from speech signals and handset distortion is compensated for based on some a priori information about the identified type in the database. Although this method works well in landline phones, it may encounter difficulty in mobile handsets because they have a large number of categories, new handset models are frequently released, and models can become obsolete in a short time. Maintaining a handset database for storing the information of all possible handset models is a great challenge and updating the compensation algorithm whenever a new handset is released is also difficult. Therefore, it is imperative to develop a channel compensation method that does not necessarily require a priori knowledge of handsets.

This section describes a blind compensation algorithm for this problem. The algorithm is designed to handle the situation in which no a priori knowledge about the channel is available (i.e., a handset model not in the handset database being used). Because the algorithm does not require a handset selector, it is suitable for a broader scale of deployment than the conventional approaches.

9.5.1 Blind Stochastic Feature Transformation

In speaker verification, it is important to ensure that channel variations are suppressed so that the interspeaker distinction can be enhanced. In particular, given a claimant's utterance recorded in an environment different from that during enrollment, one aims to transform the features of the utterance so that they become compatible with the enrollment environment. Therefore, it is not appropriate to transform the claimant's utterance either to fit the speaker model only or to fit the background model only because the former will result in an unacceptably high FAR (false acceptance rate) and the latter an excessive FRR (false rejection rate). This section describes a feature-based *blind* transformation approach to solving this problem. The transformation is blind in that it compensates the handset distortion without a priori information about the handset's characteristics. Hereafter, this transformation approach is referred to as blind stochastic feature transformation (BSFT).

Two Phases of BSFT

Figure 9.23 illustrates a speaker verification system with BSFT, whose operations are divided into two separate phases: enrollment and verification.

  1. *Enrollment Phase.* The speech of all client speakers are used to create a compact universal background model (UBM) Λ_b^M with M components. Then, for each client speaker, a compact speaker model Λ_s^M is created by adapting the UBM Λ_b^M using maximum a posteriori (MAP) adaptation [306]. Because verification decisions are based on the likelihood of the speaker model and background model, both models must be considered when the transformation parameters are computed. This can be achieved by fusing Λ_b^M and Λ_s^M to form a $2M$-component composite GMM

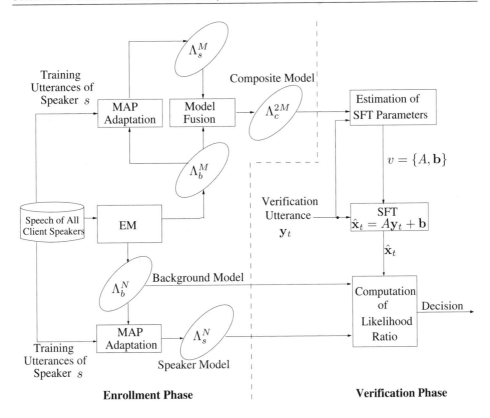

Figure 9.23. Estimation of BSFT parameters. The background model Λ_b^N, speaker model Λ_s^N, and composite model Λ_c^{2M} produced during the enrollment phase, are subsequently used for verification purposes.

Λ_c^{2M}. During the fusion process, the means and covariances remain unchanged but the value of each mixing coefficient is divided by 2. This step ensures that the output of the composite GMM represents a probability density function.

2. *Verification Phase.* Distorted features $Y = \{\mathbf{y}_1, \ldots, \mathbf{y}_T\}$ extracted from a verification utterance are used to compute the transformation parameters $\nu = \{A, b\}$. This is achieved by maximizing the likelihood of the composite GMM Λ_c^{2M} given the transformed features $\hat{X} = \{\hat{\mathbf{x}}_1, \ldots, \hat{\mathbf{x}}_T\}$:

$$\hat{\mathbf{x}}_t = f_\nu(\mathbf{y}_t) = A\mathbf{y}_t + \mathbf{b}, \quad t = 1, \ldots, T, \tag{9.5.1}$$

where A is a $D \times D$ identity matrix for zeroth-order transformation and $A = \text{diag}\{a_1, a_2, \ldots, a_D\}$ for first-order transformation, and $\mathbf{b}$ is a bias vector. The transformed vectors $\hat{X}$ are then fed to a full size speaker model Λ_s^N and a full size

UBM Λ_b^N for computing verification scores in terms of log-likelihood ratio:

$$s(\hat{X}) = \log p(\hat{X}|\Lambda_s^N) - \log p(\hat{X}|\Lambda_b^N). \tag{9.5.2}$$

The main idea of BSFT is to transform the distorted features to fit the composite GMM Λ_c^{2M}, which ensures that the transformation compensates the acoustic distortion.

A Two-Dimensional Example

Figure 9.24 illustrates the idea of BSFT in a classification problem with two-dimensional input patterns. Figure 9.24(a) plots the clean and distorted patterns of Class 1 and Class 2. The upper right (respectively lower left) clusters represent the clean (respectively distorted) patterns. The ellipses show the corresponding equal density contours. Markers ♦ and ■ represent the centers of the clean models. Figure 9.24(b) illustrates a transformation matching the distorted data of Class 2 and the GMM of Class 1 (GMM1). Because the transformation only takes GMM1 into account, while ignoring GMM2 completely, it results in a high error rate. Similarly, the transformation in Figure 9.24(c) also has a high error rate. The transformation in Figure 9.24(d) was estimated from the distorted data of Class 1 and a composite GMM formed by fusing GMM1 and GMM2. In this case, the transformation adapts the data to a region close to both GMM1 and GMM2 because it takes both GMMs into account. Therefore, instead of transforming the distorted data to a region around GMM1 or GMM2, as in Figures 9.24(b) and 9.24(c), the transformation in Figure 9.24(d) attempts to compensate for the distortion. The capability of BSFT is also demonstrated in a speaker verification task to be described next.

Experimental Evaluations

Per the discussion earlier, the experiments were divided into two phases: enrollment and verification.

1. *Enrollment Phase.* A 1,024-component UBM Λ_b^{1024} (i.e., $N = 1,024$ in Figure 9.23) was trained using the training utterances of all target speakers. The same set of data was also used to train an M-component UBM (Λ_b^M in Figure 9.23). For each target speaker, a 1,024-component speaker-dependent GMM $\Lambda_s^{1,024}$ was created by adapting $\Lambda_b^{1,024}$ using MAP adaptation [306]. Similarly, Λ_s^M was created by adapting Λ_b^M, and the two models are fused to form a composite GMM Λ_c^{2M}. The value of M was varied from 2 to 64 in the experiments.

2. *Verification Phase.* For each verification session, a feature sequence Y was extracted from the utterance of a claimant. The sequence was used to determine the BSFT parameters (A and $\mathbf{b}$ in Eq. 9.5.1) to obtain a sequence of transformed vectors $\hat{X}$. The transformed vectors were then fed to $\Lambda_s^{1,024}$ and $\Lambda_b^{1,024}$ to obtain verification scores for decision making.

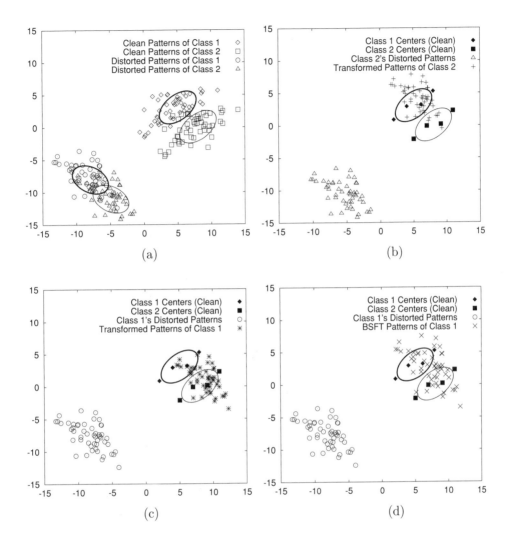

Figure 9.24. A two-class problem illustrating the idea of BSFT. (a) Scatter plots of the clean and distorted patterns corresponding to Class 1 and Class 2. The thick and thin ellipses represent the equal density contours of Class 1 and Class 2, respectively. The upper right (respectively, lower left) clusters contain the clean (respectively, distorted) patterns. (b) Distorted patterns of Class 2 were transformed to fit Class 1's clean model. (c) On the contrary, distorted patterns of Class 1 were transformed to fit Class 2's clean model. (d) Distorted data of Class 1 were transformed to fit a clean, composite model formed by the GMMs of Class 1 and Class 2 (i.e., BSFT). For clarity, only the distorted patterns before and after transformation were plotted in (b) through (d).

Speech Data and Features. The 2001 NIST speaker recognition evaluation set [151], which contains cellular phone speech of 74 male and 100 female target speakers extracted from the SwitchBoard-II Phase IV Corpus, was used in the evaluations. Each target speaker has 2 minutes of speech for training (i.e., enrollment); a total of 850 male and 1,188 female utterances are available for testing (i.e., verification). Each verification utterance has a length of between 15 and 45 seconds and is evaluated against 11 hypothesized speakers of the same sex as the speaker of the verification utterance. Out of these 11 hypothesized speakers, one is the target speaker who produced the verification utterance. Therefore, there are one target and 10 impostor trials for each verification utterance, which amount to a total of 2,038 target trials and 20,380 impostor attempts for 2,038 verification utterances.

Mel-frequency cepstral coefficients (MFCCs) [69] and their first-order derivatives were computed every 14ms using a Hamming window of 28ms. Cepstral mean subtraction (CMS) [107] was applied to the MFCCs to remove linear channel effects. The MFCCs and delta MFCCs were concatenated to form 24-dimensional feature vectors.

Performance Measures and Results. Detection error tradeoff (DET) curves and equal error rates (EERs) were used as performance measures. They were obtained by pooling all scores of both sexes from the speaker and impostor trials. In addition to DET curves and EERs, a decision cost function (DCF) was also used as a performance measure. The DCF is defined as

$$\mathrm{DCF} = C_{Miss} \times P_{Miss|Target} \times P_{Target}$$
$$+ C_{FalseAlarm} \times P_{FalseAlarm|Nontarget} \times P_{Nontarget},$$

where P_{Target} and $P_{Nontarget}$ are the prior probability of target and impostor speakers, respectively, and where C_{Miss} and $C_{FalseAlarm}$ are the costs of miss and false alarm errors, respectively. Following NIST's recommendation [288], these parameters were set as follows: $P_{Target} = 0.01$, $P_{Nontarget} = 0.99$, $C_{Miss} = 10$, and $C_{FalseAlarm} = 1$.

Table 9.16 and Figure 9.25 show the results of the baseline (CMS only), Znorm [14, 303], and BSFT with different order and number of components M in the compact GMMs. Evidently, all cases of BSFT show significant reduction in error rates when compared to the baseline. In particular, Table 9.16 shows that first-order BSFT with Znorm achieves the largest error reduction.[8] The DET curves also show that BSFT with Znorm performs better than the baseline and Znorm for all operating points.

Comparison with Other Models. It is of interest to compare BSFT with the short-time Gaussianization approach proposed in Xiang et al. [387] because both methods transform distorted features in the feature space and their transformation parameters are estimated by the EM algorithm. In short-time Gaussianization, a

---

[8]Theoretically, the larger the value of M, the better the results. However, setting M larger than 64 will result in unacceptably long verification time.

Table 9.16. EERs and minimum decision cost achieved by the baseline (CMS only), Znorm, and zero- and first-order BSFT with different number of components M in the compact GMMs.

Compensation Method	SFT Order	M	Equal Error Rate (%)	Minimum Decision Cost
Baseline	NA	NA	12.02	0.0477
BSFT	Zeroth	2	11.90	0.0473
BSFT	Zeroth	4	11.82	0.0458
BSFT	Zeroth	8	11.39	0.0449
BSFT	Zeroth	16	11.24	0.0450
BSFT	Zeroth	32	11.22	0.0450
BSFT	Zeroth	64	11.16	0.0443
BSFT	First	2	12.00	0.0506
BSFT	First	4	11.55	0.0471
BSFT	First	8	10.70	0.0464
BSFT	First	16	10.47	0.0454
BSFT	First	32	10.43	0.0446
BSFT	First	64	10.00	0.0428
Znorm	NA	NA	10.39	0.0447
BSFT+Znorm	First	64	8.18	0.0369

linear, global transformation matrix is estimated by the EM algorithm using the training data of all background speakers. The global transformation aims to decorrelate the features so that channel variations can be suppressed. The short-time Gaussianization achieves an EER of 10.84% in the NIST 2001 evaluation set [387], whereas BSFT achieves an EER of 9.26%,[9] which represents an error reduction of 14.58%. The minimum decision cost of BSFT is also lower than that of short-time Gaussianization (0.0428 versus 0.0440).

Computation Consideration. It can be argued that the inferior performance of short-time Gaussianization is due to the nonadaptive nature of its transformation parameters. However, the adaptive nature of BSFT comes with a computational price: different transformation parameters must be computed for each speaker in

---

[9]Because Xiang et al. [387] did not use Znorm, their results should be compared with the one without Znorm here.

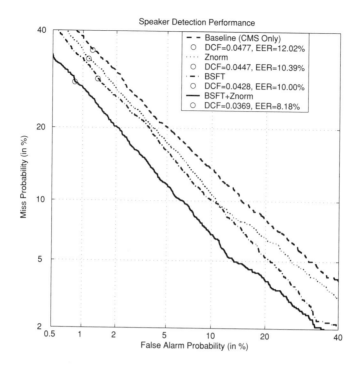

Figure 9.25. DET curves comparing speaker verification performance using CMS only (*dashed*), Znorm (*dotted*), first-order BSFT without Znorm (*dash-dot*), and first-order BSFT with Znorm (*solid*). For the BSFT, the number of components M in the compact GMMs was set to 64. The circles represent the errors at which minimum decision costs occur.

every verification session. Therefore, it is vital to have a cost-effective computation approach for BSFT. Note that the computation complexity of estimating BSFT parameters grows with the amount of adaptation data (i.e., the value of T in Eq. 9.5.1) and the number of mixture components in the GMMs (i.e., the value of M). To reduce computation time, M should be significantly smaller than N, the number of components in the full size speaker and background models. This is particularly important for the computation of BSFT parameters during the verification phase because the computation time of this phase is a significant part of the overall verification time. The evaluations suggest that a good tradeoff between performance and computation complexity can be achieved by using a suitable value of M.

9.5.2   Integration of Blind and Supervised Compensation

The blind compensation method discussed before is designed to handle handsets that are unknown to the speaker verification system. However, estimating BSFT

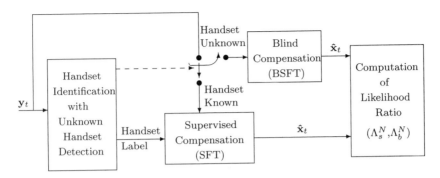

Figure 9.26. Integration of blind and supervised compensation methods.

parameters online requires additional computation during the verification phase. One possible solution to reducing computation without scarifying the advantages of blind compensation is to precompute the transformation parameters of some commonly used handsets and store them in a handset database; during verification, the most appropriate transformation parameters are selected from the database if the handset being used is detected to be one of the a priori known models. In this case, the supervised compensation approach discussed in Section 9.4.2 can be adopted. This paves the way for an integrated approach, combining blind and supervised compensation techniques. Figure 9.26 illustrates how blind and supervised compensation can be integrated into a speaker verification system. This integrated approach can enjoy the best of two worlds: (1) whenever a known handset is detected, precomputed transformation parameters can be used; and (2) should the handset being used be unknown to the system, the system can compute the transformation parameters online using blind or unsupervised techniques.

9.6   Speaker Verification Based on Articulatory Features

Most traditional speaker verification systems are based on the modeling of short-term spectral information [306]. These systems compute the ratio of the likelihood of the genuine speaker model to the likelihood of the impostor model given some spectral features extracted from a claimant. The resulting likelihood ratio is compared against a threshold for decision making. The advantage of using short-term spectral information is that promising results are obtainable from a limited amount of training data. However, spectral information is only one of many sources suitable for speaker verification. For example, in addition to the spectral information, humans make use of high-level information, such as dialect and lexical characteristics [81], to recognize speakers. Although these characteristics are closely related to the unique speech production process, they are usually ignored by most speaker verification systems. Mainly, this is because obtaining such information reliably from speech signals remains a challenging research problem.

In recent years, researchers have started to investigate the use of high-level speaker information for speaker verification [309]. Such information includes (1) prosodic features obtained from pitch and energy tracking and (2) phonetic features derived from phone sequences using speech recognizers and language models. Recent research has shown that using high-level features alone can achieve reasonably good speaker recognition performance [309]. It has also been shown that a significant improvement in speaker recognition accuracy can be obtained when the features are combined with spectral features using a simple perceptron fusion [42].

In addition to the high-level features mentioned here, properties arising from the speech production process can be used for recognizing speakers. Speech is produced by a sequential movement of articulators in the vocal tract excited by an air stream originating from the lungs. The combined effects of articulation and excitation, which give rise to the produced speech, are the origin of unique speaker characteristics [81] and the source of high-level features. Gupta et al. [122] have demonstrated that features derived from glottal excitation are useful for speaker verification; however, glottal features be derived only from voiced speech.

Combining the movement or position of articulators with glottal features for speaker verification might alleviate the shortcomings of glottal features. Specifically, articulatory features (AFs) can be used to capture the position and movement of articulators during speech production for speaker verification. AFs have been adopted as an alternative, or a supplementary feature, for automatic speech recognition (see references [88, 103, 181, 202, 312]) and confidence measures [203]. The studies have demonstrated the usefulness of AFs, either used alone or together with spectral features such as MFCCs.

In an early 2004 study, Leung et al. [201], five important speech production properties, including the manner and place of articulation, voicing of speech, lip rounding, and tongue position, were considered as features for verifying speakers. For each property, a classifier was used to determine the probability of different classes given a sequence of spectral features. The outputs of the five classifiers, which can be considered as the abstract representation of the five speech production properties, were concatenated to form the AF vectors for speaker verification. It was found that AFs are complementary to spectral features and that a significant reduction in EERs can be achieved when they are used together with spectral features in a speaker verification system.

9.6.1 Articulatory Feature Extraction

To extract AFs from speech, a set of articulatory classifiers are taught the mappings between the acoustic signals and the articulatory states. Either images (e.g., an X-ray that records the actual articulatory positions [400]) or mappings between phonemes and their corresponding articulatory properties [2, 50] can be used to train the AF classifiers. Images that track real articulatory positions can help reduce the correlation between the extracted AFs and speech data; however, collecting sufficient data to run large-scale experiments is too expensive. Consequently, it

Table 9.17. Five articulatory properties and the number of classes in each

Articulatory Properties	Class	Number of Classes
Voicing	Voiced, unvoiced	2
Front–back	Front, back, nil	3
Rounding	Rounded, not rounded, nil	3
Manner	Vowel, stop, fricative, nasal, approximant-lateral	5
Place	High, middle, low, labial, dental, coronal, palatal, velar, glottal	9

is more common to use phoneme-to-articulatory state mappings [181] to extract AFs. The average AFs classification rate can be more than 80% under matched conditions [2, 50], and the mapping approach has been adopted in many speech recognition systems [88, 181, 202, 312].

A mapping approach similar to that of Kirchhoff [181] was adopted for this work (i.e., AFs were extracted from acoustic signals using phoneme-to-articulatory mappings). Specifically, to obtain the AFs a sequence of acoustic vectors were fed to five classifiers in parallel, where each classifier represents a different articulatory property. The outputs of these classifiers (the posterior probabilities) were concatenated to form the AF vectors. As a result, the extracted AFs can be considered an intermediate representation of the acoustic signals.

In this verification system, five different articulatory properties are used, as tabulated in the first column of Table 9.17. For each property, a multi-layer perceptron (MLP) estimates the probability distributions of its predefined output classes (see second column of table). The AF extraction process is illustrated in the right-dotted box of Figure 9.27.

The inputs to these five AF–MLPs are identical, while their output numbers are equal to the number of AF classes listed in the last column of Table 9.17. To ensure a more accurate estimation of the AF values, multiple frames of MFCCs served as inputs to the AF–MLPs. More specifically, for the t-th frame of an utterance, a block of "normalized" MFCCs at frames $t - \frac{n}{2}, \ldots, t, \ldots, t + \frac{n}{2}$ (where $n = 8$ in this work) is presented to the AF–MLPs. Rather than feeding the MFCCs directly to the AF–MLPs, they are normalized to zero mean and unit variance. The normalization parameters, a mean vector μ, and a standard deviation vector σ— each with dimension the same as that of the MFCCs—are obtained globally from the training data. Given an MFCC vector $\mathbf{x}_t$, the normalization for dimension i is

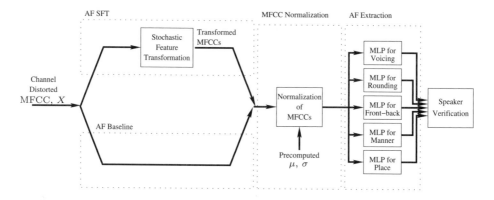

Figure 9.27. System block diagram of speaker verification using AFs extracted from MFCCs or using MFCCs after SFT. The normalization parameters, μ and σ, are determined from the clean MFCCs (i.e., senh). The thick arrows represent multidimension inputs and outputs. Each AF–MLP takes 9 consecutive, 12-dimensional normalized MFCCs as input. The number of outputs for the voicing, rounding, front–back, manner, and place MLPs are 2, 3, 3, 5, and 9, respectively.

done by applying

$$x_t^{norm}(i) = \frac{x_t(i) - \mu(i)}{\sigma(i)}, \quad i = 1, \dots, D. \tag{9.6.1}$$

The normalization aims to remove the variations of input features among different dimensions so that the determination of MLP weights will not be dominated by those input features with large magnitude.

The AF–MLPs can be trained from speech data with time-aligned phonetic labels. The alignments can be obtained from transcriptions or from Viterbi decoding using phoneme models. With the phoneme labels, articulatory classes can then be derived from the mappings between phonemes and their states of articulation [181].

9.6.2   AF-Based Speaker Verification

Figure 9.28 illustrates the procedures of phone-based speaker verification using AFs as features. The procedures can be divided into four steps. First, the most likely handset is identified by the handset selector (described in Section 9.4.2); second, the MFCCs are normalized using the parameters of the selected handset; third, the normalized MFCCs are fed to five AF–MLPs to determine the AFs; finally, the AFs are used as speaker features for speaker verification. The aim of identifying the handset before extracting the AFs is to determine the most likely handset so that handset-specific compensation can be applied to the MFCCs.

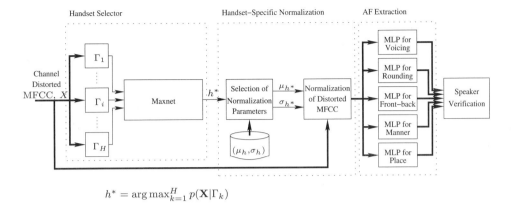

$$h^* = \arg\max_{k=1}^{H} p(\mathbf{X}|\Gamma_k)$$

Figure 9.28. Combination of handset identification, handset-specific normalization, and AF extraction for robust speaker verification. The thick arrows represent the multidimension inputs and outputs. The number of inputs and outputs of the AF–MLPs are the same as those described in Figure 9.27.

Handset-Specific Normalization

As given in Eq. 9.6.1, the means and variances of the spectral vectors estimated from the clean training data are used to normalize the MFCCs. These parameters, however, may be varied across different handsets. Therefore, the AFs of channel-distorted speech cannot be correctly determined if the means and variances of clean MFCCs are used for normalizing the channel-distorted MFCCs. To compensate for the distortion caused by different handsets, handset-dependent normalization parameters (μ_h, σ_h) are needed.

As illustrated in Figure 9.28, the normalization parameters of the identified handset, h, are used to normalize the distorted MFCCs:

$$x_t^{HS\text{-}Norm}(i) = \frac{x_t(i) - \mu_h(i)}{\sigma_h(i)}. \tag{9.6.2}$$

Eq. 9.6.2 is referred to as *HS-Norm* hereafter in this section. The handset-normalized MFCCs are then fed to the five AF–MLPs to determine the AFs. Because the MFCCs are now transformed to a range close to clean training speech, the variation of MFCCs due to handset differences can be minimized.

In addition to normalizing the MLP inputs, HS-Norm can also be applied to GMM-based verification systems to minimize the channel effect caused by different handsets. As suggested in Viikki and Laurila [361], applying a segmental means-and-variances normalization to the feature vectors regardless of the noise conditions can significantly enhance the channel robustness. Unlike the segmental feature normalization of that 1997 study [361], however, the normalization parameters in HS-Norm are handset-dependent, which makes HS-Norm more amenable to the compensation of handset-dependent channel distortion. Compared to SFT, which only shifts the feature means according to the identified channel, HS-Norm does a better feature compensation job because both feature means *and* variances are considered. Therefore, HS-Norm should be more effective than SFT in channel compensation.

Speaker Verification

As Table 9.17 shows, there are a total of 22 articulatory classes, which result in a 22-dimensional AF vector for each frame. For each testing utterance from a claimant, a sequence of 22-dimensional AF vectors $\mathbf{Y}$ were fed to a speaker model ($\mathcal{M}_s$) and a background model ($\mathcal{M}_b$) to obtain a verification score $S(\mathbf{Y})$:

$$S(\mathbf{Y}) = \log p(\mathbf{Y}|\mathcal{M}_s) - \log p(\mathbf{Y}|\mathcal{M}_b). \tag{9.6.3}$$

Similar to spectral feature-based speaker verification, speaker-independent EER is obtained from $S(\mathbf{Y})$.

9.6.3   Fusion of Spectral and Articulatory Features

Because the spectral features (MFCCs) and articulatory features focus on different aspects of speech, fusion of these features is expected to provide better performance than using individual features. Recent experimental results (see [201]) on the fusion of spectral and articulatory features for phone-based speaker verification are reported next.

Corpus

In this work, the HTIMIT corpus [304] was used for performance evaluation. HTIMIT was constructed by playing a gender-balanced subset of the TIMIT corpus through nine phone handsets and a Sennheizer head-mounted microphone. This setup introduces real handset-transducer distortion in a controlled manner but without losing the time-aligned phonetic transcriptions of the TIMIT corpus. This feature makes HTIMIT ideal for studying the handset variability in speech and speaker recognition systems [291]. It also facilitates the training of AF–MLPs by mapping the time-aligned phoneme labels to their corresponding articulatory classes.

Speaker Enrollment

A disjointed, gender-balanced speaker set and a disjointed, gender-balanced imposter set, consisting of 100 client speakers and 50 impostors, respectively, were selected from the HTIMIT corpus.

For the system that uses spectral features only (hereafter referred to as the MFCC system), twelfth-order MFCCs were computed every 14ms using a Hamming window of 28ms. For the system that uses AFs as features (hereafter referred to as AF system), 22-dimensional AF vectors were obtained from the five AF–MLPs, each with 108 input nodes (9 frames of 12-dimensional MFCCs) and 50 hidden nodes. The MLPs were trained using the Quicknet [104]. Training data includes all sentences collected using the senh microphone of all speakers in HTIMIT, excluding speakers from the speaker set and the impostor set.

For each system, a 64-center universal background model $\mathcal{M}_b$ was trained using the SA and SX utterances from all speakers in the speaker set. For each speaker in the speaker set, a speaker model $\mathcal{M}_s$ was adapted from $\mathcal{M}_b$ using MAP adaptation [306]. Only the SA and SX sentences collected using the head-mounted microphone (senh) were used for enrollment and adaptation.

Robustness Enhancement

For the MFCC system, cepstral mean subtraction (CMS), stochastic feature transformation (SFT), and handset-specific normalization (HS-Norm) were adopted to enhance the robustness to handset mismatches. Except for CMS, where the channel compensation was handset-independent, a handset selector $\{\Gamma_k\}_{k=1}^H$ was first applied to each testing utterance to identify the most likely handset. The SFT or HS-Norm corresponding to the identified handset was then applied to the distorted MFCC vectors. For each handset, the SX and SA sentences of 10 speakers, randomly chosen from the training data, were used to estimate the SFT and HS-Norm parameters.

The handset selector in the AF system was also MFCC-based, and $\{\Gamma_k\}_{k=1}^H$ are identical to those of the MFCC system. Both the verification performance of AFs extracted from MFCCs and from MFCCs with CMS were evaluated. However, most of the channel characteristics in MFCCs would be removed if CMS was applied, which reduced the accuracy of the handset selector. If required, CMS was performed after handset selection and before AF extraction.

System Fusion

Although both the AF and MFCC systems take MFCCs as input, they attempt to capture two different information sets from the speech signals. The MFCC system attempts to capture the acoustic characteristics, and the AF system attempts to capture the articulatory properties. Therefore, fusing these two systems should result in a better performance than the individual systems can provide.

In this work, utterance scores, as given in Eq. 9.6.3, obtained from the MFCC system and the AF system were linearly combined to produce the fusion scores

$$S_F(\mathbf{Y}_{mfcc}, \mathbf{Y}_{af}) = (1 - w_{af})S(\mathbf{Y}_{mfcc}) + w_{af}S(\mathbf{Y}_{af}), \qquad (9.6.4)$$

where w_{af} is a handset-dependent fusion weight. The fusion weight was determined from data used for estimating the normalization parameters and the SFT parameters of each handset type.

Results

Table 9.18 summarizes the EERs obtained from the approaches discussed before. Each EER is determined by concatenating the utterance scores of all SI sentences from 100 speakers and 50 impostors. For each approach, two sets of EERs are

Table 9.18. Average and global EERs based on different features and channel compensation approaches. The average EERs were obtained by averaging the individual EERs corresponding to nine different handsets (cb1–cb4, el1–el4, and pt1). For each global EER in the table, the utterance scores from all of the 9 mismatched handsets were merged to compute a single EER. Note that 3 + 9, 4 + 9, and 2 + 7 represent the fusion of MFCC SFT and AF HS-Norm, the fusion of MFCC HS-Norm and AF HS-Norm, and the fusion of MFCC CMS and AF CMS + HS-Norm, respectively. Fusions based on other combinations have also been performed; however, only the combinations that give the best results are listed. The handset recognition accuracy is 98.35%. *Note:* MFCC baseline, MFCC CMS, AF baseline, and AF CMS do not require the handset selector.

Row	Features	Average		Global	
		EER (%)	Rel. Red. (%)	EER (%)	Rel. Red. (%)
1	MFCC baseline	23.84	–	23.77	–
2	MFCC CMS	12.80	–	13.10	–
3	MFCC SFT	8.31	–	9.30	–
4	MFCC HS-Norm	8.15	–	8.45	–
5	AF baseline	30.93	–	30.89	–
6	AF CMS	20.73	–	20.53	–
7	AF CMS + HS-Norm	18.39	–	18.85	–
8	AF SFT	15.63	–	16.00	–
9	AF HS-Norm	14.57	–	14.66	–
10	3 + 9	7.85	5.54	8.50	8.60
11	4 + 9	7.70	5.52	8.07	4.50
12	2 + 7	11.76	8.13	12.63	3.59

obtained. The EERs and the relative reductions (Rel. Red.) under the *Average* column are the average of EERs individually obtained from the nine mismatched handsets (i.e., excluding senh). Those under the *Global* column are the global EERs obtained by merging the utterance scores of the nine mismatched handsets. The results are organized into three sections in Table 9.18, where the EERs of the MFCC systems, the AF systems, and the fusion of the two systems are listed in Rows 1 through 4, Rows 5 through 9, and Rows 10 through 12, respectively.

Evidently, for the MFCC system, MFCC SFT and MFCC HS-Norm, which adopt the handset-specific compensation, outperform MFCC baseline and MFCC CMS. In particular, MFCC HS-Norm achieves the lowest EER, which gives an average EER of 8.15% and a global EER of 8.45%. When compared to the EERs obtained from MFCC CMS, which is the simplest robustness enhancement approach, these error rates represent a relative reduction of 36.33% and 35.50% for average EER and global EER, respectively. MFCC SFT achieves a slightly less significant error reduction: a 35.08% in average EER and a 29.01% in global EER. This demonstrates that HS-Norm is better than SFT for enhancing channel robustness.

For the AF system, the verification results based on the AFs extracted from the MFCCs and from the MFCCs with CMS are named AF baseline and AF CMS in Table 9.18, respectively. AF SFT represents verification using AFs extracted from MFCCs with SFT. Because of the nature of MLP training, Eq. 9.6.1 is used to normalize the MFCCs before feeding them into the MLPs. For AF baseline, AF CMS, and AF SFT, the feature means and variances determined from the training data (i.e., senh) were used to normalize the MFCCs (or the stochastically transformed MFCCs), as shown in Figure 9.27; in AF HS-Norm and AF CMS + HS-Norm, on the other hand, handset-specific normalization was applied to the MFCCs and the MFCCs with CMS, respectively, as illustrated in Figure 9.28.

The results for AFs suggest that the lowest EERs (both average and global) can be obtained from extracting AFs from handset-normalized MFCCs rather than from the stochastically transformed MFCCs or from the CMS MFCCs. This is consistent with the results obtained from the MFCC system. During the AF extraction, the application of HS-Norm significantly reduces the EERs from 30.93% to 14.57% (average) and from 30.89% to 14.66% (global). Even if the AFs are extracted from the CMS MFCCs, applying HS-Norm can still reduce the EERs from 20.73% to 18.39% (average) and from 20.53% to 18.85% (global). This represents an error reduction of 11.29% and 8.18% for average and global EERs, respectively.

The individual MFCC and AF systems that give the lowest EER were fused together. After trying all possible combinations, the three fusions that give the lowest EERs, and their corresponding EER reduction relative to the respective MFCC systems, are summarized in Table 9.18. These include the fusion of MFCC SFT and AF HS-Norm, the fusion of MFCC HS-Norm and AF HS-Norm, and the fusion of MFCC CMS and AF CMS + HS-Norm. Of the three fusions, EERs lower than the individual systems were obtained. The relative error reductions range from 5.52% to 8.13% for the average EERs and from 3.59% to 8.60% for the global EERs. Of the three fusions, the lowest EER was obtained from the fusion of MFCC HS-Norm and AF HS-Norm, which yields 7.70% for the average EER and 8.07% for the global EER. These represent an EER reduction of 5.52% (average) and 4.50% (global). This suggests that the acoustic characteristics represented by the MFCCs and the articulatory properties represented by the AFs are partially complementary, although they are from the same source.

The global DET curves (see Martin et al. [233]) using all testing utterances from the nine mismatched handsets and the DET curves obtained from the el3 handset are shown in Figure 9.29. The curves also agree with the EERs listed in Table 9.18.

9.7 Concluding Remarks

This chapter has highlighted the key issues in speaker recognition and demonstrated how the neural models described in earlier chapters can be applied to speaker verification. The results demonstrate that the globally supervised learning of PDBNNs

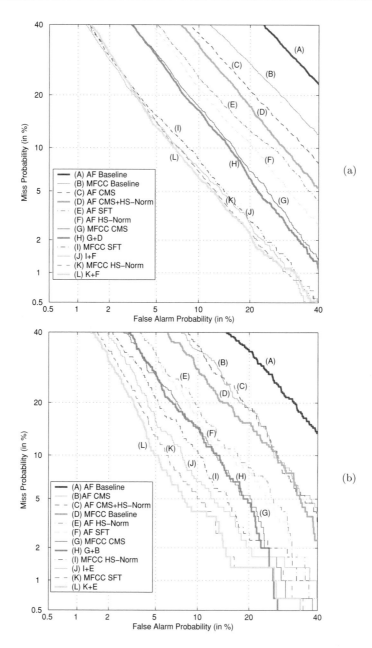

Figure 9.29. DET plots of speaker verification using the MFCCs, AFs, and fusion of the two features based on various channel compensation approaches. For ease of comparison, methods in the legend are arranged in descending order of EERs. The DET plots are generated using (a) all testing utterances from all of the 9 mismatched handsets and (b) testing utterances from the el3 handset.

can make the FAR of all speaker models very close together and that the average FAR is very small during verification; the ad hoc approach used by the EBFNs and GMMs is not able to do so. This chapter also demonstrated that PDBNNs and GMMs are more robust than EBFNs in recognizing speakers in noisy environments. Although PDBNNs and GMMs show better generalization against additive noise at different SNR, their performance is still unacceptable at low SNR. This suggests the necessity of compensation or adaptive techniques to reduce the acoustic mismatch between training and testing.

This chapter also introduced several channel and handset compensation approaches to address the problem of environmental mismatch in phone-based speaker verification systems. These techniques can be roughly categorized into three classes: feature transformation, model transformation and adaptation, and score normalization. A handset selector was combined with (1) handset-specific transformations, (2) reinforced learning, (3) stochastic feature transformation, and (4) stochastic model transformation to reduce the effect caused by the acoustic distortion. Experimental results based on 150 speakers of the HTIMIT corpus show that an environment adaptation based on MLLR, reinforced learning, and feature transformation outperforms the classic CMS, Hnorm and Tnorm approaches. The error rate achieved by the transformation technique correlates with the bit rate of the codec and, hence, reflects the perceptual quality of the coded speech. Results also demonstrate that linear and nonlinear stochastic transformation attain a comparable amount of error reduction, with nonlinear achieving a slightly better result.

The nonlinear stochastic transformation was then combined with a divergence-based handset selector with out-of-handset rejection capability for identifying unseen handsets. When speech from an unknown handset is presented, the selector will either identify the most likely handset from its handset database or reject it (consider it as unseen). Experiments have been conducted to transform utterances using the transformation parameters of the most likely handset if their corresponding handsets can be identified. On the other hand, utterances whose handsets were considered unseen were processed by CMS. Results showed this approach can reduce the average error rate and maintain the error rate of unseen handsets to values close to those obtainable by CMS. It was also found that when the unseen handset has characteristics similar to any one of the seen handsets in the handset database, the handset selector is able to select a similar handset from the database. This capability enables the verification system to maintain the error rate to values very close to those achievable by using seen handsets. On the other hand, if the unseen handset is different from all of the seen handsets, it will have a high chance of being rejected by the handset selector. The ability to reject these dissimilar, unseen handsets enables the verification system to maintain the error rate at a level achievable by the CMS method.

The chapter ends with a new speaker verification approach in which articulatory features (AFs) derived from MFCCs are fused with MFCC-based spectral features for robust speaker verification. Results have shown that AFs contain sufficient speaker-specific information for speaker verification. To increase the robustness of AFs to channel distortion, handset-specific means and variances normalization (HS-Norm) were applied to MFCCs. It was found that HS-Norm are not only useful to AF-based systems but are also applicable to MFCC-based systems. In particular, HS-Norm is able to compensate the channel effects caused by different handsets and reduce the equal error rate under handset mismatch conditions. The EERs can be further reduced when the proposed AF-based system was fused with a traditional MFCC-based system. This suggests that the speaker information captured by the MFCCs and AFs is partially complementary. The lowest global EER was 8.07%, which was obtained by fusing the scores derived from the handset-normalized MFCCs with those derived from handset-normalized AFs.

Problems

1. Suggest an appropriate statistical model for the following speaker recognition tasks:

    (a) Text-independent speaker verification.

    (b) Password-based speaker verification where the passwords are confined to spoken digits.

    (c) Phrase-prompted speaker verification.

    For each case, explain your suggestion.

2. Explain how you would use

    (a) Gaussian mixture models (GMMs)

    (b) hidden Markov models (HMMs)

    (c) radial basis function networks (RBFNs)

    (d) probabilistic decision-based neural networks (PDBNNs)

    (e) support vector machines (SVMs)

    to build a speaker verification system. Draw a block diagram for each case and explain how each of these models characterizes the impostors' speech.

3. Explain the role of the background model $\mathcal{M}_b$ in Eq. 9.4.15 in a speaker verification system. Tell what would happen if the background model were removed from the speaker verification system.

4. Derive the gradient of the second-order nonlinear stochastic feature transformation (Eqs. 9.4.10 through 9.4.12) from Eqs. 9.4.2 through 9.4.9.

5. Find the solutions of the M-step of the piecewise linear transformation (i.e., find the solutions for Eq. 9.4.13).

6. Based on Eq. 3.2.4 in Chapter 3, derive the auxiliary function $Q(\nu|\nu')$ in Eq. 9.4.5.

7. The technique used for deriving the SFT in Section 9.4.2 can also be applied to stochastic model transformation. Assume that an M-center clean speaker GMM $\Lambda_X = \{\omega_j^X, \mu_j^X, \Sigma_j^X\}_{j=1}^M$ is transformed by a function $f_\eta(\Lambda_X)$ to form another M-center GMM $\Lambda_Y = \{\omega_j^Y, \mu_j^Y, \Sigma_j^Y\}_{j=1}^M$. Consider a special case where only the means are transformed (i.e., $\omega_j^Y = \omega_j^X$ and $\Sigma_j^Y = \Sigma_j^X \; \forall \; j$), and one has

$$\mu_j^Y = \mu_j^X + b,$$

where $\eta = b$ is the transformation parameter to be estimated by the EM algorithm. The estimation of η starts with the following Q-function:

$$Q(\eta'|\eta) = E\{\log p(Y, Z|\eta', \Lambda_X)|Y, \Lambda_X, \eta\},$$

where $Y = \{\mathbf{y}_1, \ldots, \mathbf{y}_T\}$ is the sequence of distorted speech vectors, η' and η are the new and old estimates of the transformation parameter, respectively, and Z is a set containing $\{z_j(\mathbf{y}_t); t = 1, \ldots, T\}$ whose values are defined as

$$z_j(\mathbf{y}_t) = \begin{cases} 1 & \text{if } \mathbf{y}_t \text{ is generated from the } j\text{-th kernel of } \Lambda_Y \\ 0 & \text{otherwise.} \end{cases}$$

(a) Show that the Q-function can be expressed as

$$Q(\eta'|\eta) = \sum_{t=1}^T \sum_{j=1}^M h_j(\mathbf{y}_t) \log p(\mathbf{y}_t|\mu_j^X + b', \Sigma_j^X)\omega_j^X,$$

where

$$h_j(\mathbf{y}_t) = \frac{\omega_j^X p(\mathbf{y}_t|\mu_j^X + b, \Sigma_j^X)}{\sum_{i=1}^M \omega_i^X p(\mathbf{y}_t|\mu_i^X + b, \Sigma_i^X)}$$

is the probability that $\mathbf{y}_t$ is generated from the j-th kernel of Λ_Y.

(b) Show that in the M-step of the EM algorithm, the value of b that maximizes $Q(\eta'|\eta)$ is given by

$$b' = \frac{\sum_{t=1}^T \sum_{j=1}^M h_j(\mathbf{y}_t)(\Sigma_j^X)^{-1}(\mathbf{y}_t - \mu_j^X)}{\sum_{t=1}^T \sum_{j=1}^M h_j(\mathbf{y}_t)(\Sigma_j^X)^{-1}}.$$

(c) The transformation can also be expressed in a more general form:

$$\mu_j^Y = A\mu_j^X + b,$$

where A is a $D \times D$ transformation matrix and b is a bias vector. Derive the corresponding Q-function and the maximum-likelihood estimate of A and b in the M-step of each EM iteration. (*Hint*: See Leggetter and Woodland [200].)

(d) In case the variances can also be transformed as follows:

$$\Sigma_j^Y = A\Sigma_j^X A^T,$$

derive the corresponding Q-function and the maximum-likelihood solution of A and b in each M-step. (*Hint*: See Digalakis et al. [78]).

8. During the evaluation of a speaker verification system, four factors must be considered: speech quality, type of speech, length of speech, and speaker population. Detail how these factors affect the performance of a speaker verification system.

9. The CMS or CMN is a classical method of minimizing channel variation in speaker verification systems. Assume that clean speech $x(n)$ is transmitted through a channel with transfer function $H(z)$. Denote the resulting distorted speech at the other end of the channel as $y(n)$.

   (a) Express the cepstrum of $y(n)$, $c_y(n)$, in terms of the cepstrum of $x(n)$, $c_x(n)$, and the cepstrum of the impulse response of $H(z)$, $c_h(n)$.

   (b) By taking the expectation of both sides of the equation in (a) and making appropriate assumptions, derive the formula for CMS and CMN:

   $$\hat{c}_x(n) = c_y(n) - E\{c_y(n)\}, \qquad (9.7.1)$$

   where $E(\cdot)$ denotes expectation and $\hat{c}_x(n)$ denotes the recovered ceptrum. Discuss the condition under which your assumptions become valid.

   (c) Identify the similarity between Eq. 9.7.1 and zero-th order SFT (i.e., Eq. 9.4.1 with $a_i = 1 \; \forall i$).

10. You are asked to build a speaker verification system with a population size of 100. Assume that you are given both the speech of 100 client speakers as well as the speech of another 100 speakers.

    (a) Outline a method to determine a decision threshold for each of the 100 client speakers.

    (b) Discuss the advantages and disadvantages of using speaker-dependent decision thresholds in the speaker verification system.

11. Assume that the client speaker scores and the impostor scores of a speaker verification system follow the Gaussian distributions $\mathcal{N}(2,1)$ and $\mathcal{N}(0,1.5)$, respectively. Assume also that a speaker-independent decision threshold is used to make accept and reject decisions.

    (a) Calculate the EER of the system.

    (b) Calculate the FAR and FRR if the threshold is equal to 0.

(c) Sketch the DET curve of the system.

12. A pattern recognition system usually contains two distinct tasks:

(a) Preprocessing in which a subset of the potentially influential features is found

(b) Postprocessing in which optimal and classification are performed

Discuss these two tasks in the context of a speaker recognition system. Suggest the types of features, optimization algorithms, and classification schemes for each system.

Chapter 10

MULTICUE DATA FUSION

10.1 Introduction

Various research studies have suggested that no single modality can provide an adequate solution for high-security applications. These studies agree that it is vital to use multiple modalities such as visual, infrared, acoustic, chemical sensors, and so on.

The problem of combining the power of several classifiers is of great importance to various applications. In many remote-sensing, pattern recognition, and multimedia applications, it is not uncommon for different channels or sensors to facilitate the recognition of an object. In addition, for many applications with very high-dimensional feature data, fusion of multiple modalities provides some computational relief by dividing feature vectors into several lower-dimensional vectors before integrating them for final decision (i.e., the divide-and-conquer principle). Consequently, it is very important to develop intelligent and sophisticated techniques for combining information from different sensors [1].

To cope with the limitations of individual biometrics, researchers have proposed using multiple biometric traits concurrently for verification. Such systems are commonly known as multimodal verification systems [183]. By using multiple biometric traits, systems gain more immunity to intruder attack. For example, it is more difficult for an impostor to impersonate another person using both audio and visual information simultaneously. Multicue biometrics also helps improve system reliability. For instance, while background noise has a detrimental effect on the performance of voice biometrics, it does not have any influence on face biometrics. Conversely, although the performance of face recognition systems greatly depends on lighting conditions, lighting does not have any effect on voice quality (see Problem 1 for a numerical example). As a result, audio and visual (AV) biometrics has attracted a great deal of attention in recent years. However, multiple biometrics should be used with caution because catastrophic fusion may occur if the biometrics are not properly combined; such fusion occurs when the performance of an ensemble of combined classifiers is worse than any of the individual classifiers.

Biometric pattern recognition systems must be computationally efficient for real-time processing, and VLSI DSP architecture has made it economically feasible to support such intelligent processing in realtime. Neural networks (NNs) are particularly suitable for such realtime sensor fusion and recognition because they can easily adapt in response to incoming data and take special characteristics of individual sensors into consideration.

This chapter proposes and evaluates a novel neural network architecture for effective, efficient fusion of signals from multiple modalities. Taking the perspective of treating the information pertaining to each sensor as a local expert, hierarchical NNs offer a very attractive architectural solution for multisensor information fusion. A hierarchical NN comprised of many local classification experts and an embedded fusion agent (i.e., gating network) will be developed for the architecture and algorithm design. In this context, the notion of mixture-of-experts (MOE) offers an instrumental tool for combining information from multiple local experts (see Figure 6.3). For effective, efficient local experts, the decision-based neural network (DBNN) is adopted as the local expert classification module. The proposed hierarchical NN can effectively incorporate the powerful expectation-maximization (EM) algorithm for adaptive training of (1) the discriminant function in the classification modules and (2) the gating parameters in the fusion network. This chapter also shows why such a hierarchical NN is not only cost-effective in terms of computation but also functionally superior in terms of recognition performance.

In addition to the fusion of data collected from multiple sensors, it is possible to fuse the scores of multiple samples from a single sensor. This chapter details a novel approach to computing the optimal weights for fusing scores, based on the score distribution of independent samples and prior knowledge of the score statistics. Evaluations of this multisample fusion technique on speaker verification, face verification, and audio and visual (voice plus face) biometric authentication are reported.

10.2   Sensor Fusion for Biometrics

Sensor fusion is an information processing technique (see [66, 125]) through which information produced by several sources can be optimally combined. The human brain is a good example of a complex, multisensor fusion system; it receives five different signals—sight, hearing, taste, smell, and touch—from five different sensors: eyes, ears, tongue, nose, and skin. Typically, it fuses signals from these sensors for decision making and motor control. The human brain also fuses signals at different levels for different purposes. For example, humans recognize objects by both seeing and touching them; humans also communicate by watching the speaker's face and listening to his or her voice at the same time. All of these phenomena suggest that the human brain is a flexible and complicated fusion system.

Research in sensor fusion can be traced back to the early 1980s [17, 348]. Sensor fusion can be applied in many ways, such as detection of the presence of an object, recognition of an object, tracking an object, and so on. This chapter focuses on sensor fusion for verification purposes. Information can be fused at two different levels: *feature* and *decision*. Decision-level fusion can be further divided into *abstract*

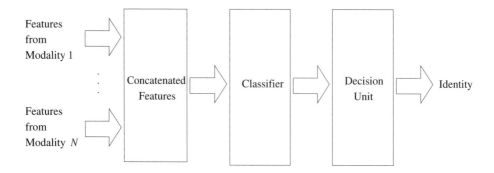

Figure 10.1. Architecture of feature-level fusion—features are concatenated before fusion takes place.

fusion and *score fusion*. These fusion techniques are discussed in the following two subsections.

10.2.1   Feature-Level Fusion

In feature-level fusion, data from different modalities are combined at the feature level before being presented to a pattern classifier [60]. One possible approach is to concatenate the feature vectors derived from different modalities [60], as illustrated in Figure 10.1. The dimensionality of the concatenated vectors, however, is sometimes too large for a reliable estimation of a classifier's parameters, a problem known as the *curse of dimensionality*. Although dimensionality reduction techniques, such as PCA or LDA, can help alleviate the problem [60, 260], these techniques rely on the condition that data from each class contain only a single cluster. Classification performance might be degraded when the data from individual classes contain multiple clusters. Moreover, systems based on feature-level fusion are not very flexible because the system needs to be retrained whenever a new sensor is added. It is also important to synchronize different sources of information in feature-level fusion, which may introduce implementation difficulty in AV fusion systems.

10.2.2   Decision-Level Fusion

Unlike feature-level fusion, decision-level fusion attempts to combine the decisions made by multiple modality-dependent classifiers (see Figure 10.2). This fusion approach solves the curse of dimensionality problem by training modality-dependent classifiers separately. Combining the outputs of the classifiers, however, is an important issue. The architecture of the classifiers can be identical but the input features are different (e.g., one uses audio data as input and the other uses video data). Alternatively, different classifiers can work on the same features and their decisions are combined. There are also systems that use a combination of these two types.

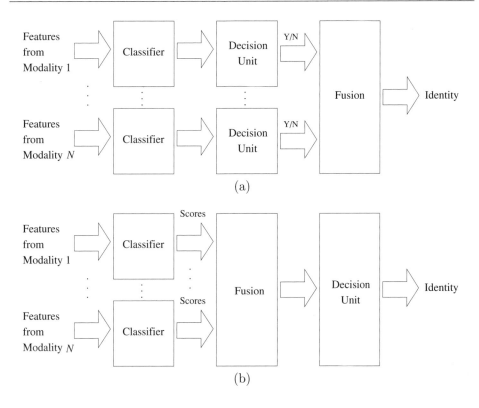

Figure 10.2. Architecture of decision-level fusion—(a) abstract fusion in which the Yes/No decisions made by the classifiers and decision units are combined; (b) score fusion in which final decisions are based on the fused scores.

The two types of decision fusions are: abstract and score. In the former, the binary decisions made by individual classifiers are combined, as shown in Figure 10.2(a); in the latter, the scores (confidence) of the classifiers are combined, as in Figure 10.2(b).

In abstract fusion, the binary decisions can be combined by majority voting or using AND and OR operators. In majority voting, the final decision is based on the number of votes made by the individual classifiers [113, 182]. However, this voting method may have difficulty making a decision when there are an even number of sensors and the decisions made by half of the classifiers do not agree with the other half.

Varshney [359] proposed using logical AND and OR operators for fusion. In the AND fusion, the final decision is not reached until all the decisions made by the classifiers agree. This type of fusion is very strict and therefore suitable only for systems that require low false acceptance. However, it has difficulty when the decisions made by different sensors are not consistent, which is a serious problem in multiclass applications. Unlike the AND fusion, the final decision in the OR fusion is made as soon as one of the classifiers makes a decision. This type of fusion is suitable only for systems that can tolerate a loose security policy (i.e., allowing high false acceptance error). The OR fusion suffers the same problem as the voting method when the decisions of individual classifiers do not agree with one other.

In score fusion, the scores of modality-specific classifiers are combined and the final score is used to make a decision (see Figure 10.2(b)). Typically, the output of modality-specific classifiers are linearly combined through a set of fusion weights [182]. The final score is obtained from

$$s = \sum_{i=1}^{K} w_i s_i, \tag{10.2.1}$$

where K is the number of modalities or experts, $\{w_i\}$ are a set of fusion weights, and $\{s_i\}$ are the scores obtained from the K modalities. This kind of fusion is also referred to as the *sum rule* [6, 182].

Scores can be interpreted as posteriori probabilities in the Bayesian framework. Assuming that scores from different modalities are statistically independent, the final score can be combined by using the *product rule* [6, 182]:

$$s = \prod_{i=1}^{K} s_i. \tag{10.2.2}$$

To account for the discriminative power and reliability of each modality, a set of weights can be introduced as follows:

$$s = \prod_{i=1}^{K} (s_i)^{w_i}. \tag{10.2.3}$$

It has been stated that the independence assumption is unrealistic in many situations. However, this challenge does hold for some applications. For example, in AV verification systems, facial and speech features are mainly independent. Therefore, fusion of audio and visual data at the score level is a possible solution to reducing verification error.

The fusion weights w_i can be nonadaptive and adaptive. Nonadaptive weights are learned from training data and kept fixed during recognition. For example, in Potamianos and Neti [284] and Sanderson and Paliwal [330], the fusion weights were estimated by minimizing the misclassification error on a held-out set; in Pigeon et al. [277], the parameters of a logistic regression model are estimated from the dispersion between the means of speakers' and impostors' scores. The nonadaptive weights, however, may not be optimal in mismatch conditions. Adaptive weights, on the other hand, are estimated from observed data during recognition—for example, according to the signal-to-noise ratio [239], degree of voicing [260], degree of mismatch between training and testing conditions [331], and amount of estimation error present in each modality [371].

Another important approach to adapting fusion weights is based on the trainable properties of neural networks. For example, in Brunelli and Falavigna [37], a person identification system based on acoustic and visual features was proposed. In particular, two classifiers based on acoustic features and three based on visual ones provide data for an integration module whose performance was evaluated. A novel technique for the integration of multiple classifiers at a hybrid rank/measurement level was introduced using HyperBF networks. This research showed that the performance of the integrated system was superior to that of the acoustic and visual subsystems.

The linear combiners described before assume that the combined scores obtained from different classes are linearly separable. In case this assumption cannot be met, the scores obtained from d experts can be considered as some d-dimensional vectors and a binary classifier (e.g., support vector machine, multi-layer perceptron, decision-tree architecture, Fisher's linear discriminant, and Bayesian classifier) can be trained from a held-out set to classify the vectors [22, 51, 96]. The experimental results showed that SVMs and Bayesian classifiers achieve about the same performance and outperform the rest of the candidate classifiers.

10.3   Hierarchical Neural Networks for Sensor Fusion

Several categories of fusion layers have been studied by Abidi and Gonzalez [1]. In the present context, the classes-in-expert hierarchical NN described in Section 6.4.2 can be naturally extended to cover the sensor-fusion network. To this end, the definition of experts needs to be properly expanded. Local experts now include not only the adaptive-trained type but also the predetermined type. For the former, the feature space (often a certain local region) represented by an expert is adaptively trained in an a posteriori fashion. The model parameters depend very much on the initial condition to which the local expert is assigned. As for the predetermined experts, each local expert has fixed model parameters because it is designated to process data in some known feature space (e.g., high- or low-frequency components). By regarding each sensor as a predetermined local expert, the classes-in-expert hierarchical NN can be made amenable to sensor-fusion applications.

More precisely, the sensor-fusion NN consists of several *classifier channels*, each of which receives an input vector from its own sensor. Its structure resembles that of the classes-in-expert NN shown in Figure 6.8. The fusion of different channels can be implemented as a (properly gated) linear combination of outputs, with the gating parameters reflecting the confidence measures of the sensor channels.

For reliable and robust biometric identification, it is critical to design fusion algorithms to effectively combine a variety of sensory inputs. The proposed hierarchical NN can effectively incorporate the divide-and-conquer hierarchical processing principle and the powerful EM algorithm for adaptive training of the classifier and gating parameters.

The EM-based fusion scheme allows quick reconfiguration of the recognition systems to cope with a diversity of audio and visual devices. Since the hierarchical NN comprises many fuzzy processing modules, the EM algorithm can serve as a basic mathematical tool for (1) training the discriminant function in the classification modules and (2) training the gating parameters in the fusion network. The EM algorithm offers two attractive attributes for intended fusion applications:

1. EM naturally accommodates model-based clustering formulation with one model corresponding to one rule used in NN.

2. EM allows the final decision to incorporate prior information. This feature could be instrumental to progressive updating of multisensor fusion in field applications.

10.3.1   PDBNNs for Multichannel Fusion

Probabilistic decision-based neural networks are amenable to multiple sensor classification. Additional sensor information can be easily incorporated to improve recognition performance. A multichannel PDBNN consists of several "classifier channels," each of which receives input vectors from either different sensors or from a portion of a higher-dimension feature vector. In this circumstance, channels are not differentiated into primary or secondary categories; therefore, "lateral" fusion of information is more appropriate. The outputs of channels are combined by proper weightings, and the weighting factor is assigned based on the *confidence* the corresponding channel has on its recognition result. Since PDBNN generates probabilistic outputs, it is natural to design the channel weightings to have probability properties. The overall configuration of a multichannel fusion network is depicted in Figure 10.3, where the score functions from two channels are combined after proper preweightings. There are two key components in a PDBNN-based fusion system: classifier module and gating network.

Classifier Module

In a PDBNN-based fusion system, each local expert needs only process information acquired by its own corresponding sensor. PDBNNs are a promising candidate for local classifiers in that they possess a modular structure: one subnet is designated to represent one object class. The decision-based learning rule is based on a minimal training principle, updating the network parameters only when necessary. PDBNNs also adopt a distributed learning principle with flexible reconfiguration capability to cope with new classes, new users, or new sensors. Moreover, it is very effective in implementing a global (i.e., interclass) supervised training scheme.

Gating Network

The score functions generated by local experts are weighted by a set of confidence indicators before they reach the top decision layer for final classification. Based on the

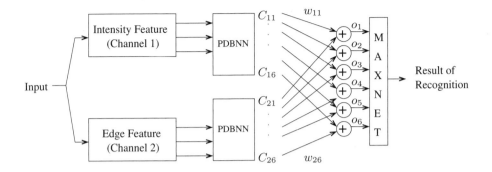

Figure 10.3. A PDBNN-based multichannel fusion network. With reference to Figure 10.4, here C_{ki} denotes the output of the i-th subnet in channel k ($=p(x|\omega_i, C_k)$) and w_{ki} ($= P(C_k|\omega_i)$) denotes the channel confidence. The final decision is based on o_i, which is the combination of the weighted outputs for subnet i; theoretically, o_i can be equated to $p(x|\omega_i)$.

input patterns, the gating network generates the confidence indicators that decide how to "fuse" the recommendations made by local experts. The gating parameters, which can be obtained via the EM algorithm, represent the confidence measures pertaining to the sensor channels. The gating network can be implemented as a softmax activation function so that a larger (respectively smaller) gating activation will mean a greater (respectively lesser) influence on the final decision.

By extending the hierarchical structure of PDBNNs, two versions of multi-channel fusion networks were developed. The first fusion scheme is called "class-dependent channel fusion" and the second is "data-dependent channel fusion." A brief description of each of these networks follows.

10.3.2   Class-Dependent Channel Fusion

In the so-called *class-dependent channel fusion* (cf. Figure 10.4(a)), the weighting factors correspond to the confidence $P(C_k|\omega_i)$ for classifier channel C_k. Here $P(C_k|\omega_i)$ represents the indicator on the confidence of channel k when the test pattern is originated from class ω_i. (By definition, $\sum_{k=1}^{K} P(C_k|\omega_i) = 1$, so it has the property of a probability function.) Suppose there are K channels in the subnet ω_i, and within each channel there are R clusters. The probability model of the PDBNN-based channel fusion network can be described as follows (see Figure 10.4(b)):

$$p(\mathbf{x}(t)|\omega_i) = \sum_{k=1}^{K} P(C_k|\omega_i)p(\mathbf{x}(t)|\omega_i, C_k), \qquad (10.3.1)$$

where $p(\mathbf{x}(t)|\omega_i, C_k)$ is the discriminant function of subnet i in channel k, and $p(\mathbf{x}(t)|\omega_i)$ is the combined discriminant function for class ω_i.

Note that $\mathbf{x}(t) = [\mathbf{x}_1^T(t), \cdots, \mathbf{x}_K^T(t)]^T$, and since $p(\mathbf{x}(t)|\omega_i, C_k)$ is conditional on C_k, only $\mathbf{x}_k(t)$ is involved in the preceding formula. More explicitly, Eq. 10.3.1 can be simplified into

$$p(\mathbf{x}(t)|\omega_i) = \sum_{k=1}^{K} P(C_k|\omega_i)p(\mathbf{x}_k(t)|\omega_i, C_k).$$

After all the parameters within the channels complete their training, the channel confidence $P(C_k|\omega_i)$ can be learned by the following steps:

1. Define $\alpha_k = P(C_k|\omega_i)$ and assign $\alpha_k = 1/K$, $\forall k = 1, \ldots, K$.

2. At step j, compute

$$h_k^{(j)}(t) = \frac{\alpha_k^{(j)} p(\mathbf{x}_k(t)|\omega_i, C_k)}{\sum_l \alpha_l^{(j)} p(\mathbf{x}_l(t)|\omega_i, C_l)} \quad \text{and} \quad \alpha_k^{(j+1)} = (1/N) \sum_{t=1}^{N} h_k^{(j)}(t). \quad (10.3.2)$$

3. Repeat Step 2 until convergence occurs.

Once the NN is trained, the fusion weights remain constant during the retrieving phase.

10.3.3   Data-Dependent Channel Fusion

A more general version of multichannel fusion is called *data-dependent channel fusion* (cf. Figure 10.5). Instead of using the likelihood of observing $\mathbf{x}(t) = [\mathbf{x}_1^T(t), \cdots, \mathbf{x}_K^T(t)]^T$ given class ω_i—that is, $p(\mathbf{x}(t)|\omega_i, C_k)$—to model the discriminant function of each cluster, the posterior probabilities of electing a class given $\mathbf{x}(t)$—that is, $p(\omega_i|\mathbf{x}(t), C_k)$—are used. For this version of a multichannel network, a new confidence $P(C_k|\mathbf{x}(t))$ is assigned, which stands for the confidence regarding channel k when the input pattern is $\mathbf{x}(t)$.

Accordingly, the probability model is also modified to become

$$P(\omega_i|\mathbf{x}(t)) = \sum_{k=1}^{K} P(C_k|\mathbf{x}(t))P(\omega_i|\mathbf{x}(t), C_k) = \sum_{k=1}^{K} P(C_k|\mathbf{x}(t))P(\omega_i|\mathbf{x}_k(t), C_k),$$

where

$$P(\omega_i|\mathbf{x}_k(t), C_k) = \frac{P(\omega_i|C_k)p(\mathbf{x}_k(t)|\omega_i, C_k)}{p(\mathbf{x}_k(t)|C_k)}$$

and the confidence $P(C_k|\mathbf{x}(t))$ can be obtained by

$$P(C_k|\mathbf{x}(t)) = \frac{P(C_k)p(\mathbf{x}(t)|C_k)}{\sum_j P(C_j)p(\mathbf{x}(t)|C_j)} = \frac{P(C_k)p(\mathbf{x}_k(t)|C_k)}{\sum_j P(C_j)p(\mathbf{x}_j(t)|C_j)},$$

where

$$p(\mathbf{x}(t)|C_k) = p(\mathbf{x}_k(t)|C_k) = \sum_i P(\omega_i|C_k)p(\mathbf{x}_k(t)|\omega_i, C_k)$$

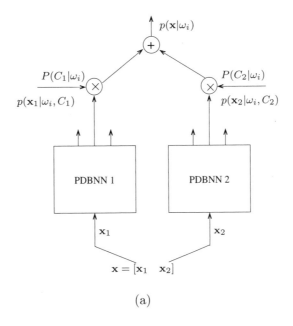

(a)

Figure 10.4. Class-dependent channel fusion. (a) The PDBNN class-dependent channel fusion scheme can be applied to applications where several sensor sources are available. Each PDBNN channel C_j receives the patterns from one sensor, and outputs from different channels (but representing the same object class) are combined by the weighting parameters $P(C_j|\omega_i)$'s. $P(C_j|\omega_i)$ is a trainable parameter; its value is fixed during the retrieval time. (b) A more detailed illustration with two clusters per channel per class. Here $p(\mathbf{x}|\omega_i, C_k)$ corresponds to C_{ki} (i.e., the output of the i-th subnet in channel k) in Figure 10.3, while $= P(C_k|\omega_i)$ corresponds to the channel confidence w_{ki} in Figure 10.3. (Figure continued on next page.)

and the channel weight $P(C_k)$ is based on the previous estimate.[1] In preparation for the next updating, the $P(C_k)$ can now be updated as follows:

$$P(C_k) = (1/N) \sum_{t=1}^{N} P(C_k|\mathbf{x}(t)). \qquad (10.3.3)$$

Unlike the class-dependent approach, the fusion weights need to be computed for each testing pattern during the retrieving phase. Notice that this data-dependent fusion scheme can be considered as the combination of PDBNN and the hierarchical mixture-of-experts (HME) [167]—compare Figures 10.4(a) and 10.5(b).

---

[1] The term $P(C_k)$ can be interpreted as the "confidence weighting" that is placed on channel k.

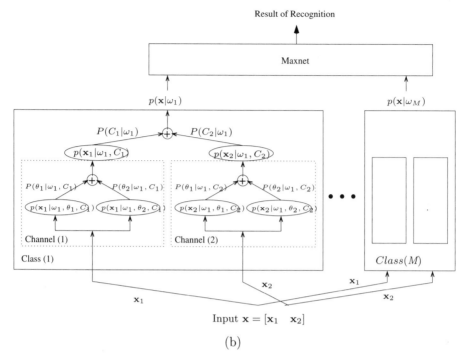

(b)

Figure 10.4. (continued)

10.4   Multisample Fusion

As discussed in the previous sections, the performance of biometric authentication systems can be improved by the fusion of multiple channels. For example, in Wark and Sridharan [371] and Jourlin et al. [169], the scores from a lip recognizer were fused with those from a speaker recognizer, and in Kittler et al. [182], a face classifier was combined with a voice classifier using a variety of combination rules. These types of systems, however, require multiple sensors, which tend to increase system cost and require extra cooperation from users (e.g., users may need to present their face and utter a sentence to support their claim). Although this requirement can be alleviated by fusing different speech features from the same utterance [330], the effectiveness of this approach relies on the degree of independence among these features.

Another approach to improving the effectiveness of biometric systems is to combine the scores of multiple input samples based on decision-fusion techniques [58, 59, 223]. Although decision fusion is mainly applied to combine the outputs of modality-dependent classifiers, it can also be applied to fuse decisions or scores from a single modality. The idea is to consider multiple samples extracted from a single modality as independent but coming from the same source. The approach is commonly referred to as multisample fusion [280].

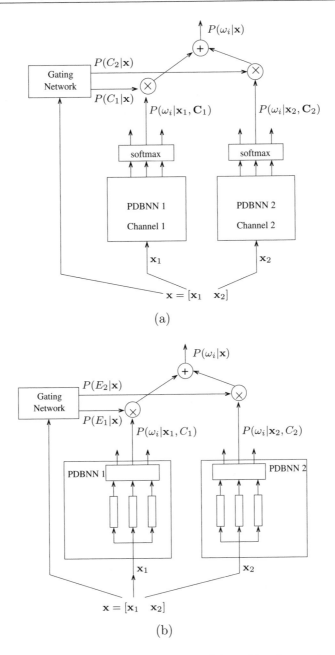

(a)

(b)

Figure 10.5. Data-dependent channel fusion. (a) In this scheme, the channel weighting parameters are functions of the input pattern $\mathbf{x}$—that is, $P(C_j|\mathbf{x})$. The outputs of PDBNNs are transformed to posterior probabilities by softmax functions. (b) This is the hybrid model of HME and PDBNN; under the global structure of the HME, the local expert network is a PDBNN classifier.

This section investigates the fusion of scores from multiple utterances to improve the performance of speaker verification from GSM-transcoded speech. The simplest way of achieving this goal is to average the scores obtained from multiple utterances, as in Poh et al. [280]. Although score averaging is a reasonable approach to combining the scores, it gives equal weight to the contribution of speech patterns from multiple utterances, which may not produce optimal fused scores. In light of this deficiency, computing the optimal fusion weights based on the score distribution of the utterances and on prior score statistics determined from enrollment data [223] was recently proposed. To further enhance the effectiveness of the proposed fusion algorithm, prior score statistics are to be adapted during verification, based on the probability that the claimant is an impostor [58]. This is called data-dependent decision fusion (DF). Because the variation of handset characteristics and the encoding/decoding process introduce substantial distortion to speech signals [397], stochastic feature transformation [226] was also applied to the feature vectors extracted from the GSM-transcoded speech before presenting them to the clean speaker model and background model.

10.4.1 Data-Dependent Decision Fusion Model

Architecture

Assume that K streams of features vectors (e.g., MFCCs [69]) can be extracted from K-independent utterances $\mathcal{U} = \{\mathcal{U}_1, \ldots, \mathcal{U}_K\}$. The observation sequence corresponding to utterance $\mathcal{U}_k$ is denoted by

$$\mathcal{O}^{(k)} = \{\mathbf{o}_t^{(k)} \in \Re^D ; t = 1, \ldots, T_k\} \qquad k = 1, \ldots, K, \tag{10.4.1}$$

where D and T_k are the dimensionality of $\mathbf{o}_t^{(k)}$ and the number of observations in $\mathcal{O}^{(k)}$, respectively. A normalized score function [215] is defined as

$$s(\mathbf{o}_t^{(k)}; \Lambda) = \log p(\mathbf{o}_t^{(k)} | \Lambda_{\omega_c}) - \log p(\mathbf{o}_t^{(k)} | \Lambda_{\omega_b}), \tag{10.4.2}$$

where $\Lambda = \{\Lambda_{\omega_c}, \Lambda_{\omega_b}\}$ contains the Gaussian mixture models (GMMs) that characterize the client speaker (ω_c) and the background speakers (ω_b), and $\log p(\mathbf{o}_t^{(k)} | \Lambda_\omega)$ is the output of GMM Λ_ω, $\omega \in \{\omega_c, \omega_b\}$, given observation $\mathbf{o}_t^{(k)}$.

The expert-in-class architecture [190] is an ideal candidate for combining normalized scores probabilistically. Specifically, frame-level fused scores are computed according to

$$s(\mathbf{o}_t^{(1)}, \ldots, \mathbf{o}_t^{(K)}; \Lambda) = s(\mathbf{O}_t; \Lambda) = \sum_{k=1}^{K} \alpha_t^{(k)} s(\mathbf{o}_t^{(k)}; \Lambda) \qquad t = 1, \ldots, T \tag{10.4.3}$$

where $\alpha_t^{(k)} \in [0, 1]$ represents the confidence (reliability) of the observation $\mathbf{o}_t^{(k)}$ and $\sum_k \alpha_t^{(k)} = 1$. The mean fused score

$$s(\mathcal{U}; \Lambda) = \frac{1}{T} \sum_{t=1}^{T} s(\mathbf{O}_t; \Lambda) \tag{10.4.4}$$

is compared against a decision threshold to make verification decisions. For notational convenience, the K utterances are assumed to contain the same number of feature vectors. If not, the tail of the longer utterances can be appended to the tail of the shorter ones to make the number of feature vectors equal. In Eq. 10.4.3, a larger (respectively smaller) fusion weight means a greater (respectively lesser) influence on the final decision. Fusion weights can be estimated using training data; alternatively, they can be determined purely from the observation data during recognition. Rather than using either training data or recognition data exclusively, a new approach in which the fusion weights depend on both training data (prior information) and recognition data is proposed.

During enrollment, the mean score of each client speaker ($\tilde{\mu}_c$) and of the background speakers ($\tilde{\mu}_b$) are determined. Then, the overall mean score

$$\tilde{\mu}_p = \frac{K_c \tilde{\mu}_c + K_b \tilde{\mu}_b}{K_c + K_b} \tag{10.4.5}$$

is used as the prior score for that client. In Eq. 10.4.5, K_c and K_b are the numbers of client speaker's utterances and background speakers' utterances, respectively. A prior variance

$$\tilde{\sigma}_p^2 = \frac{1}{K_c + K_b} \sum_{n=1}^{K_c + K_b} \left[\tilde{s}(\mathcal{O}^{(n)}; \Lambda) - \tilde{\mu}_p \right]^2 \tag{10.4.6}$$

is also to be computed, where $\tilde{s}(\mathcal{O}^{(n)}; \Lambda)$ denotes the mean score of the n-th utterance. Then, during verification, the claimant is asked to utter K utterances, and the fusion weights are computed according to

$$\alpha_t^{(k)} = \frac{\exp\{(s_t^{(k)} - \tilde{\mu}_p)^2 / 2\tilde{\sigma}_p^2\}}{\sum_{l=1}^{K} \exp\{(s_t^{(l)} - \tilde{\mu}_p)^2 / 2\tilde{\sigma}_p^2\}} \qquad k = 1, \ldots, K, \tag{10.4.7}$$

where for ease of presentation, $s_t^{(k)} \equiv s(\mathbf{o}_t^{(k)}; \Lambda)$ has been defined. Figure 10.6 depicts the architecture of the fusion model.

Theoretical Analysis

The following theoretical analysis of the fusion algorithm explains how and why it achieves better performance than the equal-weight fusion approach does.

Figure 10.7(a) illustrates the fusion weights $\alpha_t^{(1)}$ as a function of $s_t^{(1)}$ and $s_t^{(2)}$, where $s_t^{(k)} \in [-12, 12]$, $k \in \{1, 2\}$, $\tilde{\mu}_p = -2$, and $\tilde{\sigma}_p = 3.5$. A closer look at Figure 10.7(a) reveals that scores falling on the upper-right region of the dashed line L are increased by the fusion function Eq. 10.4.3. This is because in that region, for $s_t^{(1)} > s_t^{(2)}$, $\alpha_t^{(1)} \approx 1$ and $\alpha_t^{(2)} \approx 0$; moreover, for $s_t^{(1)} < s_t^{(2)}$, $\alpha_t^{(1)} \approx 0$ and $\alpha_t^{(2)} \approx 1$. Both of these conditions make Eq. 10.4.3 emphasize the larger score. Conversely, the fusion algorithm puts more emphasis on the small scores if they fall on the lower-left region of the dashed line. The effect of the fusion weights on the scores is

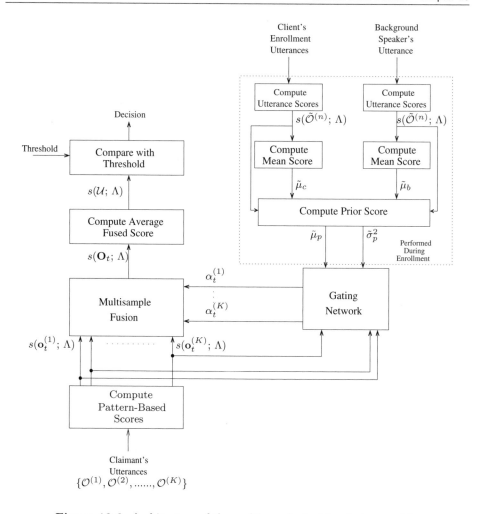

Figure 10.6. Architecture of the multisample decision fusion model.

depicted in Figure 10.7(b). Evidently, the fusion weights favor large scores if they fall on the upper-right region, whereas the fused scores are close to the small scores if they fall on the lower-left region.

The rationale behind this fusion approach is the observation that most client-speaker scores are larger than the prior score, but most of the impostor scores are smaller than the prior score. As a result, if the claimant is a client speaker, the fusion algorithm favors large scores; on the other hand, the algorithm favors small scores if the claimant is an impostor. This has the effect of reducing the overlapping area of score distribution of client speakers and impostors, thus reducing the error rate.

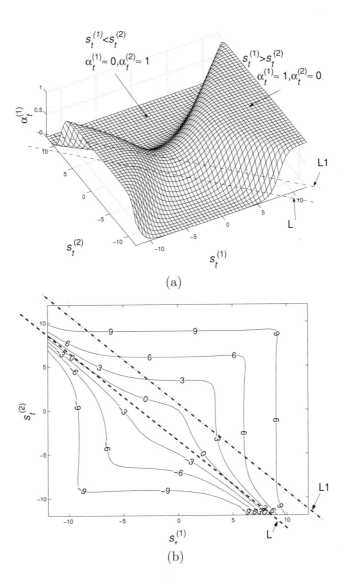

Figure 10.7. (a) Fusion weights $\alpha_t^{(1)}$ as a function of scores $s_t^{(1)}$ and $s_t^{(2)}$. (b) Contour plot of fused scores based on the fusion formula in Eq. 10.4.3 and the fusion weights in (a).

Gaussian Example

Figure 10.8 shows an example where the distributions of client speaker scores and impostor scores are assumed to be Gaussian. It is also assumed that both the client

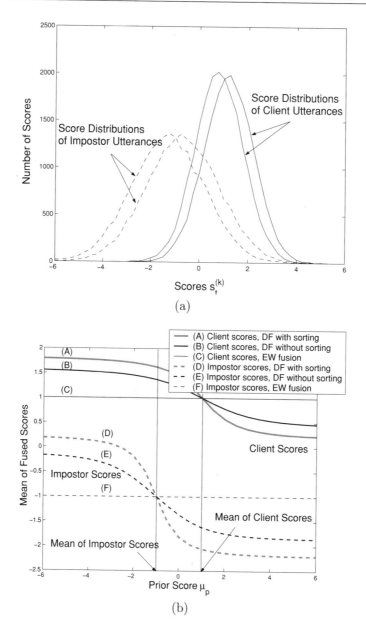

(a)

(b)

Figure 10.8. (a) Distributions of client scores and impostor scores as a result of four utterances: 2 from a client speaker and another 2 from an impostor; the means of client scores are 0.8 and 1.2, and the means of impostor scores are -1.3 and -0.7; (b) mean of fused client scores and the mean of fused impostor scores; and (c) difference between the mean of fused client scores and the mean of fused impostor scores based on EW fusion and DF with and without score sorting.

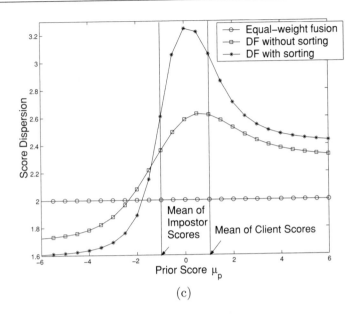

(c)

Figure 10.8. (continued)

and the impostor produce two utterances. The client speaker's mean scores for the first and second utterances are equal to 1.2 and 0.8, respectively. Likewise, the impostor's mean scores for the two utterances are equal to -1.3 and -0.7. Obviously, equal-weight (EW) fusion produces a mean speaker score of 1.0 and a mean impostor score of -1.0, resulting in a score dispersion of 2.0. These two mean scores (-1.0 and 1.0) are indicated by the two vertical lines in Figure 10.8(b). Figures 10.8(b) and 10.8(c) show that when the prior score $\tilde{\mu}_p$ is set to a value between these two means (i.e., between the vertical lines), the data-dependent fusion (DF) algorithm can produce a score dispersion larger than 2.0. As the mean of fused scores is used to make the final decision, increasing the score dispersion can decrease the speaker verification error rate.

10.4.2 Fusion of Sorted Scores

Because the fusion model described in Section 10.4.1 depends on pattern-based scores of individual utterances, the positions of scores in the score sequences can also affect the final fused scores. The scores in the score sequences can be sorted before fusion such that small scores will always be fused with large scores [59]. This is achieved by sorting half of the score sequences in ascending order and the other half in descending order. The fusion of sorted scores applies only to even numbers of utterances.

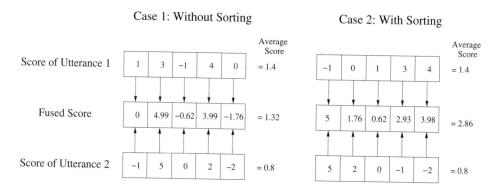

Figure 10.9. Fused scores derived from unsorted (*left*) and sorted (*right*) score sequences obtained from a client speaker. The assumption is that $\tilde{\mu}_p = 0$ and $\tilde{\sigma}_p^2 = 1$ in Eq. 10.4.7.

Theoretical Analysis

The Gaussian example can be used to extend the theoretical analysis in Section 10.4.1 to explain why the fusion of sorted scores is better than the fusion of unsorted scores.

In the analysis described in Section 10.4.1, the increase or decrease of the mean fused scores is probabilistic only because there is no guarantee that the scores of the two utterances $(s_t^{(1)}, s_t^{(2)})$ $\forall t$ fall in either the upper-right region or the lower-left region of the score space together. In many cases, some of the $(s_t^{(1)}, s_t^{(2)})$ pairs fall in the region above Line L in Figure 10.7 and others in the region below it, even though the two utterances are obtained from the same speaker. This situation is undesirable because it introduces uncertainty to the increase or decrease of the mean fused scores. This uncertainty, however, can be removed by sorting the two score sequences before fusion takes place because the scores to be fused will always lie on a straight line. For example, if the mean scores of both utterances from the client speaker are denoted as μ, and the $s_t^{(1)}$ is assumed to be sorted in ascending order and $s_t^{(2)}$ in descending order, the relationship $\mu - s_t^{(1)} \approx s_t^{(2)} - \mu$ $\forall t$ results. This is the straight line $L1$ $(s_t^{(2)} = -s_t^{(1)} + 2)$ shown in Figure 10.7, when $\mu = 1$. Evidently, Line $L1$ lies in the region where large scores are emphasized. As a result, an increase in the mean fused score can be guaranteed.

Numerical Example

A numerical example is used here to demonstrate the effect of sorting scores. Figure 10.9 shows a hypothetical situation in which scores were obtained from two client utterances. For client utterances, Eq. 10.4.7 should emphasize large scores and deemphasize small scores. However, Figure 10.9(a) illustrates the situation in which a very small score (5th score of utterance 2—i.e., -2) is fused with a relatively large score (5th score of utterance 1—i.e., 0.0). In this case, $\alpha_t^{(1)} = \frac{e^0}{e^0 + e^2} \approx 0.0$ and $\alpha_t^{(2)} = \frac{e^2}{e^0 + e^2} \approx 1.0$, which means the fifth fused score (-1.76) is dominated by the fifth score of utterance 2. This is undesirable for client utterances.

The influence of those extremely small client scores on the mean fused score can be reduced by sorting the scores of the two utterances in opposite order before fusion such that small scores will always be fused with large scores. With this arrangement, the contribution of extremely small client scores in one utterance can be compensated for by the large scores of another utterance. As a result, the mean of the fused client scores is increased.

Figure 10.9(b) shows that the mean of fused scores is increased from 1.32 to 2.86 after the scores are sorted. Likewise, if this sorting approach is applied to the scores of impostor utterances with proper prior scores μ_p (i.e., greater than the mean of impostor scores—see Figure 10.8(b)), the contribution of extremely large impostor scores in one utterance can be greatly reduced by the small scores in another utterance, which has the net effect of minimizing the mean of the fused impostor scores. Therefore, this score sorting approach can further increase the dispersion between client scores and impostor scores, as shown in Figure 10.8(c). This has the effect of lowering the error rate.

Effect on Real Speech Data

To further demonstrate the merit of sorting the scores before fusion, two client speakers (faem0 and mdac0) were selected from the handset TIMIT (HTIMIT) corpus, and the distributions of the fused speaker scores and fused impostor scores were plotted in Figure 10.10. In Eq. 10.4.7, the overall mean $\tilde{\mu}_p$ was used as the prior score. However, as the number of background speakers' utterances is much larger than that of the speaker's utterances during the training phase, the overall mean is very close to the mean score of the background speakers $\tilde{\mu}_b$—that is, $\tilde{\mu}_p \approx \tilde{\mu}_b$. Therefore, according to Section 10.4.2, the mean of fused impostor scores remains almost unchanged, but the mean of fused client scores increases significantly. Figure 10.10(a) shows that the mean of client scores increases from 0.35 to 1.08 and the mean of impostor scores decreases from -3.45 to -3.59 after sorting the score sequences. The decrease in the mean of fused impostor scores is caused by the prior score $\tilde{\mu}_p$ being greater than the mean of the unfused impostor scores—see the fourth figure in Figure 10.10(a). Thus, the dispersion between the mean client score and the mean impostor score increases from 3.80 to 4.67.

Figure 10.10(b) reveals that both the mean of client scores and the mean of impostor scores increase. This is because the means of impostor scores obtained from verification utterances are greater than the prior score $\tilde{\mu}_p$, thus resulting in the increase of the mean of fused impostor scores. However, because the increase in the mean client scores is still greater than the increase in the mean impostor scores, there is a net increase in the score dispersion. Specifically, the dispersion in Figure 10.10(b) increases from 2.14 ($= 0.24 - (-1.90)$) to 2.63 ($= 0.94 - (-1.69)$). Because

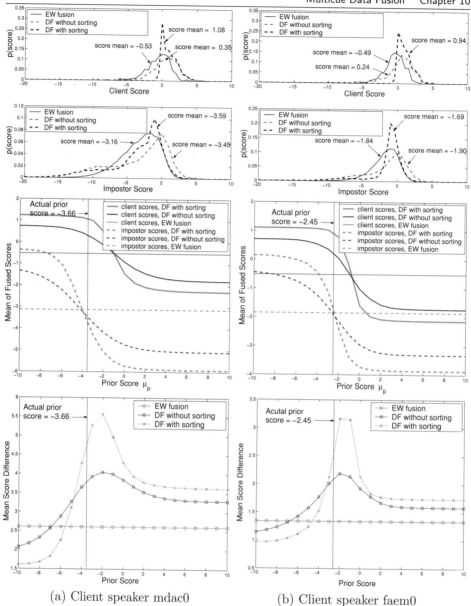

(a) Client speaker mdac0   (b) Client speaker faem0

Figure 10.10. Distribution of pattern-by-pattern client (*top*) and impostor scores (*second*), the mean of fused client and fused impostor scores (*third*), and difference between the mean of fused client and fused impostor scores (*bottom*) based on EW fusion (score averaging) and DF with and without score sorting. The means of speaker and impostor scores obtained by both fusion approaches are also shown.

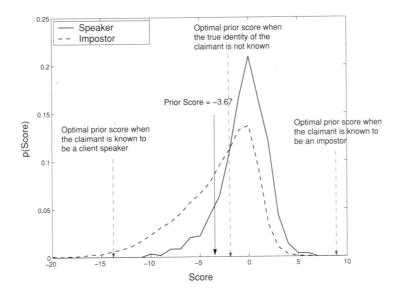

Figure 10.11. Distribution of pattern-by-pattern speaker and impostor scores corresponding to speaker mdac0. Most of the speaker scores are larger than the prior score, and only some of the impostor scores are smaller than the prior score.

verification decision is based on the mean scores, the wider the dispersion between the mean client scores and the mean impostor scores, the lower the error rate.

10.4.3   Adaptation of Prior Scores

Eq. 10.4.7 uses the overall mean $\tilde{\mu}_p$ as the prior score. However, because the number of background speakers' utterances is much larger than the number of the speaker's utterances during the training phase, the overall mean is very close to the mean score of the background speakers (i.e., $\tilde{\mu}_p \approx \tilde{\mu}_b$). This causes most of the client speaker scores to be larger than the prior score; moreover, only part of the impostor scores are smaller than the prior score. An example of this situation is illustrated in Figure 10.11 in which the distributions of the pattern-by-pattern speaker scores and impostor scores corresponding to speaker mdac0 are shown. Figures 10.7 and 10.11 show that if the claimant is a client speaker, the fusion algorithm favors large scores because most of the speaker scores in Figure 10.11 are larger than the prior score. On the other hand, the algorithm has equal preference on both small and large scores if the claimant is an impostor because the prior score is in the middle of the impostor scores' distribution.

The ultimate goal of Eq. 10.4.7 is to favor large scores when the claimant is a client speaker; on the other hand, if the claimant is an impostor, the fusion algorithm should favor small scores. In other words, the goal is to increase the separation between client speakers' scores and impostors' scores. As a result, when the claimant is a true speaker, the prior score should be smaller than all possible speaker scores so that Eq. 10.4.7 favors only larger scores; on the other hand, when the claimant is an impostor, the prior score should be larger than all possible impostor scores so that Eq. 10.4.7 favors only smaller scores. However, satisfying these two conditions simultaneously is almost impossible in practice because the true identity of the claimant is never known. Therefore, the optimal prior score should be equal to the intersection point of speaker score distribution and impostor score distribution (see Figure 10.11). At that point, the number of speaker scores smaller than the prior score plus the number of impostor scores larger than the prior score is kept to a minimum.

Adaptation Algorithm

A method to adapt the prior score during verification in order to achieve the goal mentioned before [58] follows. For each client speaker, a one-dimensional GMM score model Ω_{score} with output

$$p(\tilde{s}; \Omega_{score}) = \sum_{j=1}^{M} \pi_j p(\tilde{s}|j) = \sum_{j=1}^{M} \pi_j \mathcal{N}(\tilde{s}; \mu_j, \Sigma_j), \qquad (10.4.8)$$

$$\Omega_{score} = \{\pi_j, \mu_j, \Sigma_j\}_{j=1}^{M},$$

is trained to represent the distribution of the utterance-based background speakers' scores $\tilde{s} \equiv \frac{1}{T} \sum_{t=1}^{T} s(\mathbf{o}_t; \Lambda)$ obtained during the training phase. During verification, a claimant is asked to utter K utterances $\mathcal{U} = \{\mathcal{U}_1, \ldots, \mathcal{U}_K\}$. For the k-th utterance, the normalized likelihood that the claimant is an impostor is computed

$$\zeta = \frac{p(\tilde{s}_k; \Omega_{score})}{\max_{-\infty \leq s \leq \infty} p(s; \Omega_{score})}, \qquad (10.4.9)$$

where $\tilde{s}_k \equiv \frac{1}{T_k} \sum_{t=1}^{T_k} s(\mathbf{o}_t; \Lambda)$ is the score of the k-th utterance. The prior score is then adapted according to

$$\widehat{\mu}_p^{(k)} = \tilde{\mu}_p + f(\zeta), \qquad (10.4.10)$$

where $\widehat{\mu}_p^{(k)}$ is the adapted prior score and $f(\zeta)$ is a monotonic increasing function controlling the amount of adaption. Then, $\tilde{\mu}_p$ in Eq. 10.4.7 is replaced by $\widehat{\mu}_p^{(k)}$ to compute the fusion weights. The reason for using a monotonic function for $f(\zeta)$ is that if the normalized likelihood of a claimant is large, he or she is likely to be an impostor. As a result, the prior score should be increased by a large amount to deemphasize the high scores of this claimant. With the large scores being deemphasized, the claimant's mean fused score becomes smaller, thus increasing the chance of rejecting this potential impostor.

Notice that Eq. 10.4.10 increases the prior score rather than decreasing it because the optimal prior score is always greater than the unadapted prior score $\tilde{\mu}_p^{(k)}$. A

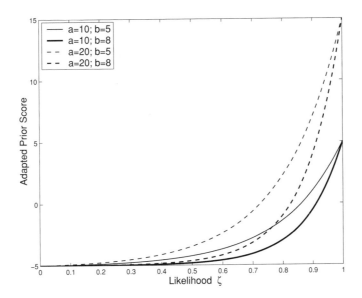

Figure 10.12. The influence of varying a and b of Eq. 10.4.10 on the adapted prior scores $\widehat{\mu}_p^{(k)}$, assuming that $\tilde{\mu}_p = -5$.

monotonic function of the form is used,

$$f(\zeta) = \tilde{\mu}_p + \frac{a(e^{b\zeta} - 1)}{e^b - 1}, \tag{10.4.11}$$

where a represents the maximum amount of adaptation and b controls the rate of increase of $f(\zeta)$ with respect to ζ. Figure 10.12 shows the effect of varying a and b on the adapted prior scores, where the larger the normalized likelihood, the greater the degree of adaption. This is reasonable because a large normalized likelihood means the claimant is likely to be an impostor; as a result, the prior score should be increased by a greater amount to deemphasize the large scores of this claimant. With the large scores being deemphasized, the claimant's mean fused score becomes smaller, thus increasing the chance of rejecting this potential impostor.

Effect of Score Adaptation on Score Distributions

To demonstrate the effect of the proposed prior score adaptation on fused score distributions, a client speaker (mdac0) from GSM-transcoded HTIMIT was arbitrarily selected and the distributions of the fused speaker scores and fused impostor scores were plotted in Figure 10.13, using EW fusion ($\alpha_t^{(1)} = \alpha_t^{(2)} = 0.5 \; \forall t$), DF without prior score adaptation (Eq. 10.4.7), and DF with prior score adaptation (Eq. 10.4.10). Evidently, the upper part of Figure 10.13 shows that the number of large client speaker scores is greater in data-dependent fusion, and the lower part of

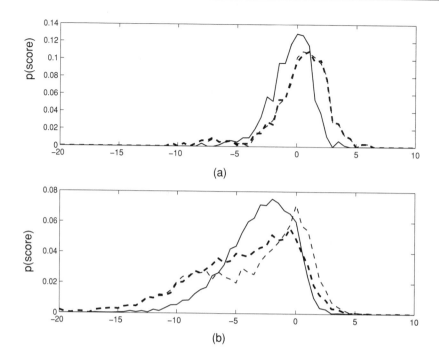

Figure 10.13. Distribution of pattern-by-pattern (a) speaker scores and (b) impostor scores based on EW fusion (score averaging) and DF. *Solid*: equal-weight fusion. *Thin dotted*: data-dependent fusion without prior score adaptation. *Thick dotted*: Data-dependent fusion with prior score adaptation.

Figure 10.13 shows that there are more small impostor scores in DF than in EW fusion. The figure also shows that data-dependent fusion with prior score adaptation outperforms the other two fusion approaches.

Further evidence demonstrating the advantage of prior score adaptation can be found in Table 10.1. In particular, Table 10.1 shows that the dispersion between the mean client score and the mean impostor score increases from 2.63 (= −0.53 − (−3.16)) to 4.04 (= 0.25 − (−3.79)) without prior score adaptation and to 4.61 (= 0.22 − (−4.39)) with prior score adaptation. Because verification decisions are based on the mean scores, the wider the dispersion between the mean client scores and the mean impostor scores, the lower the error rate. Table 10.1 also shows that with prior score adaptation, both the mean speaker score and mean impostor score are reduced. The reason for a reduction in the mean speaker score is that there is a small increase in the prior score when the claimant is a true speaker, which increases the number of scores that are smaller than the prior score. In other words, more small speaker scores are emphasized by Eq. 10.4.7, which lowers the fused speaker scores as well as the mean speaker score.

Table 10.1. The mean fused speaker and fused impostor scores of speaker mdac0. *Note:* Scores were obtained by different fusion approaches. PS stands for prior score.

	Equal Weight	Fusion without PS Adaptation	Fusion with PS Adaptation
Prior score	N/A	−3.67	Vary for different utterances
Mean speaker score	−0.53	0.25	0.22
Mean impostor score	−3.16	−3.79	−4.39
Score dispersion	2.63	4.04	4.61

10.4.4 Experiments and Results

The multisample fusion techniques described before were applied to a speaker verification task. This section details the experimental setup and discusses the results obtained.

Experiments on Multisample Fusion and Score Sorting

Data Sets and Feature Extraction. The HTIMIT corpus (see Reynolds [304]) was used in the experiments. Handset TIMIT was obtained by playing a subset of the TIMIT corpus through nine phone handsets and a Sennheizer head-mounted microphone. A GSM speech coder [91] transcoded the HTIMIT corpus and the data-dependent fusion techniques were applied to the resynthesized coded speech. In the sequel, the two corpora are denoted as HTIMIT and GSM-HTIMIT.

Speakers in HTIMIT were divided into two sets: speaker set and impostor set, as shown in Table 10.2. Similar to the speakers in HTIMIT, speakers in GSM-HTIMIT were also divided into two sets; the speaker identities (arranged alphabetically) of these sets are identical to those in HTIMIT. Twelve mel-frequency cepstrum coefficients (MFCCs) [69] were extracted from the utterances every 14ms using a 28ms Hamming window.

Enrollment Procedures. During enrollment, the SA and SX utterances from handset "senh" of the uncoded HTIMIT were used to create a 32-center GMM for each speaker. A 64-center universal background GMM [306] was also created based on the speech of 100 client speakers recorded from handset "senh." The background model will be shared among all client speakers in subsequent verification sessions.

Besides the speaker models and the background model, a two-center GMM (Λ_X in Section 9.4.2) was created using the uncoded HTIMIT utterances (from handset senh) of 10 speakers. A set of transformation parameters (b_i's in Section 9.4.1) were then estimated using this model and the handset- and codec-distorted speech (see Section 9.4.2).

Table 10.2. Speaker identities in speaker, impostor, and pseudo-impostor sets I and II

Set	Speaker Name
Speaker set (50 male, 50 female)	fadg0, faem0, ... , fdxw0; mabw0, majc0, ... , mfgk0; mjls0, mjma0, mjmd0, mjmm0, mpdf0
Impostor set (25 male, 25 female)	feac0, fear0, ... , fjem0; mfxv0, mgaw0, ... , mjlg1
Pseudo-impostor set I (10 male, 10 female)	fjen0, fjhk0, ... , fjrp1; mjpm1, mjrh0, ... , mpgr1
Pseudo-impostor set II (50 male, 50 female)	fjen0, fjhk0, ... , fmbg0; mjpm1, mjrh0, ... , msrr0

Verification Procedures.   For verification, the GSM-transcoded speech from all 10 handsets in HTIMIT was used. As a result, there were handset and coder mismatches between the speaker models and verification utterances. Stochastic feature transformation with handset identification [226, 227, 352] was used to compensate for the mismatches.

The SI sentences in the corpora were used for verification, and each claimant was asked to utter two sentences during a verification session (i.e., $K = 2$ in Eq. 10.4.1). To obtain the scores of the claimant's utterances, the utterances were first fed to a handset selector [226] to determine the set of feature transformation parameters to be used. The features were transformed and then fed to the claimed speaker model and the background model to obtain the normalized scores (see Eq. 10.4.2). Next, the data-dependent fusion algorithms were applied to fuse two independent streams of scores. Since different utterances contain different numbers of feature vectors, two utterances must have an identical number of feature vectors (length) before fusion takes place. In this work, this length was calculated according to

$$L = \lfloor \frac{L_1 + L_2}{2} \rfloor, \tag{10.4.12}$$

where L is a positive integer, and L_1 and L_2 represent the length of the first and second utterance, respectively. Figure 10.14 illustrates how the length equalization is performed; there are 105 and 138 feature vectors in the first and second utterance, respectively. According to Eq. 10.4.12, the equal length is 121 $(= \lfloor (105 + 138)/2 \rfloor)$. After finding the equal length, the remaining scores in the longer utterance (Utterance 2 in this case) are appended to the end of the shorter utterance (Utterance 1 in this case). In this example, one extra score in Utterance 2 is discarded to make the length of the two utterances equal.

To compare with EW fusion proposed by [280], the utterances' scores were also fused using equal fusion weights (i.e., $\alpha_t^{(1)} = \alpha_t^{(2)} = 0.5 \ \forall t = 1, \ldots, L$).

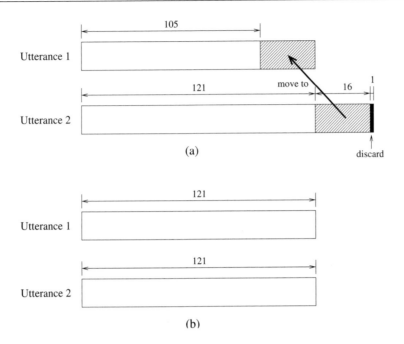

Figure 10.14. Procedure for making equal-length utterances: (a) before and (b) after length equalization. The tail of the longer utterance is appended to the tail of the shorter one.

Results and Discussion.   Figure 10.15 depicts the detection error tradeoff (DET) curves based on 100 client speakers and 50 impostors using utterances from Handset cb1 for verification. Figure 10.15 clearly shows that with feature transformation, data-dependent fusion can reduce error rates significantly, and sorting the scores before fusion can reduce the error rate even further. However, without feature transformation, the performance of DF with score sorting is not significantly better than that of EW fusion. This is caused by the mismatch between the prior scores $\tilde{\mu}_p$'s in Eq. 10.4.7 and the scores of the distorted features. Therefore, it is very important to use feature transformation to reduce the mismatch between the enrollment data and verification data.

Table 10.3 shows the speaker detection performance of 100 client speakers and 50 impostors for the EW fusion approach and the proposed fusion approach with and without sorting the score sequences. Table 10.3 also clearly shows that the proposed fusion approach outperforms the equal-weight fusion. In particular, after sorting the score sequences, the equal error rates are further reduced. Compared to the unsorted cases, an 11% error reduction has been achieved.

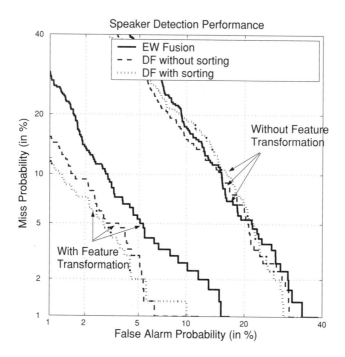

Figure 10.15. DET curves for EW fusion (score averaging) and DF with and without score sorting. The curves were obtained by using the utterances of Handset cb1 as verification speech.

Table 10.3. EERs achieved by different fusion approaches. *Note*: The utterances were obtained from 10 different handsets. Each figure is based on an average of 100 speakers, each impersonated by 50 impostors.

Fusion Method	Equal Error Rate (%)										
	cb1	cb2	cb3	cb4	el1	el2	el3	el4	pt1	senh	Ave.
No fusion (single utterance)	6.31	6.36	19.96	15.35	5.89	10.83	11.79	8.18	9.38	4.90	9.90
EW fusion	5.11	4.33	19.15	12.89	4.42	8.31	9.96	6.29	7.57	2.99	8.10
DF without sorting	4.01	3.27	15.92	10.55	3.04	6.51	8.67	4.75	7.51	2.32	6.67
DF with sorting	3.60	2.86	15.30	9.91	3.49	4.65	6.81	4.02	6.59	1.99	5.92

Experiments on Score Adaptation

Similar to the preceding experiments, HTIMIT and GSM-transcoded HTIMIT were used in the experiments. Speakers in HTIMIT were divided into four sets, as shown in Table 10.2. The pseudo-impostor sets in Table 10.2 were used to train the impostor score model (the one-dimensional GMM mentioned in Section 10.4.3) that characterizes the pseudo-impostor scores. The output of this model was used to adapt the prior score, as detailed in Section 10.4.3.

Two sets of pseudo-impostor scores were collected. For the first set, the SA and SX utterances of all speakers in the speaker set were fed to the background model and each of the speaker models to obtain the pseudo-impostor scores corresponding to the enrollment data. For the second set, the utterances of all speakers in the pseudo-impostor sets were fed to the background model and each of the speaker models to obtain the pseudo-impostor scores corresponding to unseen impostor data. These pseudo-impostor scores were averaged for each utterance. The resulting utterance-based scores were used to create a two-center one-dimensional GMM pseudo-impostor score model (Ω_{score} in Section 10.4.3).

Optimizing the Adaptation Parameters. In the experiments, the parameters a and b in Eq. 10.4.10 were set to 10 and 5, respectively. These values were empirically found to obtain reasonably good results; that is, parameters a and b would probably not be too large. If a is too large (e.g., $a = 20$ in Figure 10.12), Eq. 10.4.10 leads to a large adapted prior score $\widehat{\mu}_p^{(k)}$ even for utterances with a low normalized likelihood ζ. As a result, almost all of the claimant scores will become smaller than the prior score, which is undesirable. Moreover, a large a will result in poor verification performance when there is a lot of mismatch between the pseudo-impostor score model and the verification scores because, in a severe mismatch situation, Eq. 10.4.9 will produce an incorrect normalized likelihood. This leads to incorrect adaptation of the prior score, which in turn results in incorrect fusion weights. On the other hand, if b is too large, there will be a sharp bend in the adaptation curves, as illustrated in Figure 10.12. As a result, the prior score is adapted only when the verification utterances give a high likelihood in Eq. 10.4.9. The best range for parameter a is 5 to 20 and for parameter b, it is 5 to 8.

Results and Discussion. Figure 10.16 depicts the speaker detection performance based on 100 speakers and 50 impostors for the equal-weight fusion (score averaging) approach and the proposed fusion approach. Figure 10.16 clearly shows that with feature transformation, DF is able to reduce error rates significantly. In particular, with feature transformation, the EER achieved by data-dependent fusion with prior score adaptation is 4.01%. When compared to EW fusion, which achieves an EER of 5.11%, a relative error reduction of 22% was obtained.

In Figure 10.16, the speakers in the speaker-set were used to train the pseudo-impostor score model. The distribution of pseudo-impostor scores obtained from unseen data are of interest and are illustrated in Figure 10.17. By comparing Figure 10.17(a) with Figure 10.17(c), it is obvious that there is a mismatch between the pseudo-impostor score distribution and impostor score distribution. In particular, there are more small scores in Figure 10.17(a) than in Figure 10.17(c). This mismatch affects the adaptation of prior scores because the likelihood ζ in Eq. 10.4.9 is no longer accurate. This is evident in Table 10.4 where the EERs using different numbers of pseudo-impostors with speech extracting from different corpora

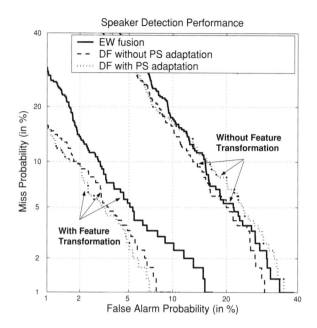

Figure 10.16. Speaker detection performance for EW fusion (score averaging) and DF.

are shown. In particular, the last column of Table 10.4 clearly shows that the improvement after adapting the prior scores is just 3.14% (from 4.14% to 4.01%).

In Table 10.4, the values inside the parentheses indicate the EERs obtained from different fusion approaches with score sorting. Each figure is based on the average of 100 speakers, each impersonated by 50 impostors. The abbreviations in the first row of the table denote the data used for creating the pseudo-impostor score model, as follows:

- PI-20 GSM: Pseudo-impostor set containing 20 speakers, with GSM-transcoded speech from Handset cb1

- PI-100 GSM: Pseudo-impostor set containing 100 speakers, with GSM-transcoded speech from Handset cb1

- PI-100 HTIMIT: Pseudo-impostor set containing 100 speakers, with HTIMIT speech from Handset senh

- SS-100 HTIMIT: Speaker set containing 100 speakers, with HTIMIT speech from Handset senh

To create a more stable and reliable pseudo-impostor score model, another set of speakers (the pseudo-impostor set) was used to train the pseudo-impostor score

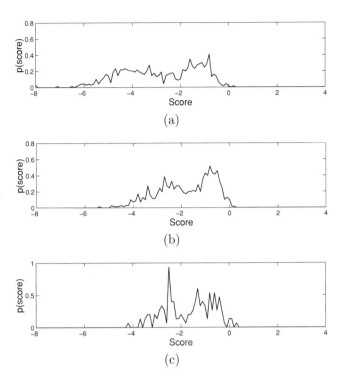

Figure 10.17. Distribution of pseudo-impostor scores for speaker fadg0 obtained by using (a) HTIMIT speech in the speaker set and (b) GSM-transcoded speech in the pseudo-impostor set. (c) Distribution of impostor scores for speaker fadg0 during verification using GSM-transcoded speech.

model. In addition, GSM-transcoded speech, instead of HTIMIT speech, was used to train the model in an attempt to create a verification environment as close to the real one as possible. The GSM-transcoded speech was transformed before calculating the scores. Figures 10.17(b) and 10.17(c) show that the score distributions are much closer after using the transformed features. The improvement after adapting the prior score is 17% (from 3.52% to 2.92%), as demonstrated in the second column of Table 10.4.

Table 10.4 shows that even for a pseudo-impostor set containing as few as 20 speakers, a lower EER (3.01%) can still be obtained by using the GSM-transcoded speech to train the pseudo-impostor score model. System performance can be further improved from 3.01% to 2.55% by applying the prior score adaptation together with score sorting.

By using the GSM-transcoded speech, the real verification environment can be modeled so that parameter a can be increased to obtain a much lower EER (from

Table 10.4. The EERs Obtained from Different Fusion Approaches. *Note:* Values inside the parentheses are the EERs obtained by using the score sorting approach.

	PI-20 GSM	PI-100 GSM	PI-100 HTIMIT	SS-100 HTIMIT
DF with PS adaptation ($a = 20$)	2.80% (2.78%)	2.57% (2.71%)	4.22% (4.23%)	4.05% (3.69%)
DF with PS adaptation ($a = 10$)	3.01% (2.55%)	2.92% (2.25%)	3.68% (3.51%)	4.01% (3.60%)
DF without PS adaptation	3.61% (2.81%)	3.52% (2.67%)	3.92% (3.80%)	4.14% (3.60%)
EW fusion	5.11%	5.11%	5.11%	5.11%
No fusion (single utterance)	6.31%	6.31%	6.31%	6.31%

3.01% to 2.80% when using 20 pseudo-impostors and from 2.92% to 2.57% when using 100 pseudo-impostors). However, the error rate increases from 3.68% to 4.22% and from 4.01% to 4.05% when there is a mismatch between the pseudo-impostor score distribution and the pseudo-impostor score model. As the fusion weights are an exponential function of the distance between the individual scores and the prior score, Eq. 10.4.7 puts more emphasis on small scores if the prior score increases. Therefore, if the claimant is an impostor, the mean impostor score becomes smaller. On the other hand, if the prior score is wrongly adapted, the mean speaker score also becomes smaller because there are more scores smaller than the prior score. Figure 10.12 shows that for the same likelihood, the increase in prior scores is larger when parameter a is increased from 10 to 20. In other words, even if inaccurate likelihood is obtained, its influence on system performance is small when parameter a is small. Therefore, if a stable and reliable pseudo-impostor score distribution is obtained, parameter a can be increased to further improve verification performance. This step is taken because if an utterance gives high likelihood in Eq. 10.4.9, the claimant is more likely to be an impostor; Eq. 10.4.7 can then be made to emphasize only small scores. On the other hand, when an utterance gives low likelihood in Eq. 10.4.9, the prior score should be increased slightly so that it is close to the optimal prior score (see Figure 10.11), because impostors' utterances can also give low likelihood. The parameter a should be kept small so that system performance is not degraded when a stable and reliable pseudo-impostor score model cannot be obtained.

Large-Scale Evaluations

Data Sets and Feature Extraction. To further demonstrate the capability of DF under practical situations, evaluations were performed using cellular phone speech:

the 2001 NIST speaker recognition evaluation set [151], which consists of 174 target speakers (clients), 74 male and 100 female. Another set of 60 speakers is also available for development. Each target speaker has 2 minutes of speech for training, and the duration of each test speech varies from 15 to 45 seconds. The set provides a total of 20,380 gender-matched verification trials. The ratio between target and impostor trials is roughly 1:10.

Twelve MFCCs and their first-order derivative were extracted from the utterances every 14ms using a Hamming window of 28ms. These coefficients were concatenated to form a 24-dimensional feature vector. Cepstral mean subtraction (CMS) [107] was applied to remove linear channel effects.

Enrollment Procedure. A 1024-component universal GMM background model [306] was created based on the speech of all 174 client speakers. A speaker-dependent GMM was then created for each client speaker by adapting the universal background model using maximum a posteriori (MAP) adaptation [306]. Finally, all 60 speakers in the development set were used to create a two-center one-dimensional GMM pseudo-impostor score model (Ω_{score} in Section 10.4.3).

Verification Procedures. For each verification session, a test sentence was first fed to the claimed speaker model and the universal background model to obtain a sequence of normalized scores. The average of these scores was then used for decision making. This traditional method (referred to as the EW approach) is the baseline. In addition to the equal-weight approach, the proposed fusion algorithm was applied to compute the weight of each score. Because only one test utterance exists for each verification session, the utterance was split into two segments and then fusion was performed. In this case, the length was calculated according to

$$ T = \lfloor \frac{L}{2} \rfloor, \tag{10.4.13} $$

where T is a positive integer and L represents the length of the test utterance.

Performance Measure. In addition to detection error tradeoff (DET) curves and equal error rates, another performance measure for the NIST evaluations is the detection cost function (DCF), which is defined as follows:

$$ \text{DCF} = C_{FA}Pr(FA|I)Pr(I) + C_{FR}Pr(FR|T)Pr(T), \tag{10.4.14} $$

where $Pr(I)$ and $Pr(T)$ are the prior probability of impostor and target speaker, respectively, and where C_{FA} and C_{FR} are the specific cost factors for false alarm and false rejection, respectively. In this work, $Pr(I) = 0.99$, $Pr(T) = 0.01$, $C_{FA} = 1$ and $C_{FR} = 10$ were set as recommended by the NIST.

Optimizing the Adaptation Parameters In the HTIMIT experiments, the parameters a and b in Eq. 10.4.10 were fixed for each speaker. However, in this experiment, a full search method and a set of training data were used to determine the best value of a and b for each client speaker. By feeding a set of pseudo-impostors into a client model, a set of mean fused pseudo-impostor scores were obtained. All 60 speakers in the development set of NIST 2001 were used as the pseudo-impostors. With different values of a and b, different sets of mean fused pseudo-impostor scores were obtained. More precisely, a particular set of a and b results in 60 mean pseudo-impostor scores. Similarly, by feeding the client speech into his or her own model and varying a and b, different mean fused client scores are obtained.

In this experiment, because there was only one utterance for training a client model, only one mean fused client score was obtained for a particular value of a and b. The value of a and b were chosen to

1. Minimize the number of mean fused pseudo-impostor scores that are greater than the mean fused client scores.

2. Maximize the difference between the mean fused client score and the largest mean fused pseudo-impostor scores provided that the mean fused client score is larger than the largest mean fused pseudo-impostor scores.

The search range of a is 0 to 10 and b is 3 to 8. The HTIMIT experiments demonstrated that a cannot be too large and b cannot be too small; otherwise, a large mismatch between the pseudo-impostor score model and the verification scores may result in poorer performance.

Results and Discussion. Figure 10.18 depicts the speaker detection performance for the DF approach with different values of a and b. According to Figure 10.12, when a and b are fixed to 10 and 3, respectively, the adapted prior scores are very large, even for small normalized likelihood ζ. This large degree of adaptation of prior scores leads to a reduction in the mean fused client scores. As a result, the miss probability (FRR) becomes larger than without adaptation when the false alarm probability is higher than 13%, as illustrated in Figure 10.18. On the other hand, when a is fixed to 10 and b is fixed to 8, there is a sharp bend in the adaptation curves (see Figure 10.12). Thus, the prior score is adapted only if the verification utterances give a high likelihood in Eq. 10.4.9, and the performance is just slightly better than without prior scores adaptation. Figure 10.18 shows that the best performance is achieved when the full search method is used to determine the value of a and b.

Figure 10.19 depicts the speaker detection performance of the EW fusion approach and the proposed fusion approach (with the parameters a and b obtained by full search). The figure clearly shows that data-dependent fusion reduces error rates. The EER achieved by DF with prior score adaptation is 10.50%. When compared to EW fusion (which achieves an EER of 12.05%), a relative error reduction of 13% was obtained. Data-dependent fusion with prior score adaptation achieves a 4% improvement as compared to the baseline in terms of minimum DCF.

Nevertheless, DF with prior score adaptation has a disadvantage. Figure 10.19 shows that when the false alarm probability is higher than 20%, the performance of data-dependent fusion with prior score adaptation degrades. This phenomenon, which is caused by the large degree of adaptation of the prior scores even for client

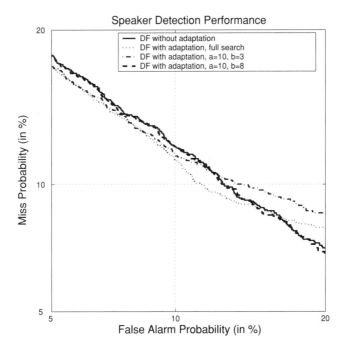

Figure 10.18. Speaker detection performance for DF with and without prior score adaptation.

speakers, decreases the mean fused client scores, resulting in the rejection of more client speakers. However, the proposed algorithm is much preferred for those systems that require low false alarm probability. At low false alarm probability, the proposed fusion algorithm can achieve a lower miss probability as compared to the EW fusion approach.

Figure 10.19 shows that splitting an utterance into two segments and fusing their pattern-based scores can achieve a performance that is better than the result of considering the utterance as only a single segment. Table 10.5 depicts the EERs and minimum detection cost obtained by splitting the verification utterances into different numbers of segments followed by the fusion algorithms. Splitting a verification utterance into two to five segments can achieve a performance better than simply taking the average of all scores (i.e., the EW approach). A much lower DCF can also be obtained by splitting the utterances into three segments—0.0442 for DF with prior score adaptation and 0.0445 for DF without prior score adaptation. These represent a 6.36% and 5.72% reduction in DCF with respect to the baseline.

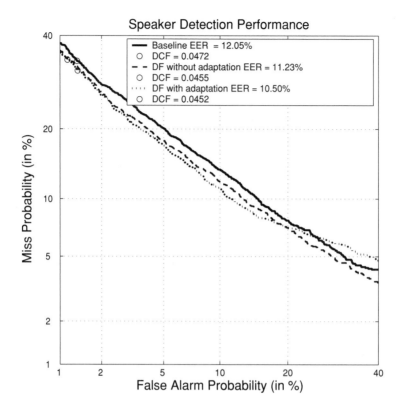

Figure 10.19. Speaker detection performance for baseline (score averaging) and DF. Adaptation parameters a and b were obtained by the full search method.

10.5   Audio and Visual Biometric Authentication

Audio and visual (AV) biometrics has long been an active area of research, primarily because of the promise it can bring to practical applications. For example, voice biometrics can suffer severe performance degradation under a noisy acoustic environment, but facial images are unaffected. Conversely, facial image quality can be severely affected in poor lighting conditions, but lighting has no effect on voice quality. These two independent and complementary information sources are ideal candidates for enhancing biometric system reliability. With the recent introduction of third-generation mobile services, the use of both voice and facial images for biometric authentication has become practical and economically viable. This section explains how the multisample fusion techniques described in Section 10.4 can be applied to AV biometric authentication.

Table 10.5. The EERs and minimum detection cost obtained by splitting the verification utterances into different numbers of segments. K stands for the number of segments define in Eq. 10.4.3.

	DF without PS Adaptation		DF with PS Adaptation	
No. of segments	EER	DCF	EER	DCF
$K = 1$ (Equal Weight)	12.05%	0.0472	12.05%	0.0472
$K = 2$	11.23%	0.0455	10.50%	0.0452
$K = 3$	10.89%	0.0445	10.50%	0.0442
$K = 4$	10.96%	0.0449	10.56%	0.0446
$K = 5$	11.10%	0.0459	10.55%	0.0450

10.5.1 AV Feature Extraction

AV Data Sets

The XM2VTSDB corpus of Luettin and Maitre [219] and Messer et al. [242] was used in this evaluation. Developed by the University of Surrey for biometric research, the corpus consists of the audio and video recordings of 295 subjects taken over a period of four months. Each subject in the corpus participated in four recording sessions, each with two utterances and two video shots. Different data sets are available, including front view images, rotating head video shots, three-dimensional VRML models, and synchronized audio and video shots of subjects. The audio recordings consist of continuous spoken digits: "zero one two three four five size seven eight nine" and "five zero six nine two eight one three seven four."

Data sets DVD003a and DVD003b, which contain the audio and video recordings of 200 client subjects and 95 impostor subjects, were used. Configuration II was adopted as specified in Luettin and Maitre [219] in the evaluation. More precisely, the database was divided into 200 clients, 70 impostors (part of the 95 impostors in DVD003b) for testing, and 25 pseudo-impostors (the remaining impostors in DVD003b) for finding decision thresholds or other system parameters. For each client, the first two sessions were used for training, and the last session was used for testing. Each client was impersonated by the 70 impostors using the audio and video data of the four sessions.

Preprocessing of Audio Files

Because the original audio files were captured in a quiet, controlled environment using a high-quality microphone, the equal error rate using the audio data alone is very low (about 0.7%); as a result, performing audio–visual fusion was unnecessary. Coder distortion and factory noise were introduced to the sound files in an attempt to simulate a more realistic acoustic environment.

The audio files in the corpus were down-sampled from 32kHz to 8kHz. Factory noise ("factory1.wav" of the NOISE92 database [142]) recorded near plate-cutting and electrical welding equipment was added to the down-sampled files at a signal-to-noise ratio of 4dB. The noisy PCM files were transcoded (encoded and then decoded) by a GSM codec [91]. Twelve MFCCs and their time derivative (delta MFCCs) were extracted from the noisy, transcoded files using a 28ms Hamming window at a rate of 71Hz.

The training sessions of 200 client speakers in the speaker set were used to create a 128-center background model. The background model was then adapted to speaker models using MAP adaptation [306]. As defined in Configuration II of XM2VTSDB, two sessions (i.e., four utterances) per speaker were used for model training. Because there is no handset variation in the audio data of XM2VTSDB, handset compensation techniques, such as SFT (Section 9.4.2) and HS-norm (Section 9.6.2), were not necessary. However, cepstral mean substraction (CMS) was performed on all MFCCs before they were used for training, testing, and evaluation because CMS is a standard technique in all speech and speaker recognition systems.

Preprocessing of Video Files

Similar to audio files, the quality of video files in the corpus was also very good, making AV fusion unnecessary (as face verification on the original video data already approaches 0% EER). As a result, distortion was introduced to the images of the video sequences using PhotoShop Version 7.0 as follows. First, each of the AVI files in the corpus was converted into a sequence of high-quality JPEG files with 720×576 pixels. Second, the frame rate was reduced to one frame per second; for each frame, the JPEG images was down-sampled to 176×144 pixels. Third, the images were blurred by setting the "Gaussian Blur" of PhotoShop to 1.0. Finally, Gaussian noise was added to the image by setting the "Gaussian Noise" of PhotoShop to 1.5. The noise-added image sequences were input to Identix's Face Verification SDK [148] to locate the head and compute the scores, which have a range of 0 to 10. The higher the score, the more likely the claimant is genuine.

10.5.2   AV Multisample Fusion

One utterance and one video shot from the claimant were obtained in a verification session. Then, the utterance and the video shot were divided into two equal-length subutterances and two equal-length subvideo shots. Feeding these subutterances and subvideo shots to the speaker verification system and the face verification system (FaceIT [148]) gives two streams of audio scores and two streams of visual scores. Multisample fusion was applied to the two audio score streams and also to the two visual score streams independently to obtain the mean of the fused audio scores—$s(\mathcal{U}; \Lambda_a)$—and the mean of the fused visual scores—$s(\mathcal{V}; \Lambda_v)$—where $\mathcal{U}$ and $\mathcal{V}$, respectively, represent utterances and video shots. These fused scores can be computed easily by applying Eqs. 10.4.2 through 10.4.7 to the audio and visual scores independently. The AV score $s(\mathcal{U}, \mathcal{V}; \Lambda_a, \Lambda_v)$ is then obtained by linearly combining the mean fused audio score and the mean fused visual score, as follows:

$$s(\mathcal{U}, \mathcal{V}; \Lambda_a, \Lambda_v) = \beta s(\mathcal{U}; \Lambda_a) + (1 - \beta)s(\mathcal{V}; \Lambda_v), \qquad (10.5.1)$$

where β is a combination weight that can be computed using training data or made dependent on the quality of audio or visual data.

Multisample Fusion of Audio Scores

For each verification trial, the sample-by-sample utterance scores (Eq. 10.4.2) obtained from a claimant were divided into two portions. Then, equal-weight and zero-sum fusion techniques were applied to fuse the scores of the two portions. The speaker-dependent fusion parameters, including the prior scores and prior variances, were obtained by feeding the utterances of the 25 pseudo-impostors to the speaker and background models. The mean fused score was then compared with a decision threshold to make a decision. Because there were only 200 client subjects with two verification utterances per subject, a client-independent decision threshold was used to increase the resolution of the error rate. Specifically, the 400 client scores (200 clients times 2 utterances per client) were lumped together and the scores compared against the 120,000 impostor scores (200 clients times impostors per client times 8 utterances per impostor) to obtain a client-independent EER and a DET plot.

Unlike systems that use client-dependent thresholds where each threshold depends only on the scale and range of the corresponding client and impostor scores, systems that use client-independent thresholds require a single scale and range for the scores of all clients and impostors. In other words, the threshold for client A can be very different from that of client B, as long as they fall into the range of their respective client and impostor score distributions.

On the other hand, systems that use client-independent thresholds must ensure that the single threshold falls into the right range of all client and impostor score distributions. This requirement can be fulfilled by "normalizing" the scores so that they fall into a predefined range. One of the many possible ways of performing normalization is to shift the mean and scale the variance of the impostor scores to zero and unity, respectively. More specifically, the score of an claimant is normalized by

$$s_{norm}(\mathbf{x}_t) = \frac{s(\mathbf{x}_t) - \mu_b}{\sigma_b},$$

where $s(\mathbf{x}_t)$ is the claimant's score (Eq. 10.4.2) because $\mathbf{x}_t$, and μ_b and σ_b, respectively, are the mean and variance of the client-dependent impostor scores. These impostor scores can be obtained during training by testing a client model against impostor observations. In this work, the impostor observations were obtained from the 25 pseudo-impostors defined in Configuration II of the XM2VTSDB corpus. This normalization technique is called "Znorm" in the literature (see [14] and [303]).

Table 10.6 shows the results of speaker verification using different types of multisample fusion techniques, as described in Section 10.4; Figure 10.20(a) plots the

Table 10.6. The EERs of speaker and face verification. *Note*: The verifications use multisample fusion and EW fusion. Rel. Red. stands for relative reduction in EER with respect to EW fusion. Fusion takes place only *within* the audio and visual scores, not between the audio and visual scores. EW + Znorm means that EW fusion was performed on Znorm scores; a similar definition applies to ZS + Znorm.

Fusion Method	Voice	Rel. Red.	Face	Rel. Red.
Equal-weight (EW)	3.99%	N/A	4.45%	N/A
EW+Znorm	3.04%	23.81%	3.55%	20.22%
Zero-sum (ZS)	2.72%	31.83%	3.91%	12.13%
ZS+Znorm	2.77%	30.58%	3.27%	26.52%

corresponding DET curves. Notice that the Znorm helps lower the EER of equal-weight fusion and that zero-sum fusion achieves the highest error reduction.

Multisample Fusion of Video Scores

Similar to the audio scores, multisample fusion was also performed on the visual scores. Table 10.6 shows the results of face verification using different types of multisample fusion techniques; Figure 10.20(b) plots the corresponding DET curves. Zero-sum fusion of Znorm scores achieves the highest error reduction.

AV Decision Fusion

Table 10.7 summarizes the results of the multisample audio and visual fusion with β set to 0.6, and Figure 10.21 plots the DET corresponding curves. Because Table 10.6 shows that Znorm can enhance the performance for both modalities, multisample fusion was performed on Znorm scores. Table 10.7 and Figure 10.21 demonstrate that ZS fusion always performs better than EW fusion.

To investigate the sensitivity of the fusion parameter β in Eq. 10.5.1, the value of β was varied and the corresponding EERs were obtained. The results are shown in Table 10.8; evidently, the EERs depend on β. This calls for further investigation of automatic techniques for determining fusion parameters. Some examples of research in this area can be found in Meier et al. [239]; Neti et al. [260]; Sanderson and Paliwal [331]; and Wark and Sridharan [371].

10.6   Concluding Remarks

This chapter has introduced and reviewed different fusion techniques for multicue biometric authentication systems. The pros and cons of fusing biometric data at different levels have been discussed. A novel multisample fusion technique, which

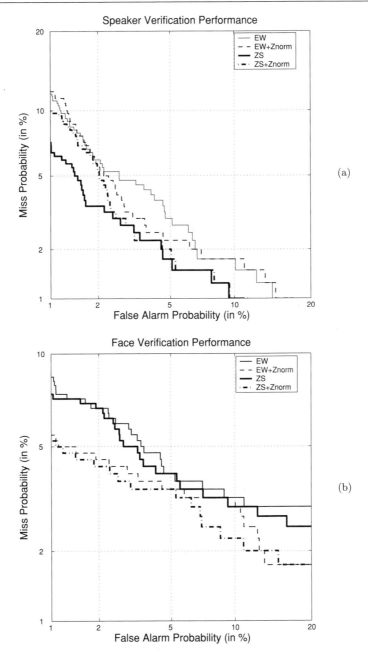

Figure 10.20. DET plots of different multisample fusion techniques in (a) speaker verification and (b) face verification. EW + Znorm means that EW fusion was performed on Znorm scores; a similar definition applies to ZS + Znorm. For clarity, the labels in the legend are arranged in descending order of EERs.

Table 10.7. EERs and voice only relative error reduction obtained by linearly combining the means of the multisample fused scores obtained from the audio and visual channels; the combination weight β in Eq. 10.5.1 is 0.6.

Voice Score Fusion Type	Face Score Fusion Type	EER (%)	Voice only Relative Error Reduction (%)
EW + Znorm	EW + Znorm	0.70%	76.97%
EW + Znorm	ZS + Znorm	0.68%	77.63%
ZS + Znorm	EW + Znorm	0.61%	77.98%
ZS + Znorm	ZS + Znorm	0.47%	83.03%

Table 10.8. EERs of multisample AV fusion for different values of combination weights β. Zero-sum fusion of Znorm scores were used to fuse the audio and visual scores.

β	0.4	0.5	0.6	0.7
EER	1.00%	0.80%	0.47%	0.59%

is general and applicable to both face and speaker recognition systems, was then proposed to fuse the scores obtained from speaker or face models. This is evident by promising experimental results using the HTIMIT corpus, NIST2001 speaker recognition benchmark test, and XM2VTSDB audio–visual database. The technique is also amenable to the fusion of AV data. It was found that error rate reduction of up to 83% can be achieved when the multisample fusion technique is applied to fuse the scores derived from speaker models *and* face models.

Problems

1. This exercise highlights the motivation of adoptng data-dependent fusion weights for a multicue biometric system. Assume that the audio and visual scores of an audio–visual biometric authentication system range from 0 to 10 (the larger the score, the more likely the claimant is a genuine user) and that the decision threshold is set to 8.0. Suppose that in a verification session, the scores of a claimant from the audio and visual channels are equal to 6.0 and 9.0, respectively. Denote the fusion weight for the audio channel as β, the fused score is $s = 6\beta + 9(1 - \beta)$.

    (a) Determine the decision (accept/reject) if the fusion weights for the audio channel and visual channel are both 0.5 (i.e., $\beta = 0.5$).

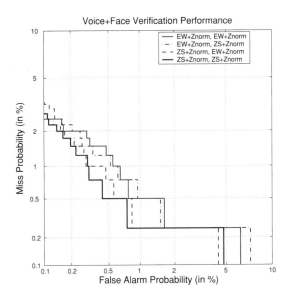

Figure 10.21. DET plots of different multisample fusion techniques for AV person authentication. EW stands for equal-weight fusion, and ZS stands for zero-sum fusion.

(b) Assume that an acoustic noise detector picks up a high level of background noise during the verification session. The system automatically reduces the contribution of the audio channel by lowering the corresponding fusion weight to 0.2 and raising the visual channel's fusion weight to 0.8 (i.e., $\beta = 0.2$). Will the claimant be accepted or rejected?

(c) Assume that the noise detector indicates a low level of background noise but a light-intensity sensor gives an unusually low reading during the verification session. The biometric system responds to this situation by lowering the visual channel's fusion weight to 0.2 and raising the audio channel's fusion weight to 0.8 (i.e., $\beta = 0.8$). Determine the decision of the system.

(d) Determine the range of fusion weights for which the claimant is acceptable (i.e., the fused score exceeds the threshold).

2. Show that when $\tilde{\sigma}_p$ in Eq. 10.4.7 tends to infinity, multisample fusion reduces to equal-weight fusion.

3. Assume that the client scores $s_{c,a}$ from the audio channel of an audio–visual biometric authentication system follow a Gaussian distribution with mean $\mu_{c,a}$ and variance $\sigma_{c,a}^2$, and the impostor scores from the audio channel follow a Gaussian distribution with mean $\mu_{i,a}$ and variance $\sigma_{i,a}^2$. Similarly, the

corresponding score statistics from the visual channel are $\mu_{c,v}$, $\sigma_{c,v}^2$, $\mu_{i,v}$, and $\sigma_{i,v}^2$. Assume also that the audio and visual scores are fused by the following formulae:

$$s_c = \beta s_{c,a} + (1 - \beta)s_{c,v}$$
$$s_i = \beta s_{i,a} + (1 - \beta)s_{i,v},$$

where $\beta \in [0, 1]$ is a fusion weight and s_c and s_i are the fused client scores and fused impostor scores, respectively.

(a) Show that the variances of the fused client scores and fused impostor scores are

$$\sigma_c^2 = \beta^2 \sigma_{c,a}^2 + (1 - \beta)^2 \sigma_{c,a}^2$$
$$\sigma_i^2 = \beta^2 \sigma_{i,a}^2 + (1 - \beta)^2 \sigma_{i,a}^2$$

State your assumption on the statistic dependence of the scores from the two channels.

(b) Hence, show that the variances of the fused scores satisfy

$$\sigma_c^2 \leq \max\{\sigma_{c,a}^2, \sigma_{c,v}^2\} \text{ and } \sigma_i^2 \leq \max\{\sigma_{i,a}^2, \sigma_{i,v}^2\}.$$

(c) Show that the fusion weight that minimizes σ_c is given by

$$\beta^* = \frac{\sigma_{c,v}^2}{\sigma_{c,a}^2 + \sigma_{c,v}^2}.$$

(d) Hence, explain why fusion of independent audio and visual scores helps increase the separation between the fused client scores and the fused impostor scores.

4. *Matlab Exercise.* Create two Gaussian distributions with means 0.8 and 1.2 and variances equal to 1.0 to simulate the score distributions of two client utterances. Similarly, create two Gaussian distributions with means -1.3 and -0.7 and variances that are equal to 2.0 to simulate the score distributions of two impostor utterances. Write a Matlab program to implement Eqs. 10.4.3 through 10.4.7.

(a) Create a plot similar to Figure 10.8(c) by computing the differences between the fused client scores and the fused impostor scores for different values of prior score $\tilde{\mu}_p$.

(b) Sort the two client score sequences in opposite order. Similarly, sort the two impostor score sequences in opposite order. Use the Matlab program to compute the score dispersion between the fused client scores and the fused impostor scores. Plot the score dispersion against the prior score $\tilde{\mu}_p$.

(c) Repeat (a) and (b) for different values of prior variance $\tilde{\sigma}_p$.

5. Repeat Problem 4 by replacing the Gaussian distributions with uniform distributions.

Appendix A

CONVERGENCE PROPERTIES OF EM

The EM algorithm is a general technique for maximum-likelihood estimation (MLE). This appendix provides proof showing that the likelihood function is guaranteed to increase during EM learning. For more convergence properties of the EM algorithm, see [74, 350, 389].

Because sample density $p(\mathbf{x}(t)|\omega)$ depends on the parameters $\mathbf{w}$, the density function can be rewritten as

$$p(\mathbf{x}(t)|\omega) = p(\mathbf{x}(t)|\omega, \mathbf{w}).$$

For simplicity, the indicator ω is omitted:

$$p(\mathbf{x}(t)|\omega, \mathbf{w}) \equiv p(\mathbf{x}(t)|\mathbf{w}).$$

Following the preceding notation, the log-likelihood to be maximized is rewritten as

$$
\begin{aligned}
l(\mathbf{w}; \mathbf{X}^+) &= \sum_{t=1}^{N} \log p(\mathbf{x}(t)|\mathbf{w}) \\
&= \log p(\mathbf{X}^+|\mathbf{w}),
\end{aligned}
\tag{A.0.1}
$$

where $\mathbf{X}^+$ is the positive training data set. The complete-data likelihood $l_c(\mathbf{w}; \mathbf{X}^+)$ is rewritten as:

$$l_c(\mathbf{w}; \mathbf{X}^+) = \log p(\mathbf{Y}|\mathbf{w}), \tag{A.0.2}$$

where $\mathbf{Y}$ is the complete data set.

Notation for the conditional density of $\mathbf{Y}$ given $\mathbf{X}^+$ and $\mathbf{w}$ is introduced next,

$$p(\mathbf{Y}|\mathbf{X}^+, \mathbf{w}) = p(\mathbf{Y}|\mathbf{w})/p(\mathbf{X}^+|\mathbf{w}) \tag{A.0.3}$$

so that the likelihood in Eq A.0.1 can be written in the useful form:

$$l(\mathbf{w}; \mathbf{X}^+) = l_c(\mathbf{w}; \mathbf{X}^+) - \log p(\mathbf{Y}|\mathbf{X}^+, \mathbf{w}). \tag{A.0.4}$$

414

In the j-th iteration of the EM algorithm, the expectation term $Q(\mathbf{w}, \mathbf{w}^{(j)})$ is

$$Q(\mathbf{w}, \mathbf{w}^{(j)}) = E[l_c(\mathbf{w}; \mathbf{X}^+)|\mathbf{X}^+, \mathbf{w}^{(j)}]. \tag{A.0.5}$$

The M-step of the EM algorithm maximizes $Q(\mathbf{w}, \mathbf{w}^{(j)})$ with respect to $\mathbf{w}$, so it guarantees that $Q(\mathbf{w}^{(j+1)}, \mathbf{w}^{(j)}) \geq Q(\mathbf{w}^{(j)}, \mathbf{w}^{(j)})$. As shown here, the EM iteration also makes $l(\mathbf{w}^{(j+1)}; \mathbf{X}^+) \geq l(\mathbf{w}^{(j)}; \mathbf{X}^+)$.

To link $Q(\mathbf{w}, \mathbf{w}^{(j)})$ with $l(\mathbf{w}; \mathbf{X}^+)$, it is convenient to write

$$H(\mathbf{w}, \mathbf{w}^{(j)}) \equiv E[\log p(\mathbf{Y}|\mathbf{X}^+, \mathbf{w})|\mathbf{X}^+, \mathbf{w}^{(j)}]. \tag{A.0.6}$$

By taking expectation on the both sides of Eq. A.0.4, the following equation is obtained:

$$E[l(\mathbf{w}; \mathbf{X}^+)] = l(\mathbf{w}; \mathbf{X}^+) = Q(\mathbf{w}, \mathbf{w}^{(j)}) - H(\mathbf{w}, \mathbf{w}^{(j)}). \tag{A.0.7}$$

After the M-step, the result is

$$l(\mathbf{w}^{(j+1)}; \mathbf{X}^+) = Q(\mathbf{w}^{(j+1)}, \mathbf{w}^{(j)}) - H(\mathbf{w}^{(j+1)}, \mathbf{w}^{(j)}). \tag{A.0.8}$$

Because $H(\mathbf{w}^{(j+1)}, \mathbf{w}^{(j)}) \leq H(\mathbf{w}^{(j)}, \mathbf{w}^{(j)})$ according to Jensen's inequality:

$$
\begin{aligned}
H(\mathbf{w}^{(j+1)}, \mathbf{w}^{(j)}) - H(\mathbf{w}^{(j)}, \mathbf{w}^{(j)}) &= E[\log \frac{p(\mathbf{Y}|\mathbf{X}^+, \mathbf{w}^{(j+1)})}{p(\mathbf{Y}|\mathbf{X}^+, \mathbf{w}^{(j)})}|\mathbf{X}^+, \mathbf{w}^{(j)}] \\
&\leq \log E[\frac{p(\mathbf{Y}|\mathbf{X}^+, \mathbf{w}^{(j+1)})}{p(\mathbf{Y}|\mathbf{X}^+, \mathbf{w}^{(j)})}|\mathbf{X}^+, \mathbf{w}^{(j)}] \\
&= \log \int [\frac{p(\mathbf{Y}|\mathbf{X}^+, \mathbf{w}^{(j+1)})}{p(\mathbf{Y}|\mathbf{X}^+, \mathbf{w}^{(j)})}]p(\mathbf{Y}|\mathbf{X}^+, \mathbf{w}^{(j)})\mu(d\mathbf{Y}) \\
&= \log \int p(\mathbf{Y}|\mathbf{X}^+, \mathbf{w}^{(j+1)})\mu(d\mathbf{Y}) \\
&= \log 1 \\
&= 0 \tag{A.0.9}
\end{aligned}
$$

Because of Eq. A.0.9 and the fact that $Q(\mathbf{w}^{(j+1)}, \mathbf{w}^{(j)}) \geq Q(\mathbf{w}^{(j)}, \mathbf{w}^{(j)})$, the log-likelihood $l(\mathbf{w}; \mathbf{X}^+)$ is guaranteed to increase iteration by iteration.

Appendix B

AVERAGE DET CURVES

To show the procedure of constructing average DET curves, three speakers are used as an example. Figure B.1 shows three dotted curves and one solid curve. Each dotted curve represents the receiver operation characteristic (ROC) of a speaker model, and the solid curve is their average. First, interpolation is applied to obtain a common set of abscissa for all dotted curves. As a result, points on Curve A have coordinates (x_1, y_{A_1}), (x_2, y_{A_2}), (x_3, y_{A_3}), ..., (x_N, y_{A_N}); points on Curve B have coordinates (x_1, y_{B_1}), (x_2, y_{B_2}), (x_3, y_{B_3}), ..., (x_N, y_{B_N}); and points on Curve C have coordinates (x_1, y_{C_1}), (x_2, y_{C_2}), (x_3, y_{C_3}), ..., (x_N, y_{C_N}). Next, the ordinates are averaged for each common abscissa value to obtain the averaged curve.

In the example shown in Figure B.1, points on the solid curve have coordinates

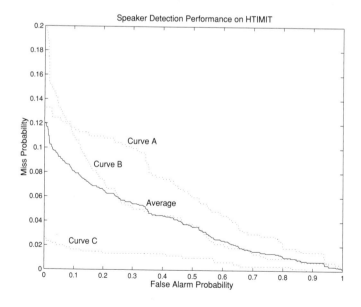

Figure B.1. ROC curves of three speakers and their average.

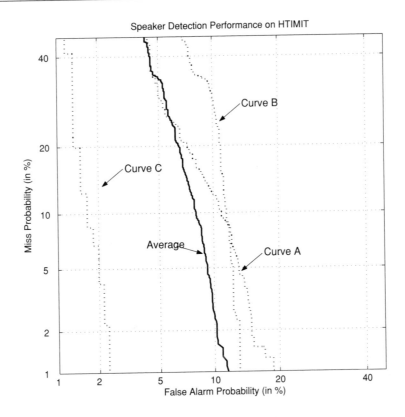

Figure B.2. DET curves of three speakers and their average.

$(x_1, \frac{y_{A_1}+y_{B_1}+y_{C_1}}{3})$, $(x_2, \frac{y_{A_2}+y_{B_2}+y_{C_2}}{3})$, $(x_3, \frac{y_{A_3}+y_{B_3}+y_{C_3}}{3})$, ..., $(x_N, \frac{y_{A_N}+y_{B_N}+y_{C_N}}{3})$. Finally, the corresponding DET curves as shown in Figure B.2 are plotted and the EER from the averaged curve is obtained; the averaged curve should be the same as the average of the EERs of the three dotted curves.

Appendix C

MATLAB PROJECTS

C.1 Matlab Project 1: GMMs and RBF Networks for Speech Pattern Recognition

C.1.1 Introduction

Gaussian mixture models (GMMs) and radial basis function (RBF) networks are two of the promising neural models for pattern classification. The goal of the following laboratory exercise is to develop pattern classification systems based on GMMs and RBF networks. The systems should be able to recognize 10 vowels. "Netlab" (*http://www.ncrg.aston.ac.uk/netlab/*) and Matlab are used to create the GMMs and RBF networks.

C.1.2 Objectives

This laboratory exercise has the following objectives.

1. Create GMMs and RBF networks to represent 10 different classes.

2. Perform pattern classification using the created networks.

3. Compare the GMM-based system against the RBF network-based system in terms of recognition accuracy, decision boundaries, training time, and recognition time.

4. Find the decision boundaries and plot them on a two-dimensional plane.

C.1.3 Procedures

GMM-Based Classifier

1. Download the "Netlab" software from *http://www.ncrg.aston.ac.uk/netlab* and save the *m*-files in your working directory. Download the training data and testing data from M.W. Mak's home page:

$$http://www.eie.polyu.edu.hk/\~mwmak/Book/gmmrbf.zip$$

This file contains two-dimensional vowel data (in the "data/" directory) that you will use in this laboratory exercise. You will also find the following files in *gmm-rbf.zip*:

- *load_pattern.m*—a function for importing the two-dimensional vowel training and test data.

- *train_gmm.m*—a function for training a GMM.

- *train_all_gmm.m*—a function for training all the GMMs.

- *train_rbf.m*—a function for training an RBF network.

- *plot_all_gmm.m*—a function for plotting the centers of all GMMs and training data.

- *plot_gmm_contour.m*—a function for plotting the equal-probability contour of a GMM.

- *plot_rbf.m*—a function for plotting RBF centers and training data.

- *VowelClassifer.m*—a program that calls the preceding functions to train 10 GMMs (or an RBF network with 10 outputs) and find the classification accuracy.

- *plot_boundary.m*—a program for plotting the decision boundaries created by 10 GMMs or an RBF network.

- *gmm_example.m*—an example program showing how to create and train a GMM-based classifier. It also suggests a method to plot the decision boundaries created by a GMM-based classifier.

Note that some of the files contain missing statements. You will need to fill in these statements in this lab exercise.

2. Run Matlab, go to "File" → "Set Path" and add the directory where "netlab" was saved.

3. Read and run the file "gmm_example.m" in *gmmrbf.zip* to see how to use Netlab to create and train a GMM.

4. Import and save the training data, *2DVowel_train_pat.dat*, to a two-dimensional array. The imported matrix should be 338×12 in size. The first two columns contain training feature vectors in a two-dimensional input space and columns 3 and 12 indicate the class to which each pattern belongs. A Matlab function *load_pattern.m* is provided to help with importing the training and test data.

5. Create 10 GMMs to represent the 10 vowels using the data from 10 different classes. The data should be separated into 10 two-dimensional arrays. Set the number of centers to 2 and the covariance type to "diag". The model can be created by using

```
model_name=gmm(data_dimension, no_of_centers, covariance_type)
```

and the model can be initialized by

$$\texttt{model\_name=gmminit(model\_name, data, options)}$$

6. Use the EM algorithm **gmmem** to train the models.

7. Plot the imported training data together with the centers after EM training.

8. Import the test data—*2DVowel_test_pat.dat*. This file is used for finding the classification rate of the GMMs you have just created. The file contains 333 data points, and again each point belongs to one of the 10 classes. The likelihood of a particular model Λ_i for a data point $\mathbf{x}_t$ (i.e., $p(\mathbf{x}_t|\Lambda_i)$) can be calculated by the function **gmmprob**. Each data point is classified to the class whose corresponding likelihood is the highest. The overall classification rate is calculated by

$$\frac{\text{Number of correctly classified points}}{\text{Total number of data points}} \times 100\%$$

9. Try different numbers of centers and different covariance types ("diag" or "full") when creating the models. Find the optimal combination that gives the highest classification rate. What is the optimal combination and what is the classification rate?

10. Plot the decision boundaries that separate the 10 classes. See *gmm_example.m* for an example.

RBF Network-Based Classifier

1. This part repeats the previous procedures but an RBF network is to be used for the classification task. Again, start by importing the training data and storing it in a 338×12 array.

2. After importing the data, separate it into two parts: one is input data, which is 338×2 in size, and the other is desired outputs, which is 338×10 in size.

3. Instead of creating 10 different RBF networks, create one RBF network. To create an RBF network, use the function **rbf**. To specify the network architecture, provide the number of inputs, the number of hidden units, and the number of output units.

4. Initialize the RBF network by calling the function **rbfsetbf**. Specify a number of option fields as in **gmm**. Before performing the classification, call the function **rbftrain** to train the RBF network. Specifying a target vector containing the class information is also necessary.

5. After training the network, import the test data and use the function **rbffwd** to perform classification. This function has two input fields: one is the RBF network, which will be used for classification, and the other is a row vector. In this exercise, the row vector has two fields: the x location and the y location. The output is again a row vector, and its size will be equal to the number of outputs that you specify in Step 3. For each test vector, class ID is determined by selecting the output whose response to the test vector is the largest.

6. Compute the classification rate of the whole test set. Try different numbers of hidden units and select the optimal one. What is the optimal number of hidden units and what is the corresponding classification rate? Compare and explain the classification performance of the RBF networks with that of the GMMs.

7. Plot the decision boundaries. Figure C.1(b) shows an example of the decision boundaries created by an RBF network. Compare the boundaries with those of the GMMs.

8. Compare the GMM-based classifier against the RBFN-based classifier in terms of classification accuracy, training time, and recognition time.

C.2 Matlab Project 2: SVMs for Pattern Classification

C.2.1 Objectives

Use linear and nonlinear support vector machines (SVMs) to classify two-dimensional data.

C.2.2 Procedures

Linear SVMs

1. Go to *http://www.cis.tugraz.at/igi/aschwaig/software.html* to download the software "svm_251" and save the *m*-files to your working directory. Some functions (e.g., `plotboundary`, `plotdata`, and `plotsv`) are included in the end of the *m*-file *demsvm1.m*. Extract these functions to form new *m*-files (e.g. *plotboundary.m*, *plotdata.m* and *plotsv.m*).

2. Open Matlab, go to "File" → "Set Path" and add the directory where "svm_251" was saved. Alternatively, add the statement "`addpath svmdir`"—where `svmdir` is the directory storing "svm_251"—to your *.m* file.

3. If your system does not have Matlab Optimization Toolbox installed, you may need to compile the files *pr_loqo.c* and *loqo.c* in Matlab's command prompt, as follows:

```
mex pr_loqo.c loqo.c
```

This will create a file named *pr_loqo.dll*; then, find the line

```
workalpha = loqo(H, f, A, eqconstr, VLB, VUB, startVal, 1);
```

in *svmtrain.m* and change `loqo` to `pr_loqo`.

4. Input the following training data. X is a set of input data, 20×2 in size. Y contains the corresponding class labels, 20×1 in size.

X	(2,7)	(3,6)	(2,5)	(3,5)	(3,3)	(2,2)	(5,1)	(6,2)	(8,1)	(6,4)	(4,8)
Y	+1	+1	+1	+1	+1	+1	+1	+1	+1	+1	-1

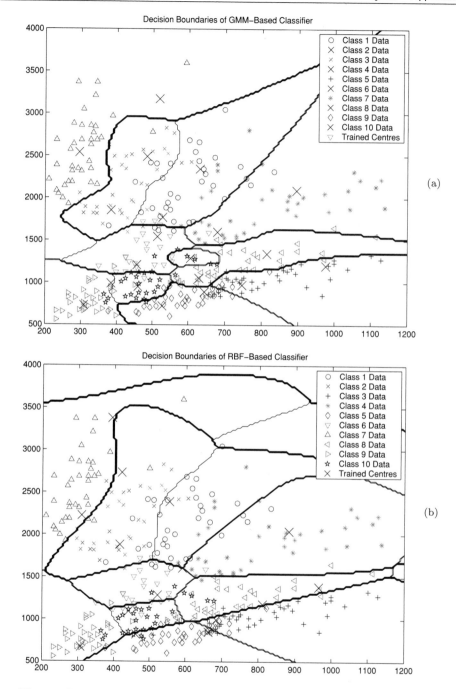

Figure C.1. Decision boundaries created by (a) 10 GMMs and (b) an RBF network, with 10 outputs for the two-dimensional vowel problems.

X	(5,8)	(9,5)	(9,9)	(9,4)	(8,9)	(8,8)	(6,9)	(7,4)	(4,4)
Y	-1	-1	-1	-1	-1	-1	-1	-1	-1

Plot a graph to show the data set using the commands `plotdata`:

```
x1ran = [0, 10]; x2ran = [0, 10]; %data range
f1 = figure; plotdata(X, Y, x1ran, x2ran);
title('Data from class +1 (squares) and class −1 (crosses)');
```

5. Create a support vector machine classifier by using the function `svm`.

$$\text{net} = \text{svm}(\text{nin}, \text{kernel}, \text{kernelpar}, \text{C}, \text{use2norm}, \text{qpsolver}, \text{qpsize})$$

Set `nin` to 2 because X contains two-dimensional data. Set `kernel` to 'linear' to use linear SVM. Set `kernelpar` to [ ] because a linear kernel does not require any parameters, set C to 100, and set `use2norm` to 0 so that 1-norm is used (standard SVM). Set `qpsolver` to 'loqo' so that the C functions in *pr_loqo.dll* will be used for performing quadratic optimization. You may leave the last parameter `qpsize` blank (i.e., the default value is to be used).

After creating an SVM, train it by using the function `svmtrain`

$$\text{net} = \text{svmtrain}(\text{net}, \text{X}, \text{Y}, \text{alpha0}, \text{dodisplay})$$

Set `alpha0` to [ ], and set `dodisplay` to 2 to show the training data.

After carrying out the preceding steps, record the number of SVs. Also, record the norm of the separating hyperplane and calculate the length of the margin from `net.normalw`.

Plot the SVM using the commands `plotboundary`, `plotdata`, and `plotsv` as follows:

```
figure; plotboundary(net, x1ran, x2ran);
plotdata(X, Y, x1ran, x2ran); plotsv(net, X, Y);
normW = norm(net.normalw);
tstring = sprintf('Linear SVM, C = %.1f, #SV = %d, normW = %.2f', ...
                  C, length(net.svind), norm);
title(tstring, 'FontSize', 11);
```

6. Vary the value of C in the function `svm` and repeat Step 5 (e.g., C = 0.1, C = 10, and C = 1000). For different values of C, plot the corresponding SVM, record the number of the support vectors, the norm of the separating hyperplanes, and the margin width. Discuss the change in the number of SVs and the margin width as a result of varying C.

Nonlinear SVMs

1. Repeat the procedures described in the preceding linear SVM, but this time replace the linear kernel with a polynomial kernel:

```
net = svm(2, 'poly', 2, C, 0, 'loqo'); % second-order polynomial kernel
net = svmtrain(net, X, Y, [ ], 2);
```

2. Vary the value of C in the function svm and repeat Step 1 (e.g. C = 0.1, C = 10, and C = 1000). For different values of C, plot the corresponding SVM and record the number of the support vectors. Discuss the change in the number of SVs and the margin width as a result of varying C. Note that you cannot obtain the margin width from net.normalw because the margin width is different in different regions of the input space.

3. Repeat Steps 1 and 2, but use a polynomial kernel of degree 3:
net = svm(2, 'poly', 3, C, 0, 'loqo'); % 2nd-order polynomial kernel
net = svmtrain(net, X, Y, [ ], 2);
Explain your observation.

4. Repeat Steps 1 and 2, but use an RBF kernel:
net = svm(2, 'rbf', 8, C, 0, 'loqo'); % RBF kernel with $2\sigma^2 = 8$
net = svmtrain(net, X, Y, [ ], 2);
Explain your observations.

The Role of Kernel Parameters

1. Edit the file *svmkernel.m* to implement the kernel function

$$K(\mathbf{x}, \mathbf{x}_i) = \left(1 + \frac{\mathbf{x} \cdot \mathbf{x}_i}{\sigma^2}\right)^p.$$

Set $p = 2$ and rerun the 20-point problem in Section C.2.2 for different values of σ. How does σ affect the shape of the decision boundary for a fixed C and p? Discuss your observations.

2. Vary the integer p but fix σ. How does p affect the shape of the decision boundary for a fixed C and σ? Discuss your observations.

3. Vary the kernel parameter σ of the RBF kernel and repeat the 20-point problem. How does σ affect the shape of the decision boundary for a fixed C? Discuss your observations.

Invariance Properties of SVMs

1. Input the following Matlab code to your *.m* file to create a noisy XOR problem:
```
shift = 0;
scale = 1;
xmean = [-1, -1; -1, 1; 1, -1; 1, 1];
xstd = 0.5;
randn('state', 0);
x = [xstd * randn(N, 1) + xmean(1, 1)  xstd * randn(N, 1) + xmean(1, 2); ...
     xstd * randn(N, 1) + xmean(2, 1)  xstd * randn(N, 1) + xmean(2, 2); ...
     xstd * randn(N, 1) + xmean(3, 1)  xstd * randn(N, 1) + xmean(3, 2); ...
     xstd * randn(N, 1) + xmean(4, 1)  xstd * randn(N, 1) + xmean(4, 2)];
x = x * scale + shift;
```

$$y = [-1 * \text{ones}(N, 1); \text{ones}(N, 1); \text{ones}(N, 1); -1 * \text{ones}(N, 1)];$$
$$\text{x1range} = [-2, 2] * \text{scale} + \text{shift};$$
$$\text{x2range} = [-2, 2] * \text{scale} + \text{shift};$$

2. Use linear SVMs, polynomial SVMs, and RBF SVMs to solve the noisy XOR problem. For each case, record the decision boundaries and the number of SVs. Note that you may need to edit the file "plotdata.m" to remove the labels of the data points on the plots. Can the linear SVM solve the XOR problem? Explain your observations.

3. *Scale Invariance.* Set `scale` = 5 to scale the training data $(\mathbf{x}_i \leftarrow 5\mathbf{x}_i)$ and repeat Step 2 for using polynomial SVMs and RBF SVMs. Note that you need to change the polynomial kernel function to $(1 + \mathbf{x}^T\mathbf{x}_i/\beta^2)^2$, where $\beta = 5$ in this case. This can be done by editing the *svmkernel.m* file. You also need to properly scale the kernel parameter of the RBF kernel, as follows:

```
net = svm(2, 'rbf', (sigma^ 0.5*scale)^ 2,C,0,'loqo');
```

where `sigma` is the kernel parameter σ in

$$K(\mathbf{x}, \mathbf{x}_i) = \exp\left\{-\frac{\|\mathbf{x} - \mathbf{x}_i\|^2}{\sigma^2}\right\}$$

for the unscaled data. You may set `sigma` = 1 to obtain reasonably good results. Are all of the SVMs scale invariant? Explain your observations.

4. Change the kernel function in *svmkernel.m* to $(\beta^2 + \mathbf{x}^T\mathbf{x}_i)^2$ and repeat Step 3. Does this kernel function lead to scale invariant SVMs? If not, why?

5. *Translation Invariance.* Set `shift` = 12 to shift the training data—$\mathbf{x}_i \leftarrow \mathbf{x}_i + [12\ 12]^T$—and repeat Step 2. Which type of SVMs is not translation invariant? Explain your observations.

6. Set `scale` = 5 and `shift` = 12 to investigate both the scale and translation invariance properties of different SVMs.

Verifying the Analytical Solutions

1. Create a 4-point XOR problem by entering the following Matlab code to your *.m* file:

```
C = 100;
shift = 0;
scale = 1;
x = [0, 0; 0, 1; 1, 0; 1, 1] * scale + shift;
y = [-1; 1; 1; -1];
x1range = [-1, 2] * scale + shift;
x2range = [-1, 2] * scale + shift;
```

2. Create a polynomial SVM to solve this XOR problem and show that the Lagrange multipliers and the bias are $\alpha_1 = 10/3$, $\alpha_2 = 8/3$, $\alpha_3 = 8/3$, $\alpha_4 = 2$, and

$b = 1$. Determine the maximum value of the Lagrangian $L(\alpha)$. You can use the following code fragment to display the multipliers and the slack variables ξ_i:

```
[svDec, svOut] = svmfwd(net, net.sv);
svSlack = 1 - y(net.svind). * svOut;
svAlpha = net.alpha(net.svind);
fprintf('svind      a_i      xi_i n');
for i = 1 : length(net.svind)
    fprintf('%d      %.4f      %fn', net.svind(i), svAlpha(i), svSlack(i));
end
```

3. Set `shift` $= -0.5$ and repeat Step 2. Compare the decision boundary between the SVM obtained in Step 2 and the one obtained in this step. What are the implications of your result? Suggest a possible method to address the translation invariance issue.

4. Repeat Steps 2 and 3, but this time use an RBF SVM to solve the four-point XOR problem. Based on your results, comment on the applicability of polynomial SVMs and RBF SVMs in real-world problems.

BIBLIOGRAPHY

[1] M. A. Abidi and R. C. Gonzalez. *Data Fusion in Robotics and Machine Intelligence*. Boston: Academic, 1992.

[2] T. Abu-Amer and J. Carson-Berndsen. Multi-linear HMM based system for articulatory feature extraction. In *Proc. ICASSP'03*, volume 2, 21–24, Hong Kong, 2003.

[3] A. Acero. *Acoustical and Environmental Robustness in Automatic Speech Recognition*. Dordrecht: Kluwer Academic, 1992.

[4] M. Afify, Y. Gong, and J. P. Haton. A general joint additive and convolutive bias compensation approach applied to noisy Lombard speech recognition. *IEEE Trans. on Speech and Audio Processing*, 6(6):524–537, 1998.

[5] H. Akaike. A new look at the statistical model identification. *IEEE Trans. on Automatic Control*, 19(6):716–723, 1974.

[6] L. A. Alexandre, A. C. Campilho, and M. Kamel. On combining classifers using sum and products rules. *Pattern Recognition Letters*, 22:1283–1289, 2001.

[7] E. Ambikairajah, M. Keane, A. Kelly, L. Kilmartin, and G. Tattersall. Predictive models for speaker verification. *Speech Communication*, 13:417–425, 1993.

[8] J. A. Anderson. *Neurocomputing—Paper Collections*. Cambridge, MA: MIT Press, 1988.

[9] G. Antonini, V. Popovici, and J. P. Thiran. Independent component analysis and support vector machine for face feature extraction. In *Proc. AVBPA'03*, 111–118, 2003.

[10] J. Ashboum. Practical implementation of biometrics based on hand geometry. In *Proc. IEE Colloquium on Image Processing for Biometric Measurement*, volume 5, 1–6, April 1994.

[11] K. T. Assaleh and R. J. Mammone. New LP-derived features for speaker identification. *IEEE Trans. on Speech and Audio Processing*, 2(4):630–638, 1994.

[12] B. S. Atal. Automatic speaker recognition based on pitch contours. *J. Acoust. Soc. Am.*, 52:1687–1972, 1972.

[13] B. S. Atal. Effectiveness of linear prediction characteristics of the speech wave for automatic speaker identification and verification. *J. Acoust. Soc. Am.*, 55(6):1304–1312, 1974.

[14] R. Auckenthaler, M. Carey, and H. Lloyd-Thomas. Score normalization for text-independent speaker verification systems. *Digital Signal Processing*, 10:42–54, 2000.

[15] J. Bala, K. DeJong, J. Huang, H. Vafaie, and H. Wechsler. Using learning to facilitate the evolution of features for recognizing visual concepts. *Evolutionary Computation*, 4(3):297–312, 1996.

[16] P. Baldi and Y. Chauvin. Neural networks for fingerprint recognition. *Neural Computation*, 5:402–418, 1993.

[17] Y. Barniv and D. Casasent. Multisensor image registration: Experimental verification. In *SPIE Proc.*, volume 292, 160–171, 1981.

[18] F. Beaufays and M. Weintraub. Model transformation for robust speaker recognition from telephone data. In *Proc. ICASSP'97*, 21–24, April 1997.

[19] R. K. Belew, J. McInerney, and N. N. Schraudolph. Evolving networks: Using the genetic algorithm with connectionist learning. In C. G. Langton, C. Taylor, J. D. Farmer, and S. Rasmussen, editors, *Artificial Life 2*, 511–547. Redwood City, CA: Addison-Wesley, 1992.

[20] P. Belhumeur, J. Hespanha, and D. Kriegman. Eigenfaces versus Fisherfaces: Recognition using class specific linear projection. *IEEE Trans. on Pattern Analysis and Machine Intelligence*, 19:711–720, 1997.

[21] S. Ben-Yacoub. Multi-modal data fusion for person authentication using SVM. In *Proc. AVBPA'99*, 1999.

[22] S. Ben-Yacoub, Y. Abdeljaoued, and E. Mayoraz. Fusion of face and speech data for person identity verification. *IEEE Trans. on Neural Networks*, 10(5):1065–1074, 1999.

[23] S. Bengio and J. Mariethoz. Learning the decision function for speaker verification. In *Proc. ICASSP'01*, volume 1, 425–428, 2001.

[24] Y. Bennani. Multi-expert and hybrid connectionist approach for pattern recognition: Speaker identification task. *International J. of Neural Systems*, 5(3):207–216, 1994.

[25] D. P. Bertsekas. *Nonlinear Programming*. Belmont, MA: Athena Scientific, 1995.

[26] S. A. Billings and G. L. Zheng. Radial basis function network configuration using genetic algorithms. *Neural Networks*, 8(6):877–890, 1995.

[27] C. M. Bishop. *Neural Networks for Pattern Recognition*. New York: Oxford University Press, 1995.

[28] V. Blanz, S. Romdhami, and T. Vetter. Face identification across different poses and illuminations with a 3D morphable model. In *Proc. 5th International Conference on Automatic Face and Gesture Recognition*, 202–207, 2002.

[29] A Bobick and A. Wilson. A state-based technique for the summarization and recognition of gesture. In *Proc. International Conference on Computer Vision*, 1995.

[30] S. F. Boll. Suppression of acoustic noise in speech using spectral subtraction. *IEEE Trans. on Acoustics, Speech, and Signal Processing*, ASSP-27(2):113–120, 1979.

[31] J. F. Bonastre, F. Bimbot, L. J. Boe, J. P. Campbell, D. A. Reynolds, and I. Magrin-Chagnolleau. Authentication by voice: A need for caution. In *Proc. Eurospeech*, 33–36, 2003.

[32] I. Booth, M. Barlow, and B. Watson. Enhancements to DTW and VQ decision algorithms for speaker recognition. *Speech Communication*, 427–433, 1993.

[33] B. E. Boser, I. M. Guyon, and V. N. Vapnik. A training algorithm for optimal margin classifiers. In D. Haussler, editor, *Proc. 5th Annual ACM Workshop on Computational Learning Theory*, 144–152, 1992.

[34] J. Bradley, C. Brislawn, and T. Hopper. The FBI wavelet/scalar quantization standard for gray-scale fingerprint image compression. Los Alamos National Lab, Technical report LA-UR-93-1659, 1993.

[35] J. S. Bridle. Probabilistic interpretation of feedforward classification network outputs, with relationships to statistical pattern recognition. In F. Fogelman and J. Herault, editors, *Neurocomputing: Algorithms, Architectures, and Applications*, 227–236. London: Springer-Verlag, 1991.

[36] F. Z. Brill, D. E. Brown, and W. N. Martin. Fast genetic selection of features for neural net classifiers. *IEEE Trans. on Neural Networks*, 3(2):324–328, 1992.

[37] R. Brunelli and D. Falavigna. Person identification using multiple cues. *IEEE Trans. on Pattern Analysis and Machine Intellegence*, 17(10):955–966, 1995.

[38] R. Brunelli and T. Poggio. Face recognition: features versus templates. *IEEE Trans. on Pattern Analysis and Machine Intelligence*, 15:1042–1052, 1993.

[39] J. Burbea and C. R. Rao. On the convexity of some divergence measures based on entropy functions. *IEEE Trans. Information Theory*, 28(3):489–495, 1982.

[40] C. J. C. Burges. A tutorial on support vector machines for pattern recognition. *Knowledge Discovery and Data Mining*, 2(2), 121–167, 1998.

[41] D. K. Burton. Text-dependent speaker verification using vector quantization source coding. *IEEE Trans. on Acoustics, Speech, and Signal Processing*, ASSP-35(2):133–143, 1987.

[42] J. P. Campbell, D. A. Reynolds, and R. B. Dunn. Fusing high- and low-level features for speaker recognition. In *Proc. Eurospeech*, 2665–2668, 2003.

[43] W.M. Campbell. Generalized linear discriminant sequence kernels for speaker recognition. In *Proc. ICASSP'02*, volume 1, 161–164, 2002.

[44] J. P. Campbell Jr. Testing with the YOHO CD-ROM voice verification corpus. In *ICASSP'95*, 341–344, 1995.

[45] J. P. Campbell Jr. Speaker recognition: A tutorial. *Proc. IEEE*, 85(9):1437–1462, 1997.

[46] F. Cardinaux, C. Sanderson, and S. Marcel. Comparison of MLP and GMM classifiers for face verification on XM2VTS. In *Proc. AVBPA'03*, 911–920, 2003.

[47] B. Carse, et al. Evolving radial basis function neural networks using genetic algorithm. In *Proc. IEEE International Conference on Evolutionary Computation*, 300–305, Dec. 1995.

[48] D. J. Chalmers. The evolution of learning: An experiment in genetic connectionism. In D. S. Touretzky, editor, *Proc. 1990 Connectionist Models Summer School*, 81–90. San Mateo, CA: M. Kaufmann, 1990.

[49] E. J. Chang and R. P. Lippmann. Using genetic algorithms to improve pattern classification performance. In R. P. Lippmann, J. E. Moody, and D. S. Touretzky, editors, *Advances in Neural Information Processing Systems 3*, 797–803. San Mateo, CA: M. Kaufmann, 1991.

[50] S. Chang, L. Shastri, and S. Greenberg. Robust phonetic feature extraction under a wide range of noise backgrounds and signal-to-noise ratios. In *Proc. Workshop on Consistent and Reliable Acoustic Cues for Sound Analysis*, Denmark, 2001 (*http://www.ee.columbia.edu/crac/program.html*).

[51] V. Chatzis, A. G. Bors, and I. Pitas. Multimodal decision-level fusion for person authentication. *IEEE Trans. on Systems, Man and Cybernetics—Part A: Systems and Humans*, 29(6):674–680, 1999.

[52] R. Chellappa, C. L. Wilson, and S. Sirohey. Human and machine recognition of faces: A survey. *Proc. IEEE*, 83(5), 705–740, 1995.

[53] K. Chen. A connectionist method for pattern classification with diverse features. *Pattern Recognition Letters*, 19(7):545–558, 1998.

[54] K. Chen, D. Xie, and H. Chi. A modified HME architecture for text-dependent speaker identification. *IEEE Trans. on Neural Networks*, 7(5):1309–1313, 1996.

[55] P. H. Chen, C. J. Lin, and B. Scholkopf. A tutorial on ν-support vector machines, 2003 (*http://www.kernel-machines.org*).

[56] Z. Chen and C. H. Kuo. A topology-based matching algorithm for fingerprint authentication. In *Proc. IEEE International Carnahan Conference on Security Technology*, 84–87, 1991.

[57] Y. Cheng, K. Liu, J. Yang, and H. Wang. A robust algebraic method for face recognition. In *Proc. 11th International Conference on Pattern Recognition*, 221–224, 1992.

[58] M. C. Cheung, M. W. Mak, and S. Y. Kung. Adaptive decision fusion for multi-sample speaker verification over GSM networks. In *Proc. Eurospeech'03*, 2969–2972, Geneva, Sept. 2003.

[59] M. C. Cheung, M. W. Mak, and S. Y. Kung. Multi-sample data-dependent fusion of sorted score sequences for biometric verification. In *Proc. ICASSP'04*, volume 5, 681–684, 2004.

[60] C. C. Chibelushi, J. S. D. Mason, and F. Deravi. Feature-level data fusion for bimodal person recognition. In *Proc. 6th International Conference on Image Processing and Its Applications*, volume 1, 399–403, 1997.

[61] P. Comon. Independent component analysis: A new concept. *Signal Processing*, 36:287–314, 1994.

[62] B. D. Costello, C. A. Gunawardena, and Y. M. Nadiadi. Automated coincident sequencing for fingerprint verification. In *IEE Colloquium on Image Processing for Biometric Measurement*, volume 3, 1–5, April 1994.

[63] R. Courant and D. Hilbert. *Methods of Mathematical Physics*, volume I and II. New York: Wiley Interscience, 1970.

[64] J. D. Cowan and D. H. Sharp. Neural nets. Mathematics Department, University of Chicago, Technical report, 1987.

[65] I. J. Cox, J. Ghosn, and P. N. Yianilos. Feature-based face recognition using mixture distance. NEC Research Institute, Technical report 95-09, 1995.

[66] B. V. Dasarathy. Sensor fusion potential exploitaiton-innovative architectures and illustrative applications. *Proc. IEEE*, 85:24–38, 1997.

[67] J. G. Daugman. High-confidence visual recognition of persons by a test of statistical independence. *IEEE Trans. on Pattern Analysis and Machine Intelligence*, 15(11):1148–1161, 1993.

[68] L. Davis. *Handbook of Genetic Algorithms.* New York: Van Nostrand Reinhold, 1991.

[69] S. B. Davis and P. Mermelstein. Comparison of parametric representations for monosyllabic word recognition in continuously spoken sentences. *IEEE Trans. on ASSP*, 28(4):357–366, 1980.

[70] I. de Falco, A. Iazzetta, P. Natale, and E. Tarantino. Evolutionary neural networks for nonlinear dynamics modeling. In A. E. Eiben, T. Bäck, M. Schoenauer, and H.-P. Schwefel, editors, *Parallel Problem Solving from Nature—PPSN V*, 593–602. New York: Springer-Verlag, 1998.

[71] H. de Garis. GenNets: Genetically programmed neural networks—using the genetic algorithm to train neural nets whose inputs and/or outputs vary in time. In *Proc. IEEE International Joint Conference on Neural Networks*, 1391–1396, 1991.

[72] K. A. de Jong and W. M. Spears. Using genetic algorithms to solve NP-complete problems. In J. D. Schaffer, editor, *Proc. Third International Conference on Genetic Algorithms*, 124–132. San Mateo, CA: M. Kaufmann, 1989.

[73] R. L. de Valois and K. K. de Valois. *Spatial Vision.* New York: Oxford University Press, 1988.

[74] A. P. Dempster, N. M. Laird, and D. B. Rubin. Maximum likelihood from incomplete data via the EM algorithm. *J. of Royal Statistical Soc., Ser. B.*, 39(1):1–38, 1977.

[75] O. Deniz, M. Castrillon, and M. Hernandez. Face recognition using independent component analysis and support vector machines. In *Proc. AVBPA '01*, 59–64, 2001.

[76] V. D. Diakoloukas and V. Diagalakis. Maximum-likelihood stochastic-transformation adaptation of hidden Markov models. *IEEE Trans. on Speech and Audio Processing*, 7(2):177–187, 1999.

[77] K. I. Diamantaras and S. Y. Kung. *Principal Component Neural Networks.* New York: Wiley, 1996.

[78] V. Digalakis, D. Ritischev, and L. Neumeyer. Speaker adaptation using constrained reestimation of Gaussian mixtures. *IEEE Trans. on Speech and Audio Processing*, 3:357–366, 1995.

[79] K. M. Dobroth, B. L. Zeigler, and D. Karis. Future directions for audio interface research: characteristics of human-human order-entry conversations. In *Proc. 8th Annual Voice I/O Systems Applications Conference*, 277–282. San Jose, CA: American Voice Input/Output Society, 1989.

[80] G. R. Doddington. *A Computer Method of Speaker Verification.* PhD thesis, Dept. of Electrical Engineering, University of Wisonsin, Madison, 1970.

[81] G. R. Doddington. Speaker recognition—identifying people by their voices. In *Proc. IEEE*, 1651–1664, 1995.

[82] G. R. Doddington. Speaker recognition based on idiolectal differences between speakers. In *Proc. Eurospeech'01*, 2521–2524, Aalborg, Sept. 2001.

[83] G. Donato, M. Bartlett, J. Hager, P. Ekman, and T. Sejnowski. Classifying facial actions. *IEEE Trans. on Pattern Analysis and Machine Intelligence*, 21:974–989, 1999.

[84] E. C. Driscoll and R. C. Fowler. A comparison of centralized versus distributed architectures in biometric access control systems. In *Proc. ICCST*, 193–198, 1989.

[85] R. O. Duda and P. E. Hart. *Pattern Classification and Scene Analysis*. New York: Wiley, 1973.

[86] R. C. Eberhart. The role of genetic algorithms in neural network query-based learning and explanation facilities. In J. D. Schaffer and D. Whitley, editors, *COGANN-92 Combinations of Genetic Algorithms and Neural Networks*, 169–183, 1992.

[87] M. J. Er, S. Wu, J. Lu, and H. L. Toh. Face recognition with radial basis function (RBF) neural networks. *IEEE Trans. on Neural Networks*, 13(3):697–710, 2002.

[88] K. Erler and L. Deng. Hidden Markov model representation of quantized articulatory features for speech recognition. *Computer Speech and Language*, 7(3):265–282, 1993.

[89] L. J. Eshelman and J. D. Schaffer. Preventing premature convergence in genetic algorithms by preventing incest. In R. K. Belew and L. B. Booker, editors, *Proc. Fourth International Conference on Genetic Algorithms*, 115–122. San Mateo, CA: M. Kaufmann, 1991.

[90] L. J. Eshelman and J. D. Schaffer. Real-coded genetic algorithms and interval-schemata. In L. D. Whitley, editor, *Foundations of Genetic Algorithms 2*, 187–202. San Mateo, CA: M. Kaufmann, 1993.

[91] European Telecommunication Standards Institute. *European Digital Telecommunications System (Phase 2)—Full Rate Speech, Part 2: Transcoding (GSM 06.10 Version 4.1.1)*, 1998.

[92] M. Fang, A. Singh, and M. Y. Chiu. A fast method for eye localization. Siemens Corporate Research, Inc., Technical report SCR-94-TR-488, 1994.

[93] G. Fant. *Title Acoustic Theory of Speech Production with Calculations Based on X-Ray Studies of Russian Articulations*. 2nd ed. The Hague: Mouton, 1970.

[94] K. R. Farrell, R. J. Mammone, and K. T. Assaleh. Speaker recognition using neural networks and conventional classifiers. *IEEE Trans. on Speech and Audio Processing*, 2(1):194–205, 1994.

[95] R. Feraund, O.J. Bernier, J. E. Viallet, and M. Collobert. A fast and accurate face detector based on neural networks. *IEEE Trans. on Pattern Analysis and Machine Intelligence*, 23(1):42–53, 2001.

[96] J. A. Fierrez, J. G. Ortega, D. R. Garcia, and J. R. Gonzalez. A comparative evaluation of fusion strategies for multimodal biometric verification. In *Proc. Audio- and Video-Based Biometric Person Authentication*, 830–836, 2003.

[97] S. Fine, J. Navratil, and R. A. Gopinath. A hybrid GMM/SVM approach to speaker identification. In *Proc. ICASSP'01*, volume 1, 417–420, 2001.

[98] J. L. Flanagan. Computers that talk and listen: Man-machine communication by voice. *Proc. IEEE*, 4:416–432, 1976.

[99] R. Fletcher. *Practical Methods of Optimization*. 2nd ed. New York: Wiley, 1987.

[100] D. B. Fogel. *Evolutionary Computation: Toward a New Philosophy of Machine Intelligence*. New York: IEEE Press, 1995.

[101] L. J. Fogel, A. J. Owens, and M. J. Walsh. *Artificial Intelligence Through Simulated Evolution*. New York: Wiley, 1966.

[102] M. E. Forsyth, A. M. Sutherland, and J. A. Jack. HMM speaker verification with sparse training data on telephone quality speech. *Speech Communication*, 411–416, 1993.

[103] J. Frankel and S. King. ASR—articulatory speech recognition. In *Proc. Euro-speech'01*, 599–602, 2001.

[104] P. Frber. Quicknet on multispert: fast parallel neural network training. ICSI, Technical report TR-97-047, 1998.

[105] K. S. Fu. *Sequential Methods in Pattern Recognition and Machine Learning*. New York: Academic, 1968.

[106] K. Fukunaga. *Introduction to Statistical Pattern Recognition*. Boston, MA: Academic, 1990.

[107] S. Furui. Cepstral analysis technique for automatic speaker verification. *IEEE Trans. on Acoustics, Speech, and Signal Processing*, ASSP-29(2):254–272, 1981.

[108] S. Furui. Comparison of speaker recognition methods using statistical features and dynamic features. *IEEE Trans. on Acoustics, Speech, and Signal Processing*, 29:342–350, 1981.

[109] S. Furui. Recent advances in speaker recognition. *Pattern Recognition Letters*, 18:859–872, 1997.

[110] M. J. F. Gales. Maximum-likelihood linear transformation for HMM-based speech recognition. *Computer Speech and Language*, 12:75–98, 1998.

[111] M. J. F. Gales and S. J. Young. Robust speech recognition in additive and convolutional noise using parallel model compensation. *Speech Communications*, 289–307, 1995.

[112] A. Garcia and R. J. Mammone. Channel-robust speaker identification using modified-mean cepstral mean normalization with frequency warping. In *Proc. ICASSP'99*, 325–328, 1999.

[113] D. Genoud, F. Bimbot, G. Gravier, and G. Chollet. Combining methods to improve speaker verification decision. In *Proc. 4th International Conference on Spoken Language Processing*, volume 3, 1756–1759, 1996.

[114] D. E. Goldberg. *Genetic Algorithms in Search, Optimization, and Machine Learning*. Reading, MA: Addison-Wesley, 1989.

[115] B. A. Golomb and T. J. Sejnowski. SEXNET: A neural network identifies sex from human faces. In D. S. Touretzky and R. Lipmann, editors, *Advances in Neural Information Proceedings Systems 3*. San Mateo, CA: M. Kaufmann, 1991.

[116] G. Golub and C. F. Van Loan. *Matrix Computations*. Baltimore: Johns Hopkins University Press, 1989.

[117] R. C. Gonzalez and P. Wintz. *Digital Image Processing*. Reading, MA: Addison-Wesley, 1977.

[118] R. M. Gray. Vector quantization. *IEEE ASSP Magazine*, 1:4–29, 1984.

[119] G. W. Greenwood. Training partially recurrent neural networks using evolutionary strategies. *IEEE Trans. on Speech and Audio Processing*, 5(2):192–194, 1997.

[120] Y. Gu and T. Thomas. A text-independent speaker verification system using support vector machines classifier. In *Proc. Eurospeech'01*, 1765–1768, 2001.

[121] S. R. Gunn and M. S. Nixon. A dual active contour for head boundary extraction. In *Proc. IEE Colloquium on Image Processing for Biometric Measurement*, volume 6, 1–4, April 1994.

[122] C. S. Gupta, S. R. M. Prasanna, and B. Yegnanarayana. Autoassociative neural network models for online speaker verification using source features from vowels. In *Proc. International Joint Conference on Neural Networks*, volume 2, 1252–1257, 2002.

[123] S. K. Gupta and M. Savic. Text-independent speaker verification based on broad phonetic segmentation of speech. *Digital Signal Processing*, 2:69–79, 1992.

[124] S. Gutta, J. R. J. Huang, P. Jonathon, and H. Wechsler. Mixture of experts for classification of gender, ethnic origin, and pose of human faces. *IEEE Trans. on Neural Networks*, 11(4):948–960, 2000.

[125] D. L. Hall and J. Llinas. An introduction to multisensor data fusion. *Proc. IEEE*, 6–23, 1997.

[126] P. W. Hallinan. Recognizing human eyes. *SPIE Proc.: Geometric Methods in Computer Vision*, 1570:214–226, 1991.

[127] S. A. Harp, T. Samad, and A. Guha. Towards the genetic synthesis of neural networks. In J. D. Schaffer, editor, *Proc. Third International Conference on Genetic Algorithms*, 360–369. San Mateo, CA: M. Kaufmann, 1989.

[128] S. Haykin. *Neural Networks: A Comprehensive Foundation*. New York: Macmillan, 1994.

[129] J. He, L. Liu, and G. Palm. A discriminative training algorithm for VQ-based speaker identification. *IEEE Trans. on Speech and Audio Processing*, 7(3):353–356, 1999.

[130] D. O. Hebb. *The Organization of Behavior*. New York: John Wiley, 1949.

[131] L. P. Heck and M. Weintraub. Handset dependent background models for robust text-independent speaker recognition. In *Proc. ICASSP'97*, volume 2, 1071–1074, 1997.

[132] B. Heisele, P. Ho, and T. Poggio. Face recognition with support vector machines: Global versus component-based approach. In *Proc. Eighth IEEE International Conference on Computer Vision*, 688–694, 2001.

[133] B. Heisele, T. Serre, M. Pontil, T. Vetter, and T. Poggio. Categorization by learning and combining object parts. In *Advances in Neural Information Processing Systems 14*, volume 2, 1239–1245, Vancouver, 2002.

[134] H. Hermansky and N. Morgan. RASTA processing of speech. *IEEE Trans. on Speech and Audio Processing*, 2(4):578–589, 1994.

[135] Sir W. J. Herschel. *The Origin of Fingerprinting*. New York: AMS Press, 1974.

[136] A. Higgins, L. Bahler, and J. Porter. Speaker verification using randomized phrase prompting. *Digital Signal Processing*, 1:89–106, 1991.

[137] J. H. Holland. *Adaptation in Natural and Artificial Systems*. Cambridge: MIT Press, 1992.

[138] K. Hornik, M. Stinchcombe, and H. White. Multilayer feedforward networks are universal approximators. *Neural Networks*, 2:359–366, 1989.

[139] *http://archive.aclu.org/news/2002/n051402b.html.*

[140] *http://home.fujifilm.com/info/share/annual/2003AR10.pdf.*

[141] *http://media.corporate-ir.net/media_files/nsd/visg/loganpilotfaq.pdf.*

[142] *http://spib.rice.edu/spib/select_noise.html.*

[143] *http://spotlight.ccir.ed.ac.uk/public_documents/technology_reports/.*

[144] *http://www.biodigest.com/BiometricDigest/BackIssues/199906.pdf.*

[145] *http://www.boston.com/news/local/articles/2003/09/03/ face_recognition_devices_failed_in_test_at_logan.*

[146] *http://www.cognitec-systems.de/press-releases/ PM0103_Cognitec_Customs_english.pdf.*

[147] *http://www.frvt.org.*

[148] *http://www.identix.com.*

[149] *http://www.igi.tugraz.at/aschwaig.*

[150] *http://www.nevengineering.us.*

[151] *http://www.nist.gov/speech/tests/index.htm.*

[152] *http://www.nist.gov/speech/tests/spk/index.htm.*

[153] *http://www.washingtonpost.com/ac2/wp-dyn/A19946-2002Jul3.*

[154] J. Huang, D. Ii, X. Shao, and H. Wechsler. Pose discrimination and eye detection using support vector machines (SVMs). In *Proc. of NATO-ASI on Face Recognition: From Theory to Applications*, 528–536, 1998.

[155] J. Huang, C. Liu, and H. Wechsler. Eye detection and face recognition using evolutionary computation. In H. Wechsler, P. J. Phillips, V. Bruce, T. Huang, and F. Soulie Fogelman, editors, *Proc. NATO-ASI on Face Recognition: From Theory to Applications*, 348–377. New York: Springer Verlag, 1998.

[156] J. Huang and H. Wechsler. Visual routines for eye location using learning and evolution. *IEEE Trans. on Evolutionary Computation*, 4(1):73–82, 2000.

[157] Q. Huo, C. Chan, and C. H. Lee. On-line adaptive learning of the continuous density hidden Markov model based on approximate recursive Bayes estimate. *IEEE Trans. on Audio and Speech Processing*, 5(2):161–172, 1997.

[158] J. N. Hwang and E. Lin. Mixture of discriminative learning experts of constant sensitivity for automated cytology screening. In *Proc. 1997 IEEE Workshop for Neural Networks for Signal Processing*, 152–161. Amelia Island, September 1997.

[159] A. Hyvarinen and E. Oja. Independent component analysis: Algorithms and applications. *Neural Networks*, 13:411–430, 2000.

[160] T. S. Jaakkola and D. Haussler. Exploiting generative models in discriminative classifiers. In M. S. Kearns, S. A. Solla, and D. A. Cohn, editors, *Advances in Neural Information Processing*, volume 11, 487–493. Cambridge: MIT Press, 1999.

[161] R. A. Jacobs. Methods for combining experts' probability assessments. *Neural Computation*, 7:867–888, 1995.

[162] R. A. Jacobs, M. I. Jordan, S. J. Nowlan, and G. E. Hinton. Adaptive mixtures of local experts. *Neural Computation*, 3:79–87, 1991.

[163] A. K. Jain. *Fundamentals of Digital Image Processing*. Englewood Cliffs, NJ: Prentice Hall, 1989.

[164] C. Z. Janikow and Z. Michalewicz. An experimental comparison of binary and floating point representations in genetic algorithms. In R. K. Belew and L. B. Booker, editors, *Proc. Fourth International Conference on Genetic Algorithms*, 31–36. San Mateo, CA: M. Kaufmann, 1991.

[165] H. Jiang and L. Deng. A Bayesian approach to the verification problem: Applications to speaker verification. *IEEE Trans. on Speech and Audio Processing*, 9(8):874–884, 2001.

[166] T. Joachims. Making large-scale SVM learning practical. In B. Schölkopf, C. Burges, and A. Smola, editors, *Advances in Kernel Methods—Support Vector Learning*. Cambridge, MA: MIT Press, 1999.

[167] M. I. Jordan and R. A. Jacobs. Hierarchies of adaptive experts. In J. Moody, S. Hanson, and R. Lipproart, editors, *Neural Information Processing Systems*, volume 4, 985–992. San Mateo, CA: M. Kaufmann, 1992.

[168] M. I. Jordan and R. A. Jacobs. Hierarchical mixtures of experts and the EM algorithm. *Neural Computation*, 6:181–214, 1994.

[169] P. Jourlin, J. Luettin, D. Genoud, and H. Wassner. Acoustic-labial speaker verification. *Pattern Recognition Letters*, 18(9):853–858, 1997.

[170] B. H. Juang and S. Katagiri. Discriminative learning for minimum error classification. *IEEE Trans. on Signal Processing*, 40(12):3043–3054, 1992.

[171] B. H. Juang, S. Y. Kung, and C. A. Kamm. *IEEE Workshops on Neural Networks for Signal Processing*. New York: IEEE Press, 1991.

[172] B. H. Juang, L. R. Rabiner, and J. G. Wilpon. On the use of bandpass liftering in speech recognition. *IEEE Trans. on Acoustics, Speech, and Signal Processing*, ASSP-35:947–954, 1987.

[173] T. Kanade. *Computer Recognition of Human Faces*. Stuttgart, Germany: Birkhauser Verlag, 1977.

[174] M. Kass, A. Witkin, and D. Terzopoulos. Snakes: Active contour models. *International J. on Computer Vision*, 1:321–331, 1987.

[175] S. Katagiri, B. H. Juang, and C. H. Lee. Pattern recognition using a family of design algorithm based upon the generalized probabilistic descent method. *Proc. IEEE*, 86(11):2345–2373, 1998.

[176] S. Katagiri, C. H. Lee, and B. H. Juang. Discriminative multilayer feedforward networks. In *IEEE Workshop Neural Networks for Signal Processing*, 11–20, Princeton, NJ, 1991.

[177] G. J. Kaufman and K. J. Breeding. Automatic recognition of human faces from profile silhouettes. *IEEE Trans. on Systems, Man, and Cybernetics*, SMC-6(2):113–121, 1976.

[178] M. D. Kelly. Visual identification of people by computer. Stanford Technical report AI-130, 1971.

[179] J. Kharroubi, D. Petrovska-Delacretaz, and G. Chollet. Combining GMM's with suport vector machines for text-independent speaker verification. In *Proc. Eurospeech'03*, volume 3, 1761, 2003.

[180] K. J. Kirchberg, O. Jesorsky, and R. W. Frischholz. Genetic model optimization for Hausdorf distance-based face localization. In *Proc. International ECCV 2002 Workshop on Biometric Authentication*, volume LNCS-2359—*Lecture Notes in Computer Science*, 103–111. Copenhagen: Springer, June 2002.

[181] K. Kirchhoff. *Robust Speech Recognition Using Articulatory Information*. PhD thesis, University of Bielefeld, 1999.

[182] J. Kittler, M. Hatef, R. P. W. Duin, and J. Matas. On combining classifiers. *IEEE Trans. on Pattern Analysis and Machine Intelligence*, 20(3):226–239, 1998.

[183] J. Kittler, G. Matas, K. Jonsson, and M. Sánchez. Combining evidence in personal identity verification systems. *Pattern Recognition Letters*, 18(9):845–852, 1997.

[184] T. Kohonen. *Self-Organization and Associative Memory*, volume 8—*Series in Information Science*. New York: Springer-Verlag, 1984.

[185] K. W. Ku, M. W. Mak, and W. C. Siu. Adding learning to cellular genetic algorithms for training recurrent neural networks. *IEEE Trans. on Neural Networks*, 10(2):239–252, 1999.

[186] S. Y. Kung. *Digital Neural Networks*. Englewood Cliffs, NJ: Prentice Hall, 1993.

[187] S. Y. Kung, K. I. Diamantaras, and J. S. Taur. Adaptive principal component extraction (APEX) and applications. *IEEE Trans. on Signal Processing*, 42(5):1202–1217, 1994.

[188] S. Y. Kung, M. Fang, S. P. Liou, and J. S. Taur. Decision-based neural network for face recognition system. In *Proc. 1995 IEEE International Conference on Image Processing*, volume I, 430–433, Washington DC, Oct. 1995.

[189] S. Y. Kung and C. Mejuto. Extraction of independent components from hybrid mixture: Kuicnet learning algorithm and applications. In *Proc. International Conference on Acoustics, Speech, and Signal Processing*, volume 2, 1209–1212, Seattle, May 1998.

[190] S. Y. Kung, J. Taur, and S. H. Lin. Synergistic modeling and applications of hierachical fuzzy neural networks. *Proc. IEEE*, 87(9):1550–1574, 1999.

[191] S. Y. Kung and J. S. Taur. Decision-based neural networks with signal/image classification applications. In *Proc. IEEE Workshop on Neural Networks for Signal Processing*, Helsingoer, Denmark, 1992.

[192] S. Y. Kung and J. S. Taur. Decision-based neural networks with signal/image classification applications. *IEEE Trans. on Neural Networks*, 6(1):170–181, 1995.

[193] T. Kuritaa and T. Takahashib. Viewpoint independent face recognition by competition of the viewpoint dependent classifiers. *Neurocomputing*, 51:181–195, 2003.

[194] M. Lades, J. Vorbruggen, J. Buhmann, J. Lange, and C. von der Malsburg. Distortion invariant object recognition in dynamic link architecture. *IEEE Trans. Computers*, 42:300–311, 1993.

[195] S. Lawrence, C. L. Giles, A. C. Tsoi, and A. D. Back. Face recognition: a convolutional neural network approach. NEC Research Institute, Technical report, 1995.

[196] Q. Le and S. Bengio. Client dependent GMM-SVM models for speaker verification. In *Proc. ICANN'03*, volume 2714, 443–451, 2003.

[197] C. H. Lee, C. H. Lin, and B. H. Juang. A study on speaker adaptation of the parameters of continuous density hidden Markov models. *IEEE Trans. on Acoustics, Speech, and Signal Processing*, 39(4):806–814, 1991.

[198] H. C. Lee and R. E. Gaensslen. *Advances in Fingerprint Technology.* New York: Elsevier, 1991.

[199] K. Lee, Y. Chung, and H. Byun. Face recognition using support vector machines with the feature set extracted by genetic algorithms. In *Proc. AVBPA'01*, 32–37, 2001.

[200] C. J. Leggetter and P. C. Woodland. Maximum likelihood linear regression for speaker adaptation of continuous density hidden Markov models. *Computer Speech and Language*, 9(2):171–185, 1995.

[201] K. Y. Leung, M. W. Mak, and S. Y. Kung. Applying articulatory features to telephone-based speaker verification. In *Proc. IEEE International Conference on Acoustics, Speech, and Signal Processing*, volume 1, 85–88, Montreal, May 2004.

[202] K. Y. Leung and M. Siu. Integration of acoustic and articulatory information with application to speech recognition. *Information Fusion*, 5(2):141–151, 2003.

[203] K. Y. Leung and M. Siu. Phone level confidence measure using articulatory features. In *Proc. ICASSP'03*, volume 1, 600–603, 2003.

[204] Y. Leung, Y. Gao, and Z. B. Xu. Degree of population diversity—A perspective on premature convergence in genetic algorithms and its Markov chain analysis. *IEEE Trans. on Neural Networks*, 8(5):1165–1176, 1997.

[205] X. Li, M. W. Mak, and S. Y. Kung. Robust speaker verification over the telephone by feature recuperation. In *Proc. International Symposium on Intelligent Multimedia, Video, and Speech Processing*, 433–436, Hong Kong, 2001.

[206] S. H. Lin. *Biometric Identification for Network Security and Access Control.* PhD thesis, Dept. of Electrical Engineering, Princeton University, Princeton, NJ, 1996.

[207] S. H. Lin, Y. Chan, and S. Y. Kung. A probabilistic decision-based neural network for locating deformable objects and its applications to surveillance system and video browsing. In *Proc. International Conference on Acoustics, Speech, and Signal Processing*, Atlanta, May 1996.

[208] S. H. Lin and S. Y. Kung. Probabilistic DBNN via expectation-maximization with multi-sensor classification applications. In *Proc. 1995 IEEE International Conference on Image Processing*, volume III, 236–239, Washington, DC, Oct. 1995.

[209] S. H. Lin, S. Y. Kung, and M. Fang. A neural network approach for face/palm recognition. In *Proc. 5th IEEE Workshop on Neural Networks for Signal Processing*, 323–332, Cambridge, MA, Aug. 1995.

[210] S. H. Lin, S. Y. Kung, and L. J. Lin. A probabilistic DBNN with applications to sensor fusion and object recognition. In *Proc. 5th IEEE Workshop on Neural Networks for Signal Processing*, 333–342, Cambridge, MA, Aug. 1995.

[211] S. H. Lin, S. Y. Kung, and L. J. Lin. Face recognition/detection by probabilistic decision-based neural network. *IEEE Trans. on Neural Networks—Special Issue on Biometric Identification*, 8(1):114–132, 1997.

[212] S. H. Lin, S. Y. Kung, L. J. Lin, and M. Fang. *Neural network for locating and recognizing a deformable object*. US Patent 5,850,470.

[213] R. P. Lippmann. An introduction to computing with neural nets. *IEEE ASSP Magazine*, 4(4-22):153, 1987.

[214] C. S. Liu, C. H. Lee, B. H. Juang, and A. E. Rosenberg. Speaker recognition based on minimum error discriminative training. In *Proc. ICASSP'94*, volume 1, 325–328, 1994.

[215] C. S. Liu, H. C. Wang, and C. H. Lee. Speaker verification using normalized log-likelihood score. *IEEE Trans. on Speech and Audio Processing*, 4(1):56–60, 1996.

[216] Y. Liu, P. Ding, and B. Xu. Using nonstandard SVM for combination of speaker verification and verbal information verification in speaker authentication system. In *Proc. ICASSP '02*, volume 1, 673–676, 2002.

[217] J. Lu, K. N. Plataniotis, and A. N. Venetsanopoulos. Face recognition using kernel direct discriminant analysis algorithms. *IEEE Trans. on Neural Networks*, 14(1):117–126, 2003.

[218] J. Luettin and S. Ben-Yacoub. Robust person verification based on speech and facial images. In *Proc. Eurospeech'99*, volume 2, 991–994, 1999.

[219] J. Luettin and G. Maitre. Evaluation protocol for the extended M2VTS database. IDIAP, Technical report, Martigny, Valais, Switzerland, 1998.

[220] M. W. Mak. Text-independent speaker verification over a telephone network by radial basis function networks. In *Proc. International Symposium on Multi-Technology Information Processing*, 145–150, National Tsing Hua University, Taiwan, 1996.

[221] M. W. Mak, W. G. Allen, and G. G. Sexton. Comparing multi-layer perceptrons and radial basis function networks in speaker recognition. *J. of Microcomputer Applications*, 16:147–159, 1993.

[222] M. W. Mak, W. G. Allen, and G. G. Sexton. Speaker identification using multi-layer perceptrons and radial basis functions networks. *Neurocomputing*, 6(1):99–118, 1994.

[223] M. W. Mak, M. C. Cheung, and S. Y. Kung. Robust speaker verification from GSM-transcoded speech based on decision fusion and feature transformation. In *Proc. IEEE International Conference on Acoustics, Speech, and Signal Processing*, 745–748, 2003.

[224] M. W. Mak and K. W. Cho. Genetic evolution of radial basis function centers for pattern classification. In *Proc. International Conference on Neural Networks*, 669–673, May 1998.

[225] M. W. Mak and S. Y. Kung. Estimation of elliptical basis function parameters by the EM algorithms with application to speaker verification. *IEEE Trans. on Neural Networks*, 11(4):961–969, 2000.

[226] M. W. Mak and S. Y. Kung. Combining stochastic feature transformation and handset identification for telephone-based speaker verification. In *Proc. ICASSP'02*, volume 1, 701–704, 2002.

[227] M. W. Mak, C. L. Tsang, and S. Y. Kung. Stochastic feature transformation with divergence-based out-of-handset rejection for robust speaker verification. *EURASIP J. on Applied Signal Processing*, 4:452–465, 2004.

[228] J. Makhoul. Linear prediction: A tutorial review. *Proc. IEEE*, 63(4):561–580, 1975.

[229] R. J. Mammone, X. Zhang, and R. P. Ramachandran. Robust speaker recognition. *IEEE Signal Processing Magazine*, 13:58–71, 1996.

[230] B. S. Manjunath, R. Chellappa, and C. von der Malsburg. A feature based approach to face recognition. In *Proc. IEEE Computer Society Conference on Computer Vision and Pattern Recognition*, 373–378, 1992.

[231] D. Mansour and B. H. Juang. A family of distortion measures based upon projection operation for robust speech recognition. *IEEE Trans. on Acoustics, Speech, and Signal Processing*, 37(11):1659–1671, 1989.

[232] J. D. Markel, B. T. Oshika, and A. H. Gray. Long-term feature averaging for speaker recognition. *IEEE Trans. ASSP*, ASSP-25:330–337, 1977.

[233] A. Martin, G. Doddington, T. Kamm, M. Ordowski, and M. Przybocki. The DET curve in assessment of detection task performance. In *Proc. Eurospeech'97*, 1895–1898, 1997.

[234] T. Matsui and S. Furui. Comparison of text-independent speaker recognition methods using VQ-distortion and discrete/continuous HMM's. *IEEE Trans. on Speech and Audio Processing*, 2(3):456–458, 1994.

[235] T. Matsui and S. Furui. Likelihood normalization for speaker verification using a phoneme- and speaker-independent model. *Speech Communications*, 17:109–116, 1995.

[236] P. McCullagh and J. A. Nelder. *Generalized Linear Models*. 2nd ed. London: Chapman and Hall, 1989.

[237] W. S. McCulloch and W. Pitts. A logical calculus of ideas immanent in nervous activity. *Bull. Mathematical Biophysics*, 5:115–133, 1943.

[238] E. McDermott. *Discriminative Training for Speech Recognition*. PhD thesis, Waseda University, Japan, 1997.

[239] U. Meier, W. Hurst, and P. Duchnowski. Adaptive bimodal sensor fusion for automatic speech reading. In *Proc. ICASSP'96*, 833–836, 1996.

[240] F. Menczer and D. Parisi. Evidence of hyperplanes in the genetic learning of neural networks. *Biological Cybernetics*, 66:283–289, 1992.

[241] J. Mercer. Functions of positive and negative type, and their connection with the theory of integral equations. *Trans. of the London Philosophical Society (A)*, 209:415–446, 1909.

[242] K. Messer, J. Matas, J. Kittler, J. Luettin, and G. Maitre. XM2VTSDB: The extended M2VTS database. In *Proc. 2nd International Conference on Audio and Video-Based Biometric Person Authentication (AVBPA'99)*, Washington, DC, 1999.

[243] Z. Michalewicz. *Genetic Algorithms + Data Structures = Evolution Programs*. New York: Springer-Verlag, 1996.

[244] S. Mika. *Kernel Fisher Discriminants*. PhD thesis, The Technical University of Berlin, Berlin, 2002.

[245] B. Miller. Vital signs of identity. *IEEE Spectrum*, 22–30, 1994.

[246] G. F. Miller, P. M. Todd, and S. U. Hegde. Designing neural networks using genetic algorithms. In J. D. Schaffer, editor, *Proc. Third International Conference on Genetic Algorithms*, 379–384. San Mateo, CA: M. Kaufmann, 1989.

[247] M. Minsky and S. Papert. *Perceptrons: An Introduction to Computational Geometry*. Cambridge, MA: The MIT Press, 1969.

[248] W. J. Mistretta and K. R. Farrell. Model adaptation methods for speaker verification. In *Proc. ICASSP'98*, volume 1, 113–116, May 1998.

[249] M. Mitchell. *An Introduction to Genetic Algorithm*. Cambridge: MIT Press, 1996.

[250] B. Moghaddam and A. Pentland. Face recognition using view-based and modular eigenspaces. *SPIE Proc.—Automatic Systems for the Identification and Inspection of Humans*, 2257, 1994.

[251] B. Moghaddam and A. Pentland. Probabilistic visual learning for object detection. In *Proc. 5th International Conference on Computer Vision*, 786–793, Cambridge, MA, June 1995.

[252] C. Mokbel. Online adaptation of HMMs to real-life conditions: A unified framework. *IEEE Trans. on Speech and Audio Processing*, 9(4):342–357, 2001.

[253] C. Mokbel, et al. Deconvolution of telephone line effects for speech recognition. *Speech Communication*, 19:185–196, 1996.

[254] D. J. Montana and L. Davis. Training feedforward neural network using genetic algorithms. In *Proc. Eleventh International Joint Conference on Artificial Intelligence*, 762–767, 1989.

[255] J. Moody and C. J. Darken. Fast learning in networks of locally tuned processing units. *Neural Computation*, 1:281–294, 1989.

[256] P. J. Moreno and P. Ho. A new SVM approach to speaker identification and verification using probabilistic distance kernels. In *Proc. Eurospeech'03*, 2965–2968, 2003.

[257] K. R. Muller, S. Mika, G. Ratsch, K. Tsuda, and B. Scholkopf. An introduction to kernel-based learning algorithms. *IEEE Trans. on Neural Networks*, 12(2):181–201, 2001.

[258] D. Naik. Pole-filtered cepstral mean subtraction. In *Proc. ICASSP'95*, volume 1, 157–160, 1995.

[259] J. M. Naik, L. P. Netsch, and G. R. Doddington. Speaker verification over long distance telephone lines. In *Proc. ICASSP'89*, volume 1, 524–527, 1989.

[260] C. Neti, et al. Audio-visual speech recognition. In *Final Workshop 2000 Report*, Center for Language and Speech Processing, The Johns Hopkins University, Baltimore, 2000.

[261] L. Neumeyer and M. Weintraub. Probabilistic optimal filtering for robust speech recognition. In *Proc. ICASSP'94*, 417–420, 1994.

[262] N. J. Nilsson. *Learning Machines: Foundations of Trainable Pattern-Classifying Systems*. New York: McGraw-Hill, 1965.

[263] M. Nixon. Eye spacing measurement for facial recognition. *SPIE Proc.*, 575:279–285, 1985.

[264] Y. Normandin. *Hidden Markov Models, Maximum Mutual Information Estimation, and the Speech Recognition Problem*. PhD thesis, Dept. of Electrical Engineering, McGill University, Montreal, 1991.

[265] J. Oglesby and J. S. Mason. Optimisation of neural models for speaker identification. In *Proc. ICASSP*, 261–264, 1990.

[266] E. Oja. A simplified neuron model as a principal component analyzer. *J. Mathematical Biology*, 15:267–273, 1982.

[267] E. Osuna, R. Freund, and E. Girosi. An improved training algorithm for support vector machines. In *Proc. IEEE Workshop on Neural Networks for Signal Processing VII*, 276–285, Amelia Island, Finland, 1997.

[268] A. S. Pandya and R. B. Macy. *Pattern Recognition with Neural Networks in C++*. Boca Raton, FL: CRC Press and IEEE Press, 1996.

[269] D. Parker. Learning logic. Center for Computational Research in Economics and Management Science, MIT, Technical Report TR-47, Cambridge, MA, 1985.

[270] E. Parzen. On estimation of a probability density function and mode. *Ann. of Mathematical Statistics*, 33:1065–1076, 1962.

[271] P. Penev and J. Atick. Local feature analysis: A general statistical theory for object representation. Computational Neuroscience Laboratory, The Rockefeller University, Technical report, 1996.

[272] A. Pentland, B. Moghaddam, and T. Starner. View-based and modular eigenspaces for face recognition. *Proc. IEEE Conference on Computer Vision and Pattern Recognition*, 84–91, 1994.

[273] E. Persoon and K. S. Fu. Shape discrimination using Fourier descriptors. *IEEE Trans. on Systems, Man, Cybernetics*, SMC-7:170–179, 1977.

[274] J. P. Phillips. Matching pursuit filters applied to face identification. *IEEE Trans. on Image Processing*, 7(8):1150–1164, 1998.

[275] J. P. Phillips, H. Wechsler, J. Huang, and P. Rauss. The FERET database and evaluation procedure for face-recognition algorithms. *Image and Vision Computing Journal*, 16(5):295–306, 1998.

[276] J. B. Pierrot, et al. A comparison of a priori threshold setting procedures for speaker verification in the CAVE project. In *Proc. ICASSP'98*, 125–128, 1998.

[277] S. Pigeon, P. Druyts, and P. Verlinde. Applying logistic regression to the fusion of the NIST'99 1-speaker submisions. *Digital Signal Processing*, 10:237–248, 2000.

[278] R. Plamondon, editor. *Progress in Automatic Signature Verification*. Series in Machine Perception and Artificial Intelligence, Vol. 13. Singapore: World Scientific, 1994.

[279] T. Poggio and F. Girosi. Networks for approximation and learning. *Proc. IEEE*, 78:1481–1497, 1990.

[280] N. Poh, S. Bengio, and J. Korczak. A multi-sample multi-source model for biometric authentication. In *Proc. IEEE 12th Workshop on Neural Networks for Signal Processing*, 375–384, 2002.

[281] V. Popovici and J. P. Thiran. Face detection using an SVM trained in eigenfaces space. In *Proc. AVBPA'03*, 190–198, 2003.

[282] A. B. Portiz. Linear predictive hidden Markov models and the speech signal. *ICASSP'82*, 2:1291–1294, 1982.

[283] V. W. Porto, D. B. Fogel, and L. J. Fogel. Alternative neural networks training methods. *IEEE Expert*, 10(3):16–22, 1995.

[284] G. Potamianos and C. Neti. Stream confidence estimate for audio-visual speech recognition. In *Proc. ICSLP'2000*, volume 3, 746–749, 2000.

[285] W. H. Press, et al. *Numerical Recipes in C*. Cambridge: Cambridge University Press, 1994.

[286] F. Preston. Automatic fingerprint matching. In *ICCST'89*, 199–202, 1989.

[287] F. J. Prokoski, R. B. Riedel, and J. S. Coffin. Identification of individuals by facial thermography. In *Proc. IEEE International Carnahan Conference on Security Technology*, 1992.

[288] M. Przybocki and A. Martin. NIST's assessment of text independent speaker recognition performance 2002. In *The Advent of Biometircs on the Internet, A COST 275 Workshop*, Rome, Nov. 2002.

[289] T. F. Quartieri, D. A. Reynolds, and G. C. O'Leary. Estimation of handset nonlinearity with application to speaker recognition. *IEEE Trans. on Speech and Audio Processing*, 8(5):567–584, 2000.

[290] T. F. Quatieri, R. B. Dunn, D. A. Reynolds, J. P. Campbell, and E. Singer. Speaker recognition using G.729 codec parameters. In *Proc. ICASSP'2000*, 89–92, 2000.

[291] T. F. Quatieri, D. A. Reynolds, and G. C. O'Leary. Estimation of handset nonlinearity with application to speaker recognition. *IEEE Trans. on Speech and Audio Processing*, 8(5):567–584, 2000.

[292] T. F. Quatieri, E. Singer, R. B. Dunn, D. A. Reynolds, and J. P. Campbell. Speaker and language recognition using speech codec parameters. In *Proc. Eurospeech'99*, volume 2, 787–790, 1999.

[293] A. Rahardja, A. Sowmya, and W. Wilson. A neural network approach to component versus holistic recognition of facial expressions in images. *SPIE Proc.: Intell. Robots and Computer Vision X: Algorithms and Technologies*, 1607:62–70, 1991.

[294] M. G. Rahim and B. H. Juang. Signal bias removal by maximum likelihood estimation for robust telephone speech recognition. *IEEE Trans. on Speech and Audio Processing*, 4(1):19–30, 1996.

[295] N. K. Ratha, A. K. Jain, and D. T. Rover. Fingerprint matching on Splash 2. Dept. of Computer Science, Michigan State University, Technical report, March 1994.

[296] I. Rechenberg. Evolution strategy: Nature's way of optimization. In *Optimization: Methods and Applications, Possibilities and Limitations*, volume 47, 106–126, *Lecture Notes in Engineering*. New York: Springer, 1989.

[297] R. A. Redner and H. F. Walker. Mixture densities, maximum likelihood and the EM algorithm. *SIAM Review*, 26(2):195–239, 1984.

[298] C. R. Reeves and S. J. Taylor. Selection of training data for neural networks by a genetic algorithm. In A. E. Eiben, T. Bäck, M. Schoenauer, and H.-P. Schwefel, editors, *Parallel Problem Solving from Nature—PPSN V*, 633–642. New York: Springer-Verlag, 1998.

[299] D. Reisfeld and Y. Yeshuran. Robust detection of facial features by generalized symmetry. In *Proc. 11th International Conference on Pattern Recognition*, 117–120, 1992.

[300] D. A. Reynolds. Experimental evaluation of features for robust speaker identification. *IEEE Trans. on Speech and Audio Processing*, 2(4):639–643, 1994.

[301] D. A. Reynolds. Large population speaker identification using clean and telephone speech. *IEEE Signal Processing Letters*, 2(3):46–48, 1995.

[302] D. A. Reynolds. Speaker identification and verification using Gaussian mixture speaker models. *Speech Communications*, 17:91–108, 1995.

[303] D. A. Reynolds. Comparison of background normalization methods for text-independent speaker verification. In *Proc. Eurospeech'97*, 963–966, 1997.

[304] D. A. Reynolds. HTIMIT and LLHDB: Speech corpora for the study of handset transducer effects. In *Proc. ICASSP'97*, volume 2, 1535–1538, 1997.

[305] D. A. Reynolds. An overview of automatic speaker recognition technology. In *Proc. ICASSP'02*, volume 4, 4072–4075, 2002.

[306] D. A. Reynolds, T. F. Quatieri, and R. B. Dunn. Speaker verification using adapted Gaussian mixture models. *Digital Signal Processing*, 10:19–41, 2000.

[307] D. A. Reynolds and R. C. Rose. Robust text-independent speaker identification using Gaussian mixture speaker models. *IEEE Trans. on Speech and Audio Processing*, 3(1):72–83, 1995.

[308] D. A. Reynolds, M. A. Zissman, T. F. Quatieri, G. C. O'Leary, and B. A. Carlson. The effects of telephone transmission degradations on speaker recognition performance. In *ICASSP'95*, 329–332, 1995.

[309] D. Reynolds, et. al. The superSID project: Exploiting high-level information for high-accuracy speaker recognition. In *Proc. International Conference on Audio, Speech, and Signal Processing*, volume 4, 784–787, Hong Kong, April 2003.

[310] J. Rice. A quality approach to biometric imaging. In *IEE Colloquium on Image Processing for Biometric Measurement*, volume 4, 1–5, April 1994.

[311] M. D. Richard and R. P. Lippmann. Neural network classifiers estimate Bayesian a posteriori probabilities. *Neural Computation*, 3:461–483, 1991.

[312] M. Richardson, J. Bilmes, and C. Diorio. Hidden-articulator Markov models: Performance improvements and robustness to noise. In *Proc. International Conference on Spoken Language Processing (ICSLP'00)*, volume 3, 131–134, 2000.

[313] J. Rissanen. Modeling by shortest data description. *Automatica*, 14:465–471, 1978.

[314] J. Rissanen. A universal prior for integers and estimation by minimum description length. *Annals of Statistics*, 11(2):416–431, 1983.

[315] G. E. Robbins, M. D. Plumbley, J. C. Hughes, F. Fallside, and R. Prager. Generation and adaptation of neural networks by evolutionary techniques (GANNET). *Neural Computers and Applications*, 1:23–31, 1993.

[316] S. Ronald. Robust encodings in genetic algorithms: A survey of encoding issues. In *Proc. of the IEEE International Conference on Evolutionary Computation*, 43–48, 1997.

[317] R. C. Rose, E. M. Hofstetter, and D. A. Reynolds. Integrated models of signal and background with application to speaker identification in noise. *IEEE Trans. on Speech and Audio Processing*, 2(2):245–257, 1994.

[318] A. E. Rosenberg, J. DeLong, C. H. Lee, B. H. Juang, and F. K. Soong. The use of cohort normalized scores for speaker verification. In *Proc. ICSLP'92*, volume 2, 599–602, 1992.

[319] A. E. Rosenberg and S. Parthasarathy. Speaker background models for connected digits password speaker verification. In *Proc. ICASSP'96*, 81–84, 1996.

[320] A. E. Rosenberg, O. Siohan, and S. Parthasarathy. Speaker verification using minimum verification error training. In *Proc. ICASSP'98*, 105–108, 1998.

[321] F. Rosenblatt. The perceptron: A probabilistic model for information storage and organization of the brain. *Psychology Review*, 65:42–99, 1958.

[322] F. Rosenblatt. *Principles of Neurodynamics: Perceptrons and the Theory of Brain Mechanisms*. Washington, DC: Spartan Books, 1961.

[323] H. A. Rowley, S. Baluja, and T. Kanade. Human face detection in visual scenes. School of Computer Science, Carnegie Mellon University, Technical report CMU-CS-95-158R, 1995.

[324] D. W. Ruck, S. K. Rogers, M. Kabrisky, M. E. Oxley, and B. W. Suther. The multilayer perceptron as an approximation to a Bayes optimal discriminant function. *IEEE Trans. on Neural Networks*, 1(4):296–298, 1990.

[325] G. Rudolph. Global optimization by means of distributed evolution strategies. In H. P. Schwefel and R. Männer, editors, *Parallel Problem Solving from Nature— PPSN I*, 209–213, Berlin: Springer-Verlag, 1991.

[326] D. E. Rumelhart, G. E. Hinton, and R. J. Williams. Learning internal representations by error propagation. In D. E. Rumelhart, J. L. McClelland, and the PDP Research Group, editors, *Parallel Distribution Processing: Explorations in the Microstruture of Cognition, Vol. 1: Foundation*, 318–362. Cambridge: MIT Press/Bradford Books, 1986.

[327] M. Sadeghi, J. Kittler, A. Kostin, and K. Messer. A comparative study of automatic face verification algorithms on the BANCA database. In *Proc. AVBPA'03*, 35–41, 2003.

[328] S. Sakamoto, R. Ishiyama, and J. Tajima. 3D model-based face recognition system with robustness against illumination changes. NEC R&D, Technical report 2002-1, 2002.

[329] F. S. Samaria and A. C. Harter. Parameterization of a stochastic model for human face identification. In *Proc. IEEE Workshop on Applications of Computer Vision*, Sarasota, FL, 1994.

[330] C. Sanderson and K. K. Paliwal. Joint cohort normalization in a multi-feature speaker verification system. In *Proc. 10th IEEE International Conference on Fuzzy Systems*, volume 1, 232–235, 2001.

[331] C. Sanderson and K. K. Paliwal. Noise compensation in a person verification system using face and multiple speech features. *Pattern Recognition*, 2:293–302, 2003.

[332] A. Sankar and C. H. Lee. A maximum-likelihood approach to stochastic matching for robust speech recognition. *IEEE Trans. on Speech and Audio Processing*, 4(3):190–202, 1996.

[333] F. S. Saramia. *Face Recognition Using Hidden Markov Model*. PhD thesis, University of Cambridge, 1994.

[334] N. Saravanan and D. B. Fogel. Evolving neural control systems. *IEEE Expert*, 10(3):23–27, 1995.

[335] J. D. Schaffer, D. Whitley, and L. J. Eshelman. Combinations of genetic algorithms and neural networks: A survey of the state of the art. In *Proc. COGANN'92*, 1–37, 1992.

[336] M. Schmidt and H. Gish. Speaker identification via support vector machines. In *Proc. ICASSP'96*, 105–108, 1996.

[337] B. Schölkopf. Statistical learning and kernel methods. Microsoft Research, Technical report MSR-TR 2000-23, 2000.

[338] H. P. Schwefel. *Evolution and Optimum Seeking*. New York: Wiley, 1995.

[339] D. Sims. Biometric recognition: Our hands, eyes, and faces give us away. *IEEE Computer Graphics and Applications*, 14–15, 1994.

[340] P. Sinha. Object recognition via image invariants: A case study. In *Investigative Ophthalmology and Visual Science*, 1735–1740, May 1994.

[341] O. Siohan, C. Chesta, and C. H. Lee. Joint maximum a posteriori adaptation of transformation and HMM parameters. *IEEE Trans. on Speech and Audio Processing*, 9(4):417–428, 2001.

[342] F. K. Soong, A. E. Rosenberg, L. R. Rabiner, and B. H. Juang. A vector quantization approach to speaker recognition. In *Proc. ICASSP'85*, 387–390, 1985.

[343] K. K. Sung and T. Poggio. Learning human face detection in cluttered scenes. *Computer Analysis of Image and Patterns*, 432–439, 1995.

[344] A. C. Surendran, C. H. Lee, and M. Rahim. Nonlinear compensation for stochastic matching. *IEEE Trans. on Speech and Audio Processing*, 7(6):643–655, 1999.

[345] J. S. Taur and S. Y. Kung. Fuzzy decision neural network. In *Proc. IEEE International Conference on Acoustics, Speech, and Signal Processing*, 577–580, Minneapolis, April 1993.

[346] J. S. Taur, S. Y. Kung, and S. H. Lin. Hierarchical fuzzy neural networks for pattern classification. In Y. H. Hu and J. N. Hwang, editors, *Handbook of Neural Network Signal Processing*. Boca Raton, FL: CRC Press, 2002.

[347] J. S. Taur and C. W. Tao. A new neuro-fuzzy classifier with application to online face detection and recognition. *J. VLSI Signal Processing Systems*, 26(3):397–409, 2000.

[348] R. R. Tenney and N. R. Sandell. Detection with distributed sensors. *IEEE Trans. on Aerospace Electronic Systems*, 17:98–101, 1981.

[349] N. Z. Tishby. On the application of mixture AR hidden Markov models to text independent speaker recognition. *IEEE Trans. on Signal Processing*, 39(3):563–570, 1991.

[350] D. M. Titterington, A. F. M. Smith, and U. E. Makov. *Statistical Analysis of Finite Mixture Distributions.* New York: Wiley, 1985.

[351] Y. Tohkura. A weighted cepstral measure for speech recognition. *IEEE Trans. on Acoustics, Speech, and Signal Processing,* ASSP-35:1414–1422, 1987.

[352] C. L. Tsang, M. W. Mak, and S. Y. Kung. Divergence-based out-of-class rejection for telephone handset identification. In *Proc. ICSLP'02,* 2329–2332, 2002.

[353] C. L. Tsang, M. W. Mak, and S. Y. Kung. Cluster-dependent feature transformation for telephone-based speaker verification. In *Proc. International Conference on Audio- and Video-Based Biometric Person Authentication (AVBPA'03),* 86–94, Surrey, U.K., June 2003.

[354] C. L. Tsang, M. W. Mak, and S. Y. Kung. Cluster-dependent feature transformation with divergence-based out-of-handset rejection for robust speaker verification. In *Proc. Fourth International Conference on Information, Communications & Signal Processing and Fourth IEEE Pacific-Rim Conference on Multimedia,* Singapore, Dec. 2003.

[355] O. Tsoi. *Voice Identification: Theory and Legal Applications.* Baltimore, Maryland: University Park Press, 1979.

[356] M. A. Turk and A. P. Pentland. Eigenfaces for recognition. *J. Cognitive Neuroscience,* 3:71–86, 1991.

[357] V. Valtchev. *Discriminative Methods in HMM-based Speech recognition.* PhD thesis, University of Cambridge, Cambridge, 1995.

[358] V. N. Vapnik. *The Nature of Statistical Learning Theory.* New York: Springer-Verlag, 1995.

[359] P. K. Varshney. *Handbook of Multisensor Data Fusion.* New York: Springer-Verlag, 1997.

[360] R. Vergin and D. O'Shaughnessy. On the use of some divergence measures in speaker recognition. In *Proc. ICASSP'99,* 309–312, 1999.

[361] O. Viikki and K. Laurila. Noise robust HMM-based speech recognition using segmental cepstral feature vector normalization. In *Proc. ESCA-NATO Workshop on Robust Speech Recognition for Unknown Communication Channels,* 107–110, France, 1997.

[362] E. Vonk, L. C. Jain, and R. P. Johnson. *Automatic Generation of Neural Network Architecture Using Evolutionary Computation.* Singapore: World Scientific, 1997.

[363] M. Wall. GAlib: A C++ library of genetic algorithm components. Mechanical Engineering Dept., MIT, Technical report, 1996.

[364] V. Wan and W. M. Campbell. Support vector machines for speaker verification and identification. In *Proc. IEEE Workshop on Neural Networks for Signal Processing X*, volume 2, 775–784, 2000.

[365] V. Wan and S. Renals. SVMSVM: Support vector machine speaker verification methodology. In *Proc. ICASSP'03*, volume II, 221–224, 2003.

[366] Y. Wang, S. H. Lin, H. Li, and S. Y. Kung. Data mapping by probabilistic modular networks and information-theoretic criteria. *IEEE Trans. on Signal Processing*, 46(12):3378–3397, 1998.

[367] Y. Wang, L. Luo, M. T. Freedman, and S. Y. Kung. Probabilistic principal component subspaces: A hierarchical finite mixture model for data visualization. *IEEE Trans. on Neural Networks—Special Issue on Neural Networks for Data Mining and Knowledge Discovery*, 11(3):625–636, 2000.

[368] Y. Wang, A. Reibman, F. Juang, T. Chen, and S. Y. Kung. *Proc. IEEE Workshops on Multimedia Signal Processing*. Princeton: IEEE Press, 1997.

[369] Y. J. Wang. Database mapping by mixture-of-experts in computer-aided diagnosis. Private communication.

[370] Z. Wang, Y. Wang, J. Lu, S. Y. Kung, J. Zhang, R. Lee, J. Xuan, J. Khan, and R. Clarke. Discriminatory mining of gene expression microarray data. *J. VLSI Signal Processing Systems—Special Issue on Signal Processing and Neural Networks for Bioinformatics*, 35:255–272, 2003.

[371] T. Wark and S. Sridharan. Adaptive fusion of speech and lip information for robust speaker identification. *Digital Signal Processing*, 11:169–186, 2001.

[372] S. R. Waterhouse, D. MacKay, and A. J. Robinson. Bayesian methods for mixtures of experts. In *Advances in Neural Information Processing 8*, 351–357. Denver, 1995.

[373] S. R. Waterhouse and A. J. Robinson. Classification using hierarchical mixtures of experts. Engineering Department, Cambridge University, Technical report, 1994.

[374] F. Weber, L. Manganaro, B. Peskin, and E. Shriberg. Using prosodic and lexical information for speaker identification. In *Proc. ICASSP'02*, volume 1, 141–144, 2002.

[375] Y. Weiss and E. H. Adelson. Motion estimation and segmentation using a recurrent mixture of experts architecture. In *Proc. 5th IEEE Workshop on Neural Networks for Signal Processing*, 293–302, Cambridge, MA, Aug. 1995.

[376] J. Weng, T. S. Huang, and N. Ahuja. Learning recognition and segmentation of 3D objects from 2D images. In *Proc. IEEE International Conference on Computer Vision*, 121–128, 1993.

[377] P. J. Werbos. *Beyond Regression: New Tools for Prediction and Analysis in the Behavior Science.* PhD thesis, Harvard University, Cambridge, 1974.

[378] B. A. Whitehead and T. D. Choate. Evolving space-filling curves to distribute radial basis functions over an input space. *IEEE Trans. on Neural Networks,* 5(1):15–23, 1994.

[379] B. A. Whitehead and T. D. Choate. Cooperative-competitive genetic evolution of radial basis function centers and widths for time series prediction. *IEEE Trans. on Neural Networks,* 7(4):869–880, 1996.

[380] D. Whitley. Genetic algorithms and neural networks. In G. Winter, J. Periaux, M. Galan, and P. Cuesta, editors, *Genetic Algorithms in Engineering and Computer Science,* 191–201. New York: Wiley, 1995.

[381] D. Whitley, T. Starkweather, and C. Bogart. Genetic algorithms and neural networks: Optimizing connections and connectivity. *Parallel Computing,* 14:347–361, 1990.

[382] G. Widrow and M. E. Hoff. Adaptive switching circuit. In *IRE Western Electronic Show and Convention: Convention Record,* 96–104, 1960.

[383] A. Wieland. Evolving neural network controllers for unstable systems. In *Proc. of the International Joint Conference on Neural Networks,* 667–673, 1991.

[384] L. Wiskott and C. Von der Malsburg. Face recognition by dynamic link matching. Institut für Neuroinformatik, Ruhr-Universität Bochum, Germany, Technical report IR-INI 96-05, 1996.

[385] S. L. Wood, G. Y. Qu, and L. W. Roloff. Detection and labeling of retinal vessels for longitidunal studies. In *Proc. IEEE International Conference on Image Processing,* volume III, 164–167, Washington, DC, Oct. 1995.

[386] C. Wu, C. Liu, H. Y. Shum, Y. Q. Xu, and Z. Zhang. Automatic eyeglasses removal from face images. *IEEE Trans. on Pattern Analysis and Machine Intelligence,* 26(3):322–336, 2004.

[387] B. Xiang, U. Chaudhari, J. Navratil, G. Ramaswamy, and R. Gopinath. Short-time Gaussianization for robust speaker verification. In *Proc. ICASSP'02,* volume 1, 681–684, 2002.

[388] D. Xin and Z. Wu. Speaker recognition using continuous density support vector machines. *Electronics Letters,* 37(17):1099–1101, 2001.

[389] L. Xu and M. I. Jordan. On convergence properties of the EM algorithms for Gaussian mixtures. MIT, AI Lab, Technical report 1520, 1995.

[390] L. Xu, M. I. Jordan, and G. E. Hinton. A modified gating network for the mixture of experts architecture. In *World Congress on Neural Networks*, volume 2, 405–410, San Diego, 1994.

[391] F. Yang and M. Paindavoine. Implementation of an RBF neural network on embedded systems: Real-time face tracking and identity verification. *IEEE Trans. on Neural Networks*, 14(5):1162–1175, 2003.

[392] B. L. Yeo and B. Liu. A unified approach to temporal segmentation of motion JPEG and MPEG compressed video. In *Proc. International Conference on Multimedia Computing and Systems*, 81–88, 1995.

[393] K. K. Yiu, M. W. Mak, and S. Y. Kung. Channel distortion compensation based on the measurement of handset's frequency responses. In *Proc. International Symposium on Intelligent Multimedia, Video and Speech Processing*, 197–200, Hong Kong, 2001.

[394] K. K. Yiu, M. W. Mak, and S. Y. Kung. Environment adaptation for robust speaker verification. In *Proc. Eurospeech'03*, 2973–2976, 2003.

[395] S. Young. A review of large-vocabulary continuous-speech recognition. *IEEE Signal Processing Magazine*, 13(5):45–57, 1996.

[396] K. C. Yow and R. Cipolla. Finding initial estimates of human face locations. In *Proc. 2nd Asian Conference on Computer Vision*, volume 3, 514–518, Singapore, 1995.

[397] W. M. Yu, M. W. Mak, and S. Y. Kung. Speaker verification from coded telephone speech using stochastic feature transformation and handset identification. In *Proc. Pacific-Rim Conference on Multimedia 2002*, 598–606, Taiwan, 2002.

[398] W. M. Yu, M. W. Mak, C. H. Sit, and S. Y. Kung. Speaker verification based on G.729 and G.723.1 coder parameters and handset mismatch compensation. In *Proc. Eurospeech'03*, 1681–1684, 2003.

[399] A. Yuille, D. Cohen, and P. Hallinan. Feature extraction from faces using deformable templates. In *Proc. IEEE Computer Society Conference on Computer Vision and Pattern Recognition*, 104–109, 1989.

[400] J. Zacks and T. R. Thomas. A new neural network for articulatory speech recognition and its application to vowel identification. *Computer Speech, and Language*, 8:189–209, 1994.

[401] B. L. Zhang and Y. Guo. Face recognition by auto-associative radial basis function network. In *Proc. AVBPA'01*, 52–58, 2001.

[402] H. J. Zhang and S. W. Smoliar. Developing power tools for video indexing and retrieval. *SPIE Proc.—Storage and Retrieval for Image and Video Databases II*, 2185:140–149, 1994.

[403] W. D. Zhang, M. W. Mak, C. K. Li, and M. X. He. A priori threshold determination for phrase-prompted speaker verification. In *Proc. Eurospeech'99*, volume 2, 1023–1026, 1999.

[404] X. Y. Zhang, C. L. Myers, and S. Y. Kung. Cross-weighted Fisher discriminant analysis for visualization of DNA microarray data. In *Proc. IEEE Conference on Acoustics, Speech, and Signal Processing*, Montreal, May 2004.

[405] Y. Zhao. An EM algorithm for linear distortion channel estimation based on observations from a mixture of Gaussian sources. *IEEE Trans. on Speech and Audio Processing*, 7(4):400–413, 1999.

[406] M. S. Zilovic, R. P. Ramachandran, and R. J. Mammone. Speaker identification based on the use of robust cepstral features obtained from pole-zero transfer functions. *IEEE Trans. on Speech and Audio Processing*, 6(3):260–267, 1998.

INDEX